## Uncertainty Analysis for Engineers and Scientists

Build the skills for determining appropriate error limits for quantities that matter with this essential toolkit. Understand how to handle a complete project and how uncertainty enters into various steps. This book provides a systematic, worksheet-based process to determine error limits on measured quantities, and all likely sources of uncertainty are explored, measured, or estimated. It also features instructions on how to carry out error analysis using Excel and MATLAB ®, making previously tedious calculations easy. Whether you are new to the sciences or an experienced engineer, this useful resource provides a practical approach to performing error analysis. This book is suitable as a text for a junior- or senior-level laboratory course in aerospace, chemical, and mechanical engineering, and for professionals.

FAITH A. MORRISON is Professor of Chemical Engineering at Michigan Technological University. She is the author of *An Introduction to Fluid Mechanics* (Cambridge University Press, 2013). She is a fellow of the Society of Rheology, member of the Dean's Teaching Showcase (2019), and was a board member of the American Institute of Physics.

# Uncertainty Analysis for Engineers and Scientists
## A Practical Guide

FAITH A. MORRISON
*Michigan Technological University*

CAMBRIDGE
UNIVERSITY PRESS

# CAMBRIDGE
## UNIVERSITY PRESS

University Printing House, Cambridge CB2 8BS, United Kingdom

One Liberty Plaza, 20th Floor, New York, NY 10006, USA

477 Williamstown Road, Port Melbourne, VIC 3207, Australia

314-321, 3rd Floor, Plot 3, Splendor Forum, Jasola District Centre, New Delhi – 110025, India

79 Anson Road, #06–04/06, Singapore 079906

Cambridge University Press is part of the University of Cambridge.

It furthers the University's mission by disseminating knowledge in the pursuit of education, learning, and research at the highest international levels of excellence.

www.cambridge.org
Information on this title: www.cambridge.org/9781108478359
DOI: 10.1017/9781108777513

First published 2021

*A catalogue record for this publication is available from the British Library.*

Library of Congress Cataloging-in-Publication Data
Names: Morrison, Faith A., author.
Title: Uncertainty analysis for engineers and scientists : a practical
guide / Faith A. Morrison, Michigan Technological University.
Description: Cambridge, UK ; New York, NY : Cambridge University Press,
2021. — Includes bibliographical references and index.
Identifiers: LCCN 2020019649 (print) — LCCN 2020019650 (ebook) |
ISBN 9781108478359 (hardback) | ISBN 9781108745741 (paperback) |
ISBN 9781108777513 (epub)
Subjects: LCSH: Uncertainty (Information theory)–Mathematical models. |
Error analysis (Mathematics)
Classification: LCC Q375 .M67 2021 (print) | LCC Q375 (ebook) | DDC 003/.54–dc23
LC record available at https://lccn.loc.gov/2020019649
LC ebook record available at https://lccn.loc.gov/2020019650

ISBN 978-1-108-47835-9 Hardback
ISBN 978-1-108-74574-1 Paperback

# Contents

# Preface

This is a new, short book on uncertainty analysis, targeted both to be a guide for practicing engineers and scientists and as a contribution to the undergraduate textbook market.

Hard numbers from measurements are the backbone of both scientific knowledge and engineering practice. Measurements must be interpreted in context, however: no measured value is infinitely precise, and measurements are certainly not always accurate. This book introduces basic aspects of uncertainty analysis as applicable to the physical sciences and engineering. We focus on one objective: enabling readers to determine appropriate uncertainty limits for measured and calculated quantities. Better conclusions can be drawn from results when they are reported with the appropriate uncertainty limits.

Uncertainty analysis is part of STEM education (science, technology, engineering, and mathematics) to varying degrees: early science education introduces the concept of significant figures; some disciplines teach error propagation in laboratory courses; and many STEM practitioners study statistics. In our advanced technological society, however, all practicing scientists and engineers need refined skills in assessing the reliability of measurements. More broadly, the public needs literacy in measurement of uncertainty, as the reliability of measurements impacts regulations, guilt or innocence in court, budget priorities, and many other issues in the public sphere. To compare "apples to apples" in political or scientific discussions that hinge on measurement accuracy or sufficiency, we need tools like the error-analysis methods emphasized in this book.

We have argued that error analysis is essential, but unfortunately error analysis can seem tedious and involved and, perhaps, of limited practical value. This book addresses that impression head-on. Reading formal texts on error analysis, one might infer that uncertainty is determined with complex statistical formulas that require prolonged study. This impression sets up a barrier to the most important aspects of error analysis: understanding the quality and meaning of day-to-day measurements and using measurements to draw valid, quantitative conclusions on an ongoing basis. With this book, the reader is given the tools, the justification for using those tools, and clear guidance on the common ways to apply classic error analysis.

This book is not a compilation of formulas with no context. The text engages the reader in the issue of the quality of numbers that come from instruments and calculators. The presence of 16 digits in a calculator obscures the fact that the actual precision of the answer is probably quite small and may contain only one or two (and perhaps zero) significant digits. We also guide the reader to ask questions about calculations, such as how to identify the dominant source of uncertainty in the result. We show the reader how to use some modern tools (Microsoft Excel and MATLAB) to make error calculations, allowing us to run trial scenarios to see how uncertainty affects final answers.

Practicing engineers and scientists can be passionate about practical expertise, such as knowing which kind of meter or device is reliable in a given circumstance, or understanding that, in the appropriate context, almost any number can be "essentially zero" or "below our ability to measure." That kind of expertise is often hard-won through experience, but it can be learned systematically with the study of error analysis. We see in this book how to interrogate our devices to determine the extent to which we can rely on them.

The book is divided into six chapters:

1. Introduction and Definitions
2. Quick Start: Replicate Error Basics
3. Reading Error
4. Calibration Error
5. Error Propagation
6. Model Fitting

The first chapter sets the stage for the discussion of error analysis and introduces the three categories of error presented in the subsequent chapters. **Replicate error** is classic random error, and Chapter 2 presents a brief introduction of the statistical topics that feed into error analysis, including the properties of continuous stochastic variables and their probability density distributions, and the definition of standard error, which allows us to combine independent contributing errors. In addition to replicate error we discuss **reading error** (Chapter 3) and **calibration error** (Chapter 4), including how to estimate errors when necessary (Type B errors). The book contributes worksheets associated with the three error types (Appendix A), which guide the user to ask appropriate questions. When error estimates are used, the worksheets record and archive the decisions made, a practice that allows improvements to be incorporated as we learn more about our systems.

Chapter 5 presents error propagation and an error propagation worksheet. The equation for error propagation (combination of variances) is relatively straightforward, but to the uninitiated, the presence of second partial derivatives and confusing nomenclature makes the calculation seem harder than it is. In some texts, use of the full equation is replaced with several equations that apply to specific circumstances (for example, special equations for when two variables are added or subtracted or multiplied). In this book we prefer to use the worksheet and the full equation for two reasons: (1) it builds familiarity with the variance-combination equation in the mind of the user and (2) it is easily evaluated with software. Programming an error propagation into computer software allows us to readily identify the dominant error and also allows us to run scenarios with different values of contributing error. The computer approach thus enables the teaching of critical thinking about errors.

Chapter 6 focuses on ordinary least-squares regression and the errors associated with model predictions and with a model's parameters. Many undergraduates and practitioners use plotting functions and curve-fitting options, but with a bit of additional knowledge they can expand their use to full error analysis. We discuss 95% confidence error limits and 95% confidence prediction intervals and show how these can improve our understanding of correlations in our datasets.

In summary, the book puts error-analysis methods into the hands of students and practitioners in a way that encourages their use. The four worksheets (one for each type of error plus one for error propagation) become tools that guide the determination of error limits, and they form an archive of estimates made and errors determined. The book is short and is designed to become a go-to guide for assessing the quality of measurements and associated calculations.

Many people have helped in various aspects of this book. The junior chemical engineering laboratory facility at Michigan Tech that was the inspiration for this book was built and maintained for many years by David Caspary and Timothy Gasperich and they were instrumental in its success. Jim ('58) and Sally Brozzo provided the essential funding and moral support for the lab. Dr. Nam Kim was the original instructor. The following individuals were my teaching assistants over the years: Austin Bryan, Cho Hui Lim, Matt Chye, Danial Lopez Gaxiola, Ben Conard, Jithendar Guija, Shubham Borole, Hwiyong Lee, Kaela Leonard, Neeva Benipal, Rachael Martin, Erik Leslie, and Yang Zhang. David Chadderton taught the lab during my 2012 sabbatical and did a great job. All these individuals impacted the material presented here.

I wish to express my gratitude to and appreciation of my first department chair, Dr. Edward R. Fisher, who had the vision to develop Michigan Tech's world-class unit operations and junior laboratories and who created an environment in which working hard for our students was fun and productive.

I offer deep thanks to the many people who have given me support over the years. My husband Dr. Tomas Co is and has been unwavering in his support and encouragement with this and all my book-writing projects. For this project, he was also my MATLAB tutor. I would like to thank my uncle, Mike Palmer, for encouraging me to write books. Thank you to the whole Morrison/Palmer/Co family and to all the members of my family and "family of choice," who provide support and encouragement and are central to my life. Love you Betty and Dar.

To my Michigan Tech students over the years, thank you for your engagement, optimism, and willingness to work hard. I'm proud of you, and I celebrate your learning and contributions.

# 1

# Introduction and Definitions

This book provides an introduction to basic aspects of uncertainty analysis as used in the physical sciences and engineering. We focus on one objective: enabling readers to determine appropriate uncertainty limits for measured and calculated quantities. Better conclusions can be drawn from results when they are reported with the appropriate uncertainty limits.

The topic of data uncertainty is often called *error analysis*, a phrase that has the unfortunate connotation that some error has been made in the measurement or calculation. When something has been done incorrectly, we call that a blunder or a mistake. Known mistakes should always be corrected. Due to limitations in both the precision and accuracy of experiments, however, uncertainty is present even when a measurement has been carried out correctly. In addition, there may be hard-to-eliminate random effects that prevent a measurement from being exactly reproducible. Our goal in this book is to present a system for recognizing, quantifying, and reporting uncertainty in measurements and calculations.

It is straightforward to use replication and statistics to assess uncertainty when only *random errors* are present in a measurement – errors that are equally likely to increase or decrease the observed value. We discuss random error and the use of statistics in error analysis in Chapter 2. While replication techniques are powerful, they cannot directly quantify many *nonrandom* errors, which are very common and which often dominate the uncertainty. It is much harder to account for nonrandom or *systematic* errors because they originate from a wide variety of sources and their effects often must be estimated. In dealing with systematic errors it helps if we, too, are systematic, so that we can keep track of both our assumptions and how well our assumptions compare with reality. By following an orderly system, we zero in on good error estimates and build confidence in our understanding of the quality and meaning of our measurements.

1

In this book we focus on three categories of error: random (Chapter 2), reading (Chapter 3), and calibration (Chapter 4). For all three types we provide worksheets (Appendix A) that guide our choices about the likely error magnitudes. If, as is often the case, an experimental result is to be used in a follow-on calculation, we must propagate the errors from each variable to obtain the uncertainty in the calculated result. This process is explained in Chapter 5, and again a worksheet is provided. The worksheet format for error propagation is easily translated to a computer for repeated calculations. In Chapter 6, we provide an introduction to determining error measures when using software to produce curve fits (Microsoft Excel's LINEST or MATLAB).

Through the use of the error-analysis methods discussed in this text, one can determine plausible error limits for experimental results. The techniques are transparent, with the worksheets keeping track of assumptions made as well as the relative impacts of the various assumptions. Keeping track of assumptions facilitates the ever-so-important process of revisiting error estimates, when, for example, seeking a more precise answer, seeking to improve a measurement process, or trying to improve the error estimates previously made.

Having stated our goals, we now devote the rest of this chapter to discussing how four important concepts relate to our error-analysis system: precision and accuracy; significant figures; error limits; and types of uncertainty or error. In Chapter 2 we turn to the statistics of random errors – the statistics of random processes form the basis of all error analyses.

## 1.1 Precision and Accuracy

In common speech the words "precision" and "accuracy" are nearly syn-onyms, but in terms of experimentally determined quantities, we use these words to describe two different types of uncertainty. *Accuracy* describes how close a measurement is to its true value. If your weight is 65 kg and your bathroom scale says that you weigh 65 kg, the scale is accurate. If, however, your scale reports 75 kg as your weight, it fails to reflect your true weight and the scale is inaccurate. To assess accuracy, we must know the true value of a measured quantity.

> To assess accuracy, we must know
> the true value of a measured quantity.

*Precision* is also a measure of the quality of a measurement, but precision makes no reference to the true value. A measurement is precise if it may be distinguished from another, similar measurement of a quantity. For example, a bathroom scale that reports weight to one decimal place as 75.3 kg is more precise than a scale that reports no decimal places, 75 kg. Precise numbers have more digits associated with them. With a precise bathroom scale, we can distinguish between something that weighs 75.2 kg and something that weighs 75.4 kg. These two weights would be indistinguishable on a scale that reads only to the nearest kilogram.

From this discussion we see that accuracy and precision refer to very different things: a three-digit weight is precise, but knowing three digits does not ensure accuracy, since the highly precise scale may be functioning poorly and thus reporting an incorrect weight. Measurement precision and accuracy both affect how certain we are in a quantity. In this book, precision is associated with random and reading error (discussed in Chapters 2 and 3, respectively), while accuracy is addressed in calibration error (discussed in Chapter 4).

## 1.2 Significant Figures

An important convention used to communicate the quality of a result is the number of *significant figures* reported. The significant-figures convention (sig figs) is a short-hand adopted by the scientific and engineering communities to communicate an estimate of the quality of a measured or calculated value. Under the sig-figs convention, we do not report digits beyond the known precision of a number. The rules of the sig-figs convention are provided in Appendix B.

The rules of significant figures allow the reporting of all certain digits and, in addition, one uncertain digit. The idea behind this is that we likely know a number no better than give-or-take "1" in the last digit. For example, if we say we weigh 65 kg, that probably means we know we weigh in the 60s of kilograms, but our best estimate of the exact value is no better than the range 64–66 kg (65 ± 1 kg). As we discuss in this book, determining the true uncertainty in a measurement is more complicated than just making the "±1 in the last digit" assumption (it must involve knowing how the quantity was measured, for example). The sig-figs convention is therefore only a rough, optimistic indication of certainty. The limited goal of the sig-figs convention is to prevent ourselves reporting results to more precision than is justified.

> The significant-figures convention is
> a rough, optimistic indication of certainty.

A person making a calculation often faces the difficulty of assigning the correct number of significant figures to a result. This occurs, for example, when extra digits are generated by arithmetic. For example, if we use a precise bathroom scale to weigh 12 apples as 2.3 kg (two significant figures in both numbers), we can use a calculator to determine that the average mass of an apple is

$$\text{average apple mass} = \frac{2.3 \text{ kg}}{12} = 0.19166666666667 \text{ kg}$$

The calculator shows a large number of digits to accurately convey the arithmetic result. It would be inappropriate, however, to report the average mass of these apples to 14 digits. The rules of significant figures guide us to the most optimistic number of digits to report. As discussed in Appendix B, there are rules for how significant figures propagate through calculations, depending on whether we are adding/subtracting or multiplying/dividing numbers in the calculation. The sig-fig rules are based on the error-propagation methods discussed in Chapter 5. In the case of the average mass of an apple, we assign the answer as 0.19 kg/apple (two sig figs). This is because the sig-figs rule when multiplying/dividing is that the result may have no more sig figs than the least number of sig figs in the numbers in the calculation – in this case, two (see Appendix B).[1]

Learning and using the significant-figures rules allows us to write results more correctly. The principal advantage of following the sig-figs convention is that it is easy, and it prevents us from inadvertently reporting significantly more precision than we should. However, because the sig-figs convention is approximate and may still greatly overestimate precision and accuracy (see Example 1.1), we cannot rely solely on sig-figs rules for scientific and engineering results. When important decisions depend on our calculations, we must perform rigorous error analysis to determine their reliability – this is the topic of this book.

---

[1] In intermediate calculations we retain all digits to avoid round-off error. It is only the final reported result that is subjected to the sig-figs truncation. For greater arithmetic accuracy, intermediate calculations made by calculators and computers automatically employ all available digits. If the number obtained is itself an intermediate calculation, all digits should be employed in downstream calculations.

**Example 1.1: Temperature displays and significant figures.** *A thermocouple is used to register temperature in an experiment to determine the melting point of a compound. If the compound is pure, the literature indicates it will melt at 46.2 ± 0.1°C. A sample of the chemical is melted on a hot plate and a thermocouple placed in the sample reads 45.2°C when the compound melts. Is the compound pure?*

**Solution:** The tenths place of temperature is displayed by the electronics connected to the thermocouple. Assuming the display follows the rules of significant figures, this optimistically implies that the device is "good" to ±0.1°C. Since the highest observed melting temperature is therefore 45.2 + 0.1 = 45.3°C, we might conclude that the compound is not pure, since the observed melting point is less than the lowest literature value obtained for the pure compound, 46.2 − 0.1 = 46.1°C.

Unfortunately, this would not be a justified conclusion. As we discuss in Chapter 4, the most accurate thermocouples have a calibration error that reduces their reliability to no better than ±1.1°C. The manufacturers of thermocouple-based temperature indicators provide extra digits on the display to allow users to avoid round-off error in follow-on calculations; it is the responsibility of the user to assign the appropriate uncertainty to a reading, taking the device's calibration into account.

In the current example and using calibration error limits of ±1.1°C (see also Example 4.6), the highest possible observed melting temperature is 45.2 + 1.1 = 46.3°C, which is within the expected range from the literature. It does not mean that the substance is pure, however; rather, we can say that our observations are consistent with the compound being pure or that we cannot rule out that the compound is pure. To actually test for purity with melting point, we would need to use a temperature-measuring device known to be accurate to within the accuracy of the literature value, ±0.1°C.

## 1.3 Error Limits

This book explains accepted techniques for performing rigorous error analysis. We use *error limits* to report the results of our error analyses. Error limits are a specified range within which we expect to find the true value of a quantity. For example, if we measure the length of 100 sticks and find an average stick length of 55 cm, and if we further find that most of the time (with a 95% level of confidence) the stick lengths are in the range 45–65 cm, then we might report our estimate of the length of a stick as 55 ± 10 cm. The number 55 cm

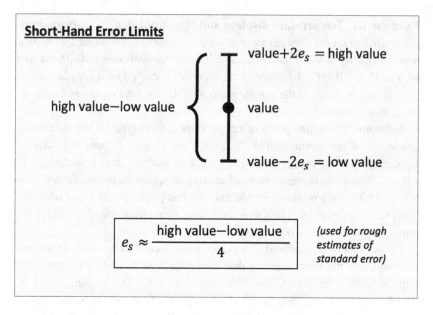

Figure 1.1 When errors are distributed according to the *normal* distribution (see Chapter 2 and Appendix E), then at a 95% level of confidence the true value will be within a range of about plus or minus two standard errors, $e_s$; thus, the span from the smallest probable value to the highest probable value is four standard errors. If we determine a highest and lowest reasonable values for a quantity, we can estimate a plausible standard error as one-fourth of that range.

is our best estimate of the length of a stick, but the uncertainty in the length is such that the length of any chosen stick might be as low as 45 cm and as high as 65 cm.

For the error limits in this example, we explained that the stated range represented the values observed with a 95% level of confidence. This type of error range is a form of 95% confidence interval (discussed in depth in Section 2.3.2), which is associated with an error range of plus or minus two *standard errors*. The standard error $e_s$ is a statistic that we discuss more in Chapter 2. For now we can think of it as a kind of regularized error associated with a quantity. We need to regularize or standardize because uncertainty comes from many sources, and to combine errors from different sources the errors must be expressed in the same way. We discuss the accepted way to standardize errors in Chapter 2. Note that if with 95% confidence the true value is within plus or minus two standard errors of the best estimate of the number, then the total range between the lowest and highest probable values of a number (at a 95% level of confidence) is four standard errors (Figure 1.1). If we estimate a worst-case high value of a quantity and a worst-case low value, subtract the two and divide by four, we obtain a rough estimate of the standard error $e_s$. We use this rule of thumb in situations where we are obliged to estimate error amounts from little information (see the example that follows and Chapter 4).

**Example 1.2: Reader's weight, with uncertainty limits.** *What do you weigh? Please give your answer with appropriate uncertainty limits. What is the standard error of your estimate?*

**Solution:** For a quantity, we are often able to estimate an optimistic (high) number and a pessimistic (low) number. As we have seen, it is conventional to use a 95% level of confidence on stochastic quantities, which corresponds to error limits that are about $\pm 2e_s$. For such a range, there is a span of $4e_s$ between the high value and the low value. This provides a method for estimating $e_s$.

$$\begin{array}{c} \text{Short-hand estimate} \\ \text{for standard error} \end{array} \qquad \boxed{e_s \approx \frac{\text{high value} - \text{low value}}{4}} \qquad (1.1)$$

A person might answer the question about weight by saying he weighs at most 185 lb$_f$ and at least 180 lb$_f$. Then we would write his weight as

$$e_s = \frac{185 - 180}{4} = 1.25 \text{ lb}_f$$

$$\text{average weight} = \frac{185 + 180}{2} \text{ lb}_f$$

$$\text{weight} = \text{average weight} \pm 2e_s$$

$$= 182.5 \pm 2.5 \text{ lb}_f$$

The standard error for this estimate is $e_s = 1.25$ lb$_f$.

Some scientists and engineers report error ranges using other than a 95% level of confidence: 68% or $\pm$ one standard error and 99% or $\pm$ three standard errors are sometimes encountered (These percentages assume that the errors are distributed via the normal distribution; see Appendix E). It is essential to make clear in your communications which type of error limit you are using. The 95% level of confidence is the most common convention; in this text we use exclusively a 95% level of confidence ($\approx \pm$ two standard errors).

## 1.4 Types of Uncertainty or Error

Measurement uncertainty has many sources. We find it convenient to divide measurement uncertainty into three categories – random error, reading error, and calibration error – and each of these errors has its own chapter in this book. We briefly introduce these categories here.

Figure 1.2 An electronic display (left) gives the reading from the scale underneath the water tank on the right. Displays from which we read data limit the precision of the measurement. This limitation contributes to reading error.

*Random* or *stochastic errors* are generated by (usually) unidentified sources that randomly affect a measurement. In measuring an outside air temperature, for example, a small breeze or variations in sunlight intensity could cause random fluctuations in temperature. By definition, random error is equally likely to increase or decrease an observed value. The mathematical definition of a *random process* allows us to develop techniques to quantify random effects (Chapter 2).

A second type of error we commonly encounter is *reading error*. Reading error is a type of uncertainty that is related to the precision of a device's display (Figure 1.2). A weighing scale that reads only to the nearest gram, for example, systematically misrepresents the true mass of an object by ignoring small mass differences of less than 0.5 g and by over-reporting by a small amount when it rounds up a signal. Another component of reading error is needle or display-digit fluctuation: we can estimate a reading from a fluctuating signal, but there is a loss of precision in the process. In Chapter 3 we discuss the statistics of reading error and explain how to estimate this effect.

The third category of error we consider is *calibration error*. While reading error addresses issues of precision, issues of accuracy are addressed through calibration. Instruments are calibrated by testing them against a known, accurate standard. For example, a particular torque transducer may be certified by its manufacturer to operate over the range from 0.01 to 200 mN·m. The manufacturer would typically certify the level of accuracy of the instrument by, for example, specifying that over a chosen performance range the instrument is calibrated to be accurate to within $\pm 2\%$ of the value indicated by the device. In Chapter 4 we outline the issues to consider when assessing uncertainty due to calibration limitations.

As discussed further in Chapter 2, all three error types – random, reading, and calibration – may be present simultaneously, which leads to a complicated situation. If one of the three error sources dominates, then only that error matters, and the error limits on the measurement (at a 95% level of confidence) are plus or minus twice the dominant error in standard form. The task in that case is to determine the dominant error in standard form. If more than one error source matters, we must combine all effects when reporting the overall uncertainty. To combine independent errors we put them in their standard form and combine them according to the statistics of independent stochastic events (Chapter 2). It turns out that standard errors combine in quadrature [52].

Independent errors
combine in quadrature
$$e_{s,cmbd}^2 = e_{s,random}^2 + e_{s,reading}^2 + e_{s,cal}^2 \quad (1.2)$$

where $e_{s,cmbd}$ is the combined standard error, and $e_{s,random}$, $e_{s,reading}$, and $e_{s,cal}$ are standard random, reading, and calibration error of a measurement, respectively.

Summarizing, uncertainty in individual measurements comes from three types of error:

1. **Random** errors due to a variety of influences (these are equally likely to increase or decrease the observed value)
2. **Reading** errors due to limitations of precision in measuring devices (systematic)
3. **Calibration** errors due to limitations in the accuracy of the calibration of measuring devices (systematic)

For a given measurement, each of these independent error sources should be evaluated (Chapters 2, 3, and 4) and the results combined in quadrature (Equation 1.2).

## 1.5 Summary

### 1.5.1 Organization of the Text

The focus of this chapter has been to establish the footings on which to build our error-analysis system. The project is organized into six chapters as outlined here.

- In this first chapter, we have defined terms and categorized the uncertainties inherent in measured quantities. We identify three sources of measurement uncertainty: random error, reading error, and calibration error. These three sources combine to yield a combined error associated with a measured quantity.

$$\text{Independent errors combine in quadrature} \qquad e^2_{s,cmbd} = e^2_{s,random} + e^2_{s,reading} + e^2_{s,cal}$$

- In Chapter 2 we discuss random errors, which may be analyzed through random statistics. From the statistics of stochastic processes we learn a method to standardize different types of errors: measurement error is standardized by making an analogy between making a measurement and taking a statistical sample. We also present a technique widely used for expressing uncertainty, the 95% confidence interval.
- In Chapter 3 we discuss reading error, which is a systematic error produced by the finite precision of the reading display of a measuring device or method. Sources of reading error include the limited number of digits in an electronic display, display fluctuations, and the limit to the fineness of subdivisions on a knob or analog display.
- In Chapter 4 we discuss calibration error, a systematic error attributable to the finite accuracy of a measuring device or method, as determined by its calibration. The accuracy of the calibration of a device is known by the investigator who performed the calibration, and thus the instrument manufacturer is the go-to source for calibration accuracy. If the manufacturer's calibration numbers are not known or are difficult to find, we suggest how to estimate the calibration error using rules of thumb or other short-cut techniques (at our own risk). Also in Chapter 4 we begin two sets of linked examples that follow a calibration process and the use of a calibration curve. These examples are discussed and advanced in Chapters 4–6, as mapped out in Section 1.5.2.
- In Chapter 5 we discuss how error propagates through calculations; this discussion is inspired by how stochastic variables combine.
- The final chapter of the book is dedicated to one family of error-propagation calculations, those associated with ordinary least squares curve fitting. In Chapter 6 we discuss the process of fitting models to experimental data and of determining uncertainty in quantities derived from model parameters. These are computer calculations – we use Microsoft Excel's LINEST and MATLAB in our discussion.

Five appendices are included:

- Appendix A contains worksheets that guide the determination of random, reading, and calibration errors in standard form. There is also an error-propagation worksheet, which structures error-propagation calculations in a form that is compatible with spreadsheet software.
- Appendix B presents the rules and techniques of the significant-figures convention.
- Appendix C has a summary of Microsoft Excel functions that are helpful when performing measurement-error calculations.
- Appendix D lists the contents of the MATLAB functions that are referenced in the text. Also included is a table that shows equivalent Excel and MATLAB commands.
- Appendix E elaborates on some statistical concepts, including both the normal probability distribution and the Student's $t$ distribution, which is the sampling distribution of the normal distribution (Section 2.2.3). Sampling distributions are used to standardize errors, as discussed in Chapter 2.
- Appendix F describes empirical models that are often used to represent and interpret data.

### 1.5.2 Linked Examples Roadmap

Throughout the later chapters of this text, we consider two sets of examples associated with interpreting raw data signals from a differential-pressure (DP) meter. This set of ten interrelated problems (summarized and organized in Figure 1.3) illustrates several aspects of error analysis, among them the use of calibration, error propagation, and model fitting. Here, as a roadmap through those linked examples, we take the opportunity to catalog the issues that are addressed; as a whole, the linked examples show the power and practical application of the error-analysis techniques discussed in the text. We recommend that readers refer to this section and Figure 1.3 as they work through Chapters 4, 5, and 6.

The topics of the linked example sets are measurements from a DP meter, a device used to determine the relative pressure between two locations in a flow loop. In Example 4.2 we begin the discussion by introducing the task of interpreting raw DP-meter readings associated with the flow of water through a flow network. The DP meter is capable of sensing a pressure difference and creating a signal of electric current (4–20 mA) that is proportional to pressure difference between two points. The data presented in Example 4.2 represent

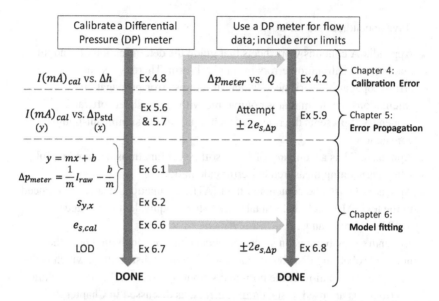

Figure 1.3 Ten linked examples show how the techniques of error analysis of this text are applied to a practical example, the calibration and use of a differential-pressure (DP) meter. Also addressed is the lowest value that can be accurately measured by a device, the limit of determination (LOD).

fluid pressure drops $\Delta p$ along a 6-ft-long tube section, recorded as a function of water volumetric flow rate $Q$.

When the DP-meter calibration curve is known, as it is in Example 4.2, it is straightforward to transform the raw data signals to true values of $\Delta p$. When the $\Delta p$ results are used or reported, we need to indicate the uncertainty associated with the final values, and this information is not obtainable from the calibration curve alone. The task of determining error limits on data obtained with a calibration curve is left unfinished in Example 4.2, to be addressed in subsequent chapters, using the tools of the text (Examples 5.9 and 6.8; see Figure 1.3).

The error limits $\pm 2e_{s,cal}$ associated with measurement values determined from a calibration curve depend on the details of the calibration process itself. The calibration of the DP meter is taken up in a separate set of examples that begins with Example 4.8. The operation of the DP meter is described in that example, and we introduce the differential-pressure *calibration standards*, which in this case are obtained with a manometer. In Example 4.8 we are provided with raw calibration data of DP-meter signal $(I, mA)_{cal}$ versus manometer fluid-height-differences $\Delta h$, but we need error-propagation techniques to advance the calculation of calibration error $e_{s,cal}$; thus, further progress on DP-meter calibration is delayed until after the Chapter 5 discussion of error propagation.

In Chapter 5 we discuss how error propagates through mathematical operations. An early example in Chapter 5 is taking manometer fluid-height data (we use the calibration data introduced in Example 4.8) and assessing the uncertainty for the final calculated values of differential pressure; this is performed in Examples 5.6 and 5.7. We also create a plot of DP-meter signal $y = (I, \mathrm{mA})_{cal}$ versus pressure differences from the calibration standards, $x = (\Delta p, \mathrm{psi})_{std}$, and a best-fit model is determined with Excel or MATLAB (Figure 5.10). The best-fit model can be inverted to form a calibration curve $(\Delta p, \mathrm{psi})_{meter}$ versus $(I, \mathrm{mA})_{raw}$, first introduced in Equation 5.39.

Having established the error-propagation process, we return to the Example 4.2 data to attempt to determine error limits through error propagation (Example 5.9). We make progress, but we lack error measures for the coefficients of the calibration curve, and we need these errors for the error propagation. We determine the coefficients and their associated errors from a least-squares regression calculation that is explained in Chapter 6. Thus, although the Example 4.2 project has been advanced by the Example 5.6, 5.7, and 5.9 discussions, we cannot finish the flow-data problem until the topic of fitting models to data is addressed in Chapter 6 (the Example 4.2 problem continues and concludes in Example 6.8).

Chapter 6 addresses how to use the ordinary least-squares method to determine best-fit model parameters for a dataset. In Example 6.1 we apply the ordinary least-squares technique to the DP-meter calibration dataset first introduced in Example 4.8, which consists of raw calibration data of DP-meter signal versus manometer fluid-height differences. The least-squares technique allows us to find a slope and an intercept that represent the data well (Equation 6.16). We follow up the curve fit by calculating the calibration curve for the DP meter (Equation 6.23), which we obtain by inverting the best-fit curve to allow differential-pressure values to be obtained from DP-meter readings. The calibration equation thus determined is that used in Example 4.2.

In Figure 6.5 we use the error-propagation worksheet to guide an error propagation that leads to good estimates of uncertainty for the slope (and later, intercept) obtained with the DP-meter calibration data. To carry out the error propagation we need the quantity $s_{y,x}$, the standard deviation of $y$-values within an ordinary least-squares fit (Example 6.2). Once $s_{y,x}$ is known, we are able to determine $s_m^2$ and $s_b^2$, the variances of the slope and intercept, respectively (Equations 6.32 and 6.33).

The value of $s_{y,x}$ for the DP-meter calibration dataset is needed in Example 6.6 when we calculate the calibration error for the DP meter. We obtain the calibration error by assessing the error in predicted values of $\Delta p$ for all possible values registered by the DP meter. In Example 6.6, in addition to addressing the uncertainty in the calibration curve of the DP meter, we present the derivation of the general case of uncertainty in values obtained from calibration. In Example 6.7 we show how to calculate the *limit of determination* (LOD) for the DP meter. The LOD is defined as the lowest value that can accurately be measured by a device. Our recommendation is to disregard any data that have more than 25% error; this rule of thumb sets the limit of determination (see Section 4.3).

> The limit of determination (LOD) is defined as the lowest value that can accurately be measured by a device.

The standard calibration error for the DP meter $e_{s,cal}$, determined in Example 6.6, is the last piece of information we need to complete the series of examples that began with Example 4.2. The final resolution of the issue of error limits for the pressure-drop/flow-rate data of Example 4.2 is presented in Example 6.8. The outcome of the calculations is a plot of $\Delta p$ versus flow-rate with appropriate error limits on measured values of $\Delta p$ (Figure 6.17).

The two sets of linked examples outlined here and organized in Figure 1.3 give a flavor of how the methods of this book may be used to structure error calculations to accompany data measurements. The chapters of this text present the methods, and the justification of the methods, helping a reader learn how to apply and benefit from the techniques. We recommend that the reader return to this section and to Figure 1.3 while working through the linked examples of Chapters 4–6.

### 1.5.3 Introductory Examples

We finish this chapter with a few examples that draw on the concepts introduced here. Performing error analysis requires that we learn how to pose appropriate questions about our measurements and calculations. We cannot simply memorize procedures and answers; instead, we must train our minds to identify sources of uncertainty and learn to think about numbers as less-than-perfect representations of quantities. The dilemmas posed in the examples here illustrate the first steps along a path of developing this important thinking skill.

Plausible answers are provided at the end of the chapter. Your own reasoning may be as valid as the answers provided.

**Example 1.3: Precision versus accuracy** *For each situation described here, is the issue presented one of precision or accuracy or both? Justify your answer.*

1. *A racer in a downhill ski race beats the current world record by 0.9 s; the officials refuse to certify the record. What could justify this action?*
2. *Alpine skier Olu clocks a time of 1:34.3 (m:s) on a downhill course while a second skier Hannu also clocks 1:34.3. The course officials are blamed for the tie. What could justify this claim?*
3. *On her driving vacation Jean Louise uses cruise control to limit her driving speed to the speed limit; nevertheless, she is pulled over by the State Police and given a ticket. The citation indicates she was traveling 10 mph (miles per hour) over the speed limit. Jean Louise contests the ticket. On what scientific basis could this case be decided?*
4. *Feng has a global positioning system (GPS) navigation device. When he compares the car's speed indicated by the GPS to that indicated on the car's dashboard, they are often different by a few kph (kilometer per hour). Which should he trust?*
5. *First responders use the 911 emergency system to dispatch help to those in need. Emma called 911 from the seventh floor of a building and heard the emergency vehicles arrive but was confused when they took a long time to locate her, even though she knows they can locate her phone by tracking its GPS signal. What do you think could have caused the delay?*

**Solution:** We provide these situations to allow you to reflect on the role of precision or accuracy in each circumstance; plausible interpretations appear at the end of the chapter.

**Example 1.4: Significant figures practice** *For the following numbers, carry out the arithmetic and assign the correct number of significant figures. The significant figures rules are given in Appendix B.*

1. $132 + 43.2 =$
2. $(67)(143.6) =$
3. $1.000 - (0.53 * 4.0) =$
4. $\ln\left(3.45 \times 10^{-3}\right) =$
5. $\frac{453}{21} =$
6. $10^{-3.9} =$
7. *The area of a circle of radius 0.25 m =*

**Solution:** The answers are at the end of the chapter.

Although a calculator returns a large number of digits, the finite precision of the numbers that enter into a calculation limits the significance of the digits in the answer. The rules of significant figures are a best-case scenario of the number of digits that should be reported in the final answer. In Chapter 5 we show how to determine and display a number's uncertainty with more precision than is possible with significant figures.

## 1.6 Problems

1. Which of these two numbers is more precise: $1.3 \pm 0.2$ m; $23.01 \pm 0.02$ m? Which is more accurate? Explain your answers.
2. Density $\rho$ of a liquid was determined by weighing a fixed volume. The mass of 1.3 liters of the fluid was found to be 2.3260 kg. What is the density? Give your answer with the correct number of significant figures.
3. A recycling center is set up to weigh vehicles and their contents. A truck loaded with metal for recycling weighs 15,321 lb$_\mathrm{f}$. The truck is then emptied and re-weighed and found to weigh 15,299 lb$_\mathrm{f}$. How much metal was unloaded? Give the weight to the correct number of significant figures.
4. An airline only permits a checked bag to weigh up to 50 lb$_\mathrm{f}$. Before leaving home, Helen weighs herself holding the bag (174.1 lb$_\mathrm{f}$) and not holding the bag (124.0 lb$_\mathrm{f}$). What is the weight of the suitcase? Give the weight to the correct number of significant figures.
5. In Chapters 2 and 5 we discuss viscosity data obtained with Cannon–Fenske viscometers (see Example 2.19). The equation for viscosity $\tilde{\mu}$ is

$$\tilde{\mu} = \rho \tilde{\alpha} \Delta t_{eff}$$

where $\rho$ is fluid density, $\tilde{\alpha}$ is a calibration constant, and $\Delta t_{eff}$ is efflux time in the viscometer. For $\rho = 1.124$ g/cm$^3$, $\tilde{\alpha} = 0.01481$ mm$^2$/s$^2$, and $\Delta t_{eff} = 158.4$ s, what is the viscosity? Report the correct number of significant figures based on the data provided.
6. A bag of eleven apples weighs 1.6 kg. What is the mass of an apple? Report the correct number of significant figures in your answer.
7. Bricks weigh 2.9 kg each, on average. How much would nine bricks weigh? Report the correct number of significant figures.

8. The specific gravity of mercury is 13.6 (dimensionless). How much would 2 ml of mercury weigh? Report the correct number of significant figures.

9. A reservoir initially holds 500.0 ml of solvent. A technician decants 492.3 ml of solvent from the reservoir. What is the remaining volume in the reservoir? Report the correct number of significant figures.

10. What is the logarithm (base 10) of 520? Report the correct number of significant figures.

11. What is $10^{3.2}$? Report the correct number of significant figures.

12. Vance uses a bathroom scale to measure the weight of some luggage for his trip abroad. He uses two different methods. In the first method, he places the suitcase on the bathroom scale and measures the weight. In the second method he steps onto the scale and gets one reading and then steps onto the scale without the suitcase for a second reading; he then subtracts the two readings. Which method is more preferable? Discuss your answer.

13. For fourteen days in a row Milton fills a 500.0 ml volumetric flask with water, weighs it, and then empties the flask and stores it. The mass of the full flask is always a little different each day. What could be contributing to this variation?

14. Over the years, travelers made many repeated trips from Hancock, Michigan, to Minneapolis, Minnesota. The longest it took was eight and a half hours. The shortest time for the 365-mile trip was five and three quarters hours. What is a good best estimate for the time the trip will take and what are the error limits on this estimate? Justify your answers.

15. The scale at the gym is used to settle a dispute about who weighs more. Chris weighs 172 lb$_f$ and Lee weighs 170 lb$_f$. The dispute then turns to the issue of whether the scale is accurate enough to settle the dispute. What is your take on the dispute?

16. In the United States, recipes for cooking baked goods specify the quantity of each ingredient by volume (teaspoon, cup, half cup, etc.). In Europe, recipes are specified by mass (grams, kilograms, etc.). Which way is more accurate? Explain your choice.

17. Using the methods described in this book, we determine that the random error, reading error, and calibration error on a temperature indicator are:

$$e_{s,random} = \frac{s}{\sqrt{n}} = 0.23°C$$
$$e_{s,reading} = \text{(negligible)}$$
$$e_{s,cal} = 0.55°C$$

Does any one type of error dominate for this device? If yes, which one? What is the combined error?

18. We measure room temperature to be 20.5°C using a digital temperature indicator using thermocouples to sense temperature. The standard calibration error for the thermocouple is determined to be $e_{s,cal} = 0.55°C$. The readings of room temperature are unvarying, and thus the replicate error is zero. How big would the reading error have to be to influence the combined error for this indicator?

19. In Section 1.5.2, we introduce a set of linked examples that thread their way through the text. Find these examples and add their page numbers to Figure 1.3 (also given on the inside cover of the book). These notations will assist you as you use this book. What are the two overall tasks discussed in the linked examples, as organized by the major threads in Figure 1.3? Which is typically performed by the manufacturer, and which by the instrument user?

20. What is the limit of determination (LOD) of a device? When does the LOD have an effect on the reported accuracy of results from the device? Explain your answer.

## Answers to Example 1.3

1. Perhaps the official thinks that such a large difference cannot be correct and that the timing apparatus is insufficiently accurate to register a record.

2. The two numbers are identical; a timing method with more precision would have solved the problem.

3. The officer says one number and Jean Louise asserts that her car is more accurate than the officer's equipment. She is unlikely to win unless she can test the officer's apparatus for its accuracy. For a driver, having faulty equipment (inaccurate) is not an acceptable excuse for breaking the law.

4. The two devices were calibrated independently, and it is not clear which would be more accurate. Perhaps if we knew the error limits on both measurements, we would find that they agree within the expected uncertainty of the devices.

5. The vertical resolution may be the problem here, as early GPS systems did not have enough vertical precision to determine the correct floor of an emergency when the call came from a multistory building.

## Answers to Example 1.4

(1)  175.2
(2)  $9.6 \times 10^3$
(3)  $-1.1$
(4)  5.669
(5)  22
(6)  $1.3 \times 10^{-4}$
(7)  $0.20 \text{ m}^2$

# 2

# Quick Start

## Replicate Error Basics

Our first topic is random error, a subject intimately tied to statistics.

When we make an experimental determination of a quantity, one of the questions we ask about our result is, if someone else came along and did the same measurement on similar equipment, would they get the same value as we did? We would like to think that they would, but there are many slight, random differences between what is done in any two laboratories and in how two similar apparatuses perform, so we accept that a colleague's answer might be a little different from our answer. A non-laboratory example of this would be weighing oneself on the same kind of scale at home and at the gym—these two numbers might differ by a kilogram or two. Even if you weigh yourself repeatedly throughout the day on the same bathroom scale, you may see some variation due to what you have eaten recently, whether you have exercised, or if your clothing is a bit different for each measurement. Quantities that have this characteristic of variability are called *stochastic variables*.

To identify a good value for a measured variable that is subject to a variety of influences, we turn to statistics. If effects are random, statistics tells us the probability distribution of the effect happening (random statistics), and we can rigorously express both a best estimate for the quantity and error limits on the best estimate [5, 38]. The *mean* of replicated measurements expresses the expected value of the measured quantity, and the *variance* of replicated measurements can be used to quantify the effect of random events on the reproducibility of the measurements, allowing us to construct error limits. We discuss these topics now. Additional background on the statistics of stochastic variables may be found in the literature [5, 38].

## 2.1 Introduction

When we repeatedly determine a quantity from measurements of some sort, the measured numbers often vary a bit, preventing us from knowing that number with absolute precision. Consider, for example, the time it takes to go from your home to your workplace. You may know, roughly, that it takes 30 min, but that number changes a bit from day to day and may vary with what time of day you make the trip, with the type of weather encountered, and with traffic conditions.

To determine a good estimate of the time it takes to make the trip from your home to your workplace, you might measure it several times and take the average of your measurements. Repeated measurements of a stochastic variable are called *replicates*. From replicates we can calculate an average or mean; the mean of a set of replicates is a good estimate of the value of the variable. In replicate analysis we use the following terms:

$x$  stochastic or random variable (2.1)

$x_i$  an observation of $x$ (collectively, the sample set) (2.2)

$n$  number of observations in a sample set (sample size) (2.3)

$\bar{x}$  mean value of the observations of $x$ (2.4)

$s^2$  variance of the observations of $x$ (2.5)

$s$  standard deviation of the observations of $x$ (2.6)

We define these terms in the paragraphs that follow.

Repeated measurements of a quantity such as commuting time may be thought of as observations of a stochastic variable. When we identify a quantity as a stochastic variable, we are saying that the value is subject to influences that are random. These random influences serve both to increase the variable (heavy traffic due to a visiting dignitary slows you down and increases commuting time) and to decrease the variable (leaving earlier in the morning before there is too much traffic decreases your commuting time). Because the influences are random, they average out, leaving a mean value that stays constant throughout the random effects. The definition of the mean of $n$ observations $x_i$ of a stochastic variable $x$ is

$$\text{Sample mean} \quad \bar{x} \equiv \left( \frac{x_1 + x_2 + \cdots + x_n}{n} \right) \tag{2.7}$$

$$= \frac{1}{n} \sum_{i=1}^{n} x_i \tag{2.8}$$

This formula is the familiar arithmetic average. In the spreadsheet program Microsoft Excel,[1] the mean of a range of numbers is calculated with the function AVERAGE(range); all Excel functions mentioned in the text are listed for reference in Appendix C. In the MATLAB computing environment,[2] the mean is calculated with the built-in function mean(array); all MATLAB functions or commands mentioned in the text are listed for reference in Appendix D. Appendix D also contains a table comparing equivalent Excel and MATLAB commands.

The list of terms given earlier includes two quantities that assess the variability of replicates: the sample variance $s^2$ and the sample standard deviation $s$. The definitions of these are [52]:

$$\text{Sample variance } s^2 \equiv \left( \frac{\sum_{i=1}^{n} (x_i - \bar{x})^2}{n - 1} \right) \tag{2.9}$$

$$\text{Sample standard deviation } s = \sqrt{s^2} \tag{2.10}$$

Looking at the definition of variance in Equation 2.9, we see that it is a modified average of the squared differences between the individual measurements $x_i$ and the sample mean $\bar{x}$. The use of *squared* differences ensures that both positive and negative deviations count as deviations and do not cancel out when the sum is taken. The sample variance is not quite the average of squared differences – the average of the squared differences would have $n$ in the denominator instead of $(n-1)$ – but this difference is not significant for our purposes. The variance turns out to be a very useful measure of variability of stochastic quantities. The presence of $(n-1)$ in the denominator of the equation defining sample variance (Equation 2.9) rather than $n$ is called for by statistical reasoning.[3] Sample variance and its square root, sample standard deviation, are widely used to express the variability or spread among observations $x_i$ of stochastic variables.

In Excel, the variance of a sample set is calculated with the function VAR.S(range) and the standard deviation of a sample set with STDEV.S(range) or SQRT(VAR.S(range)); in MATLAB, these commands are $var$(array) and

---

[1] We used Microsoft Excel 2013, www.microsoft.com, Redmond, WA.
[2] We used MATLAB r2018a, www.mathworks.com/products/matlab.html, Natick, MA.
[3] In short, we use $s^2$ as an estimate of the population variance $\sigma^2$, but when we use $n$ instead of $(n-1)$ in Equation 2.9, the expected value of $s^2$ is not $\sigma^2$ but rather the quantity $\left( \sigma^2 - \sigma^2/n \right)$. Knowing this, and with a little algebra, we arrive at the conclusion that, if we use the definition of sample variance in Equation 2.9, the expected value of sample variance $s^2$ will be $\sigma^2$ as we desire. This is called Bessel's correction; see [38, 52].

| Replicated Variable, $Y$: | | | | | Units: | | |
|---|---|---|---|---|---|---|---|
| Measured values $Y_1, Y_2, ..., Y_n$ | Sample Mean, $\bar{Y}$ | Sample Variance, $s^2$ | Sample Standard Deviation, $s$ | Standard Error, $e_s = \frac{s}{\sqrt{n}}$ | 95% Confidence Interval based on $n$ replicates (*Student's t* distribution) | | |
| $Y_1$ | | | | | $n = 1$ | n/a | (include units) |
| $Y_2$ | | | | | $n = 2$ | $\pm 12.7 e_s$ | $\pm$ |
| $Y_3$ | | | | | $n = 3$ | $\pm 4.30 e_s$ | |
| $Y_4$ | | | | | $n = 4$ | $\pm 3.18 e_s$ | |
| $Y_5$ | | | | | $n = 5$ | $\pm 2.78 e_s$ | |
| $Y_6$ | | | | | $n = 6$ | $\pm 2.57 e_3$ | |
| $Y_7$ | | | | | $n \geq 7$ | $\pm 2 e_s$ | |
| | | | | | $\infty$ | $\pm 1.96 e_s$ | |

$$\bar{Y} \equiv \frac{1}{n} \sum_{i=1}^{n} Y_i \qquad s^2 \equiv \frac{1}{(n-1)} \sum_{i=1}^{n} (Y_i - \bar{Y})^2$$

Figure 2.1 When $n$ data replicates are available for a measured variable, the average of the measurements is a good estimate for the value of the measured variable, and the sample variance $s^2$ and sample standard deviation $s$ are associated with the variability of the measurements. As we see later in the chapter, these quantities allow us to determine standard replicate error $e_{s,random} = s/\sqrt{n}$, as well as error limits for the quantity of interest. The replicate error worksheet in Appendix A organizes this calculation.

$std$(array).[4] In Example 2.1 we show a calculation of sample mean, variance, and standard deviation using spreadsheet software. A worksheet in Appendix A (an excerpt is shown in Figure 2.1) organizes these types of calculations.

**Example 2.1: A good estimate of the time to commute from home to workplace.** *Over the course of a year, Eun Young takes 10 measurements of her commuting time under all kinds of conditions (Table 2.1). Calculate a good estimate of her time to commute. Calculate also the variance and the standard deviation of the dataset. What is your estimate of Eun Young's typical commuting time? What is your estimate of Eun Young's commuting time tomorrow?*

---

[4] The sample variance $s^2$ and the sample standard deviation $s$ are different from the population variance $\sigma^2$ and population standard deviation $\sigma$, respectively. We calculate the sample variance from $n$ measurements, which is just a sample of the entire population of all possible measurements. We cannot take all possible measurements, so we cannot know the population variance, but $s^2$ is a good estimate of the population variance in most cases. These issues are considered further in the statistics literature [5, 38].

Table 2.1. *Data for Example 2.1: ten replicate measurements of commuting time.*

| Index, $i$ | Commuting time, min |
|---|---|
| 1 | 23 |
| 2 | 45 |
| 3 | 32 |
| 4 | 15 |
| 5 | 25 |
| 6 | 28 |
| 7 | 18 |
| 8 | 50 |
| 9 | 26 |
| 10 | 19 |

| Replicated Variable, $Y$: | | | | | | | | Units: | |
|---|---|---|---|---|---|---|---|---|---|
| Measured values $Y_1, Y_2, ..., Y_n$ | Sample Mean, $\bar{Y}$ | Sample Variance, $s^2$ | Sample Standard Deviation, $s$ | Standard Error, $e_s = \frac{s}{\sqrt{n}}$ | 95% Confidence Interval based on $n$ replicates (*Student's* $t$ distribution) | | | | |
| $Y_1$  23 min | 28 min | 131 min | 11 min | ? | $n = 1$ | n/a | (include units) | | |
| $Y_2$  45 | | | | | $n = 2$ | ±12.7$e_s$ | ± | ? | |
| $Y_3$  32 | | | | | $n = 3$ | ±4.30$e_s$ | | | |
| $Y_4$  15 | | | | | $n = 4$ | ±3.18$e_s$ | | | |
| $Y_5$  25 | | | | | $n = 5$ | ±2.78$e_s$ | | | |
| $Y_6$  28 | | | | | $n = 6$ | ±2.57$e_s$ | | | |
| $Y_7$  ... | | | | | $n \geq 7$ | ±2$e_s$ | | | |
| | | | | | $\infty$ | ±1.96$e_s$ | | | |

$$\bar{Y} \equiv \frac{1}{n} \sum_{i=1}^{n} Y_i \qquad s^2 \equiv \frac{1}{(n-1)} \sum_{i=1}^{n} (Y_i - \bar{Y})^2$$

Figure 2.2 The replicate error worksheet can organize the calculations for the commuting-time example. We have not yet introduced the replicate standard error $e_{s,random}$; see the discussion surrounding Equation 2.26.

**Solution:** The data vary from 15 to 50 min, indicating that some factors significantly influence the time it takes to make this trip. We are asked to calculate the sample mean (Equation 2.8), variance (Equation 2.9), and standard deviation (Equation 2.10) of commuting time, $x$. Using the Excel functions AVERAGE(), VAR.S(), and STDEV.S(), and the data in Table 2.1, we calculate (see Figures 2.2 and 2.3):

**AVERAGE(B4:B13)**

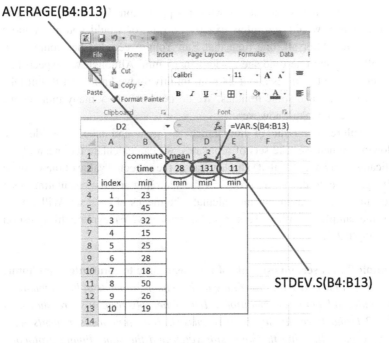

STDEV.S(B4:B13)

Figure 2.3  We use the spreadsheet software Excel to carry out the calculation of the mean and standard deviation of the data provided in Table 2.1.

$$\bar{x} = 28 \text{ min}$$
$$s^2 = 131 \text{ min}^2$$
$$s = 11 \text{ min}$$

Note the units on the quantities.

The meaning of the statistic $\bar{x}$ or "mean commuting time" is straightforward: on average, for the data available, the commute took 28 min. The other two statistics, $s^2$ and $s$, have been calculated with the help of the formulas and Excel, but as yet we do not know the meaning of these numbers. We know them simply as measures of the variability of the data; we address this topic next. Based on the calculations just performed, we estimate that, typically, it would take about 28 min (give or take) for Eun Young to make the commute.

In Example 2.1 we calculated the mean of a sample set, which is also the *expected value* of the variable we are measuring, the commuting time. At any given time, all things being equal, we expect that if Eun Young made her commuting trip, it would take about 28 min.

Perhaps you feel some unease with this prediction. If you were going to make that trip, would you allow exactly 28 min, or would you allow more or less time? After all, one time that Eun Young recorded her commute it took 50 min, and one time she made it in 15 min. Although we expect that, on average, the trip takes about 28 min, intuitively we know that it will take 28 min, give or take some minutes. We do not know how many minutes we should "give and take," however.

To explore this issue of how much to "give and take," consider the following. When we take sets of replicates, we can calculate the mean of the replicates $\bar{x}$ and may use that sample mean as the expected value of the variable $x$. Imagine we do this entire process – the sampling and the calculating – six times: take $n$ measurements and calculate the means of each set. Will we get the same sample mean for all six sets of $n$ replicates? We explore this question in Example 2.2.

**Example 2.2: A second estimate of the mean time to commute from home.**
*Over the course of a year, Eun Young took 20 measurements of her commuting time under all kinds of conditions. Ten of her observations are shown in Table 2.1 and have already been discussed. Ten other measurements were recorded as well, using the same stopwatch and the same timing protocols; this second sample of 10 commuting times is shown in Table 2.2. Calculate a good estimate of Eun Young's time-to-commute from the second set of 10 observations. Calculate also the variance and the standard deviation of the*

Table 2.2. *Data for Example 2.2: ten replicate measurements of commuting time (second dataset).*

| Index, $i$ | Commuting time, min |
|---|---|
| 1 | 27 |
| 2 | 40 |
| 3 | 25 |
| 4 | 22 |
| 5 | 45 |
| 6 | 22 |
| 7 | 28 |
| 8 | 35 |
| 9 | 32 |
| 10 | 41 |

*second dataset. Compare your results to those obtained from the first dataset and discuss any differences.*

**Solution:** The procedure for calculating the sample mean, variance, and standard deviation for this second set is identical to that used with the first set. The results are:

$$\bar{x} = 32 \text{ min}$$
$$s^2 = 68 \text{ min}^2$$
$$s = 8 \text{ min}$$

These numbers are different from those we obtained from the first set: the mean was 28 min in the first set, and it is 32 min in the second set. The variance is twice as big in the first set ($131 \text{ min}^2$) as in the second set ($68 \text{ min}^2$). It seems that taking the second dataset has only made things less clear, rather than more clear.

Although the numbers for the estimates of sample mean and standard deviation differ between the two datasets, this is not automatically cause for concern. Remember that we are sampling a population (all possible commuting times), and observations vary. It is because of this type of variation from sample to sample that the field of statistics has developed good methods (described here) to allow us to draw appropriate conclusions from data samples. The results in both Examples 2.1 and 2.2 are reasonable estimates of the mean commuting time, given the number and reproducibility of measurements used in calculating $\bar{x}$ and $s$ each time. What we lack, as yet, is a way to express any single estimate of the mean commuting time $\bar{x}$ with its error limits. In addition, we would benefit from being able to express and defend how confident we are in the estimate that we calculate.

Before we acquired the second dataset, we thought we knew the average commuting time to be 28 min; then, the second dataset gave an average that was 4 min longer. Example 2.2 showed us that we need to exercise care when interpreting results drawn from a single dataset. If we obtain a second dataset, we may well get (indeed, will most likely get) different values for the mean and the standard deviation.

One positive aspect of the two measurements discussed in Examples 2.1 and 2.2 is that the two calculated means are not too different. If we took a third, fourth, fifth, and sixth sets of data, we expect we would get different numbers for $\bar{x}$ and $s$ each time, but, as we saw earlier, we would expect the means of all sets to be reasonably close to each other.

The near reproducibility of sample means is a phenomenon that is well understood [5, 38]. When we repeatedly measure a quantity that is subject to stochastic variations, we do not get the same value every time. However, if we repeatedly sample the variable and calculate the average for each sample of size $n$, the values of the sample mean we obtain will be grouped around the true value of the mean (which we call $\mu$) and will be symmetrically distributed around the true value. We can use these facts to develop error limits around the estimate of the sample mean.

We began with the question of determining a good estimate of Eun Young's commuting time, and this question has led us into the topic of how to quantify things that vary a bit each time we observe them. The uncertainty problem is not unique to commuting time: whenever we make an experimental measurement, we encounter stochastic variations and other sources of uncertainty.

Taking stock of the situation thus far:

1. In the course of our work, we often find that there is a stochastic variable $x$ that interests us: in Examples 2.1 and 2.2, the variable is the commuting time; in a future example, it is the measured value of density for a liquid; or it could be any other measured quantity.

2. Due to random effects, individual observations $x_i$ of the variable $x$ are not identical.

3. If we average a set of $n$ observations $x_i$ of the stoichiometric variable $x$, we can obtain the average value $\bar{x}$ and the standard deviation $s$ of the sample set. The average of the sample set is a good estimate of the value of the variable, and the standard deviation is an indication of the magnitude of stochastic effects observed in the sample set.

4. If we have one such sample of size $n$, mean $\bar{x}$, and standard deviation $s$, we would like to estimate how close (the "give and take") the sample mean $\bar{x}$ is to the true value of the mean of the distribution of $x$. The true value of the mean is given the symbol $\mu$. A mathematical way of expressing our question is to ask, what is a range around a sample mean $\bar{x}$ within which there is a high probability that we will capture $\mu$, the true value of the mean of $x$?

$$\mu = \text{estimate} \pm (\text{error limits})$$

$$\mu = \bar{x} \pm (\text{error limits})$$

5. The wider the error limits placed on a quantity, the higher the probability that we will capture the true value between the limits. Unfortunately, while expanding the error limits increases certainty, it also decreases the usefulness of the answer – it is not so helpful to say that the commuting time is somewhere between zero and a million hours. We need to establish

a reasonable middle ground between answers that are highly precise but uncertain (narrow error limits) and imprecise but certain (wide error limits).

We have taken important steps toward our goal of quantifying uncertainty. To understand and communicate quantities determined from measurements, we classify them as stochastic variables and apply the methods of statistics. We use statistical reasoning based on sampling to obtain a good estimate of the measured quantity, and statistics also allows us to obtain error limits and the associated probability of capturing the true value of the quantity within the error limits. Our goal in this chapter is to explain the basics of how all this is accomplished.[5]

## 2.2 Data Sampling

We have a single purpose, which is to identify appropriate error limits for a quantity we measure. Quantities we measure are continuous stochastic variables.

$$\begin{matrix} \text{Goal: to determine a} \\ \text{plausible value} \\ \text{of a measured quantity} \\ \text{in the form} \end{matrix} \qquad \text{Answer} = (\text{value}) \pm (\text{error limits}) \qquad (2.11)$$

We seek to address this purpose by taking samples of the stochastic variable of interest. To explain the role of sampling in uncertainty analysis, we begin with a discussion of the mathematics of continuous stochastic variables. Taking an experimental data point is, in a statistical sense, "making an observation of" or "sampling" a *continuous stochastic variable*.

A key tool that characterizes a continuous stochastic variable is its *probability density function (pdf)*. The pdf of a stochastic variable is a function that encodes the nature of the variable – what values the variable takes on and with what frequency or probability. A stochastic variable has its own inherent, underlying pdf called the population distribution. As we discuss in this section (see Equation 2.12), the probability of a continuous stochastic variable taking on a value within a range is expressed as an integral of its pdf across that range.

---

[5] Once the basics are established, we finish the consideration of Eun Young's commuting time in Example 2.7, where we determine the error range for mean commuting time based on the 95% confidence level.

We *sample* stochastic variables to learn about their probability density functions. When we sample a stochastic variable and calculate a statistic such as the sample mean or the sample variance, the value obtained for the chosen statistic likely will be different every time we draw a sample, as we saw in the previous section. The chosen statistic (i.e., the sample mean $\bar{x}$ or the sample variance $s^2$) is itself a continuous stochastic variable, and it has its own probability density function separate from that of the underlying distribution associated with $x$. It turns out that the pdf of the statistic "sample mean" allows us to quantify the probabilities we need to establish sensible error limits for measured data.

> The pdf of the statistic *sample mean*
> allows us to quantify the probabilities we need
> to establish sensible error limits for measured data.

We discuss this in Section 2.2.3.

Finally, the probability density function of sample means is quite reasonably taken to be a well-known distribution called the *Student's t distribution*. We discuss why this is the case and show how Excel and MATLAB can facilitate determinations of error limits with the Student's $t$ distribution.

To recap, this section contains (1) an introduction to the topic of continuous stochastic variables and their pdfs; (2) information on how to determine probabilities from probability density functions; and (3) discussion of determining error limits for stochastic variables using sampling and the Student's $t$ distribution. These topics advance our goal of learning to quantify uncertainty in experimental measurements. For some readers, it may serve your purposes to skip ahead to Section 2.3, which shows how the methods discussed here are applied to practical problems. After reviewing the examples, a reader who has skipped ahead may wish to return here to explore why, when, and how these methods work.

### 2.2.1  Continuous Stochastic Variables

When flipping a coin, what is the probability (Pr) that the result comes up heads? This is a first question in the study of probability. The answer is $\mathrm{Pr} = \frac{1}{2}$, since there are two possible states of the system – heads and tails – and each is equally likely. Thus, we expect that half the time, on average, the result of a coin toss will yield heads and half the time the result will be tails. The outcome of a coin toss is a discrete random variable. When a stochastic variable is *discrete*, meaning the variable has a finite set of outcomes, it is straightforward to calculate probabilities: probability of an outcome equals the number of ways the outcome may be produced divided by the total number of possible outcomes.

When measuring commuting time (Example 2.1), what is the probability that it will take Eun Young 29.6241 min to make her commute? This probability is not at all obvious. In addition, perhaps we can agree that it would be highly unlikely that she would ever take exactly 29.6241 min to make her commute. Commuting time is a *continuous* variable, and this type of variable requires a different type of probability question.

> Getting a useful answer
> requires first identifying
> a useful question.

What is the probability that it will take Eun Young between 20 and 30 min to make her commute? This probability is also not obvious, but it seems like a more appropriate question for a variable such as commuting time. Based on the data we have seen in Examples 2.1 and 2.2, it seems likely that the probability would be pretty high that Eun Young's commute would take between 20 and 30 min. We guess that this probability (Pr) is greater than 0.5 and perhaps as high as Pr = 0.7 or 0.8 (80% chance). We know the probability would not be Pr = 1 (100%), since when we measured the commuting time in Example 2.1 one trip took 50 min. At this point it is not so clear how to be more rigorous in estimating these probabilities.

In the preceding discussion, we explored the difference between establishing probabilities with discreet stochastic variables, such as the outcome of a coin toss, and with continuous stochastic variables, such as commuting time. For discrete stochastic variables, we establish probabilities by counting up all the possible outcomes and calculating the number of ways of achieving each outcome. For continuous stochastic variables, we cannot follow that procedure. We cannot count the number of possible commuting times between 20 and 30 min and we cannot count the number of ways of having Eun Young take 29.6241 min to make her commute. Continuous stochastic variables require a different approach than that used for discrete variables. The approach we use, based on calculus, is to define a pdf and to calculate probabilities by integrating the pdf across a range of values. We discuss how and why this works in the next section.

## 2.2.2 Probability Density Functions

Commuting time is a continuous stochastic variable. Quantities such as commuting time are *continuous* because, in contrast to discrete quantities (the number of people in a room, for example), values of commuting time are not limited to integer values but can take on any decimal number. The continuous nature of experimental variables affects how we quantify the likelihood of observing different values of the variable.

When sampling a continuous stochastic variable, the probability of observing any *specific* outcome is very small, basically zero [38]. For example, in the commuting-time measurement, if we ask about the likelihood of observing a commuting time of exactly 29.6241 min, the answer is zero. Any other precise value is highly unlikely as well; we can even say that between 20 and 30 min there are an infinite number of unlikely values.

Yet our experience tells us that observing some commuting time in this interval is likely. Individual values are unlikely, but when we aggregate over intervals (integrate), the probability becomes finite. The choice of interval matters as well. The interval between 20 and 30 min captures much of the data we know about for Eun Young's commuting time, but this interval is special: not all intervals are equally likely to contain observed commuting times. For example, it is unlikely to observe a commuting time between 100 and 110 min. Thus, probability changes when we ask about different intervals. In addition, the breadth of the interval changes the probability. At one extreme, if we choose a very narrow interval, we find the probability is zero. If we broaden our interval, we are more likely to capture observed commuting times. If we use an extremely broad interval, the probability of observing a value of commuting time becomes nearly certain. If we choose finite-sized limits throughout the domain of possible values, we obtain different, finite probabilities.

An effective approach to the challenge of calculating probabilities for continuous variables is to think of probability in terms of how it adds up over various intervals. We define the *probability density function* $f(x)$ to calculate the probability Pr of the variable $x$ taking on values in an interval between limits $a$ and $b$ (Figure 2.4):

Definition of $f$:
probability is expressed
as an integral of a probability
density function (pdf)
(continuous stochastic variable)

$$\Pr[a \leq x \leq b] \equiv \int_a^b f(x')dx' \qquad (2.12)$$

## Probability Density Function (pdf), $f(x)$

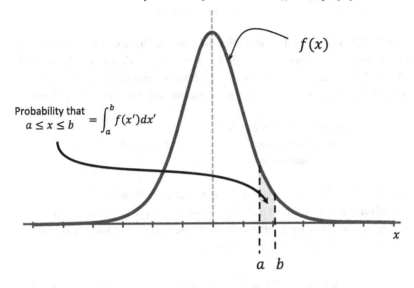

Figure 2.4 For continuous probability distributions, we cannot evaluate the probability of observing a particular value such as an average commuting time of 29.6241 min, but we can calculate the probability that the variable of interest is bracketed in a range of values. (What is the probability that the commuting time is between 28 and 30 min?) The probability is calculated as the area under the curve of the probability density function, between the two values that form the range of interest.

The quantity $f(x')dx'$ is the probability that $x$ takes on a value between $x'$ and $x' + dx'$. The integral represents the result of adding up all the probabilities of observations between $a$ and $b$. If we let $a = -\infty$ and $b = \infty$, then the probability is 1. For all other intervals $[a,b]$, the probability is less than 1 and is calculated by integrating the pdf $f(x)$ between $a$ and $b$.[6]

In the remaining sections and chapters of this text, Equation 2.12 is the key tool for determining error limits for experimental data. We seek error limits $[a,b]$ on the expected value of $x$ so that the probability is high that the true value of a quantity we measure is found in that range. We now explain how that error-limit range $[a,b]$ is found.

To build familiarity with probability density functions and probability calculations, it is helpful at this point to work out a specific probability example

---

[6] We use the notation $x'$ in the integral to make clear that $x'$ is a dummy variable of integration that disappears once the definite integral is carried out. This distinguishes temporary variable $x'$ from the variable $x$, which has meaning outside of the integral.

for a case when the pdf is known. After the example, we turn to the question
of how one determines a pdf for a system of interest.

**Example 2.3: Likelihood of duration of commuting time.** *If we know the
pdf of George's commuting time is the function given in Equation 2.13 (plotted
in Figure 2.5), what is the probability that his commute takes between 25 and
35 min? What is the probability that the commute will take more than 35 min?*

$$
\begin{array}{c}
\text{Probability density function} \\
\text{(pdf) of George's} \\
\text{commuting time}
\end{array}
\qquad
f(x) = Ae^{-\frac{(x-B)^2}{C}}
\qquad (2.13)
$$

where $A = (1/\sqrt{50\pi})\ min^{-1}$, $B = 29\ min$, and $C = 50\ min^2$.

**Solution:** The definition of pdf in Equation 2.12 (repeated here) allows us to
calculate probabilities for continuous stochastic variables if the pdf is known,
as it is in the case of George's commuting time.

$$
\Pr[a \leq x \leq b] \equiv \int_a^b f(x')dx'
\qquad \text{(Equation 2.12)}
$$

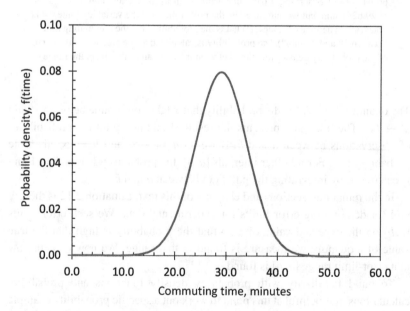

Figure 2.5 The probability density function that characterizes George's commute.
Having the pdf makes probability calculations straightforward; later we see how
to make probability estimates for variables without knowing the underlying pdf.

For the question posed in this example, we obtain the requested probability by integrating the pdf (Equation 2.13) between 25 and 35 min.

$$\Pr[25 \leq x \leq 35 \text{ min}] = \int_{25}^{35} Ae^{-\frac{(x'-B)^2}{C}} dx' \tag{2.14}$$

We have everything we need to finish the problem; the rest is mathematics.

The integral in Equation 2.14 is a deceptively simple one, and it does not have a closed-form solution. This integral is sufficiently common in mathematics that it has been defined as a function all its own, called the *error function*, erf $(u)$. Like sin $u$ and ln $u$, the function erf $u$ comes preprogrammed in mathematical software.

$$\text{Error function (defined):} \quad \boxed{\text{erf}(u) \equiv \frac{2}{\sqrt{\pi}} \int e^{-u^2} du} \tag{2.15}$$

$$\int e^{-u^2} du = \frac{\sqrt{\pi}}{2} \text{erf}(u) \tag{2.16}$$

We carry out the integral in Equation 2.14 in terms of the error function, which we subsequently evaluate in Excel [in both MATLAB and Excel the command for the error function is ERF()]:[7]

$$\Pr[25 \leq x \leq 35 \text{ min}] = \int_{25}^{35} Ae^{-\frac{(x'-B)^2}{C}} dx' = A\sqrt{C} \int_{25}^{35} e^{-\left(\frac{x'-B}{\sqrt{C}}\right)^2} \left(\frac{1}{\sqrt{C}} dx'\right)$$

$$= \left(A\sqrt{C}\right) \left(\frac{\sqrt{\pi}}{2}\right) \text{erf} \left(\frac{x'-B}{\sqrt{C}}\right) \Big|_{x'=25}^{x'=35}$$

$$= \frac{1}{2} \left[\text{erf} \left(\frac{(35-29)}{\sqrt{50}}\right) - \text{erf} \left(\frac{(25-29)}{\sqrt{50}}\right)\right]$$

$$= \frac{1}{2}(0.76986 - (-0.57629))$$

$$= 0.673075 = 67\%$$

We obtain the result that about two thirds of the time George needs between 25 and 35 min to make his commute.

For the second question, to determine how often the commute will be more than 35 min, we integrate the pdf from 35 to $\infty$; the answer is that there is a 12% probability of a commute 35 min or longer (this calculation is left to the reader; Problem 2.5).

---

[7] Hint: In the calculation shown, let $u \equiv \frac{(x'-B)}{\sqrt{C}}$ and thus $du = \frac{dx'}{\sqrt{C}}$.

As we saw in Example 2.3, it is straightforward to calculate probabilities for continuous stochastic variables when the pdf of the underlying distribution is known. With this ability, we can answer some interesting questions about the variable. The problem now becomes, how do we determine the probability density function for a continuous stochastic variable of interest? For Eun Young's commuting time, for instance (Examples 2.1 and 2.2), how do we obtain $f(x)$?

Determining a pdf is a modeling question. As with throwing a die, if we know the details of a process and can reason out when different outcomes occur, we can, in principle, reason out the pdf. To do this, we research the process and sort out what affects the variable, and we build a mathematical model.

For complicated processes such as commuting time, a large number of factors impact the duration of the commute – the weather, ongoing road construction, accidents, seasonal events. Unfortunately, there are too many factors to allow us to model this process accurately. This is the case with many stochastic variables. Because of complexity, the most accurate way to determine the pdf of a real quantity turns out to be to measure it, rather than to model it. If we have patience and resources, a reasonably accurate version of the pdf is straightforward to measure: we make a very large number of observations over a wide range of conditions and cast the results in the form of a pdf.

Although measuring the pdf is straightforward if we are patient and well financed, measuring a pdf is rarely easy. Measuring the pdf for commuting time is a substantial project: to accurately determine the pdf, we must ask Eun Young to time her commute for years under a variety of conditions. Before embarking on this measurement, it would be reasonable to ask, are we justified in making this effort? If we just want to know how to plan a future commute, can we do something useful that is less time-consuming than measuring the pdf?

For casual concerns about one person's commuting time, there is little justification for undertaking the difficult and complex task of measuring the pdf. Many realistic questions we might ask about Eun Young's commuting time could be addressed by taking a guess at the probable time, then adding some extra time to the estimate to protect against circumstances that cause the commute to be on the longer side. For questions relating to commuting time, we probably do not really need to know the best value and its error limits – a worst-case estimate is sufficient (see, for example, Problem 2.30). We characterize this approach as relying on an expert's opinion and incorporating a safety factor; the effort to determine the pdf is not warranted.

Although worst-case thinking has its place in decision making, for scientific and engineering work a worst-case estimate of a quantity of interest is often insufficient. In science, our measurements are usually part of a broader project in which we hope to make discoveries, learn scientific truths, or build reliable devices, processes, or models. We may measure density, for example, as part of the calibration of a differential-pressure meter. The accuracy of the calibrated meter depends substantially on the accuracy of the density measurement for the fluid used during calibration, and we need to know the extent to which we can rely on our measurements. Many technological applications of measurements are like this – dependent on instrument and measurement accuracy. Not infrequently, accuracy can be a matter of life and death (for example, when building structures, designing safety processes, and manufacturing health-related devices) or can be what determines the success/failure of a project (cost estimates, instrument sizing, investments). In such cases we cannot use a worst-case value and talk ourselves out of the need to establish credible error limits on our numbers.

The objections raised here seem to argue that we have no choice but to measure the pdf for stochastic variables that we study for scientific and engineering purposes. This sounds like a great deal of work (and it is), but there is some good news that will considerably reduce this burden.

First, it turns out that we can quite often reasonably assume that the stochastic effects in measurements are *normally distributed*; that is, their pdf has the shape of the normal distribution (Figure 2.6):

$$
\begin{array}{c}
\text{Normal} \\
\text{probability} \\
\text{distribution} \\
\text{(pdf)}
\end{array}
\qquad
\boxed{\;f(x) = \frac{1}{\sqrt{2\sigma^2\pi}} e^{-\frac{(x-\mu)^2}{2\sigma^2}}\;}
\qquad (2.17)
$$

The normal distribution (see the literature [38] and Section E.2 in the appendix and for more details) is a symmetric distribution with two parameters: a mean $\mu$, which specifies the location of the center of the distribution, and the standard deviation $\sigma$, which specifies the spread of the distribution. With the assumption that the random effects are normally distributed (that is, they follow Equation 2.17), we reduce the problem of determining the pdf to determining the two parameters $\mu$ and $\sigma$.

Second, often the questions we have about a stochastic variable can be answered by examining a *sample* of the values of the stochastic variable. We introduce sampling to get around the difficulty of determining the underlying pdf of the variable. The approach is this: we take data replicates – three, four, or

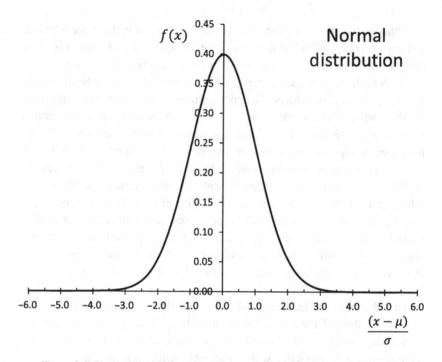

Figure 2.6 The normal distribution, famous as the bell curve, is a symmetric probability density distribution characterized by two parameters, $\mu$, which is its center, and $\sigma$, which characterizes its spread. The pdf has been plotted here versus a dimensionless version of $x$, translated to the mean $(x - \mu)$ and scaled by the standard deviation of the distribution. When $\mu = 0$ and $\sigma = 1$, this is called the standard normal distribution. For the normal distribution, 68% of the probability is located within $\pm\sigma$ of the mean, and 95% of the probability is located within $\pm 1.96\sigma$ or $\approx \pm 2\sigma$ of the mean.

more observations – and we ask, what is a good value (including error limits) of the stochastic variable based on this sample set? The sample is not a perfect representation of the variable, but thanks to considerable study, the field of statistics can tell us a great deal about how the characteristics of samples are related to the characteristics of the true distribution – all without us having to know in detail the underlying pdf of the variable. We can put this statistical knowledge to good use when determining error limits.

In the next section we show how we use a probability density function called the *sampling distribution of the sample mean* $\bar{x}$ to determine a good value (including error limits) for a stochastic variable based on finite data samples.

### 2.2.3 Sampling the Normal: Student's *t* Distribution

We return to our purpose, which is to identify error limits for a quantity we measure.

Goal: to determine a
plausible value
of a measured quantity
in the form

$$\text{Answer} = (\text{value}) \pm (\text{error limits}) \quad (2.18)$$

We seek to address this purpose by taking samples of the measured quantity, which is a stochastic variable. We take replicate measurements (a sample of size $n$) of the quantity, from which we wish to estimate a good value of the measured quantity. To determine the appropriate error range about that value, we also ask: what is the size of the error range we need to choose (Figure 2.7) to create an interval that has a good chance (we define a "good chance" as 95% probability) of including the true value of the variable?

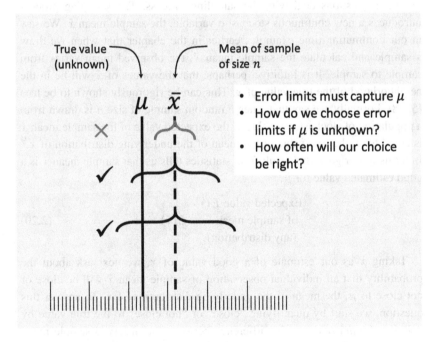

Figure 2.7 We measure $\bar{x}$ and seek to determine error limits that capture the true value of the variable, $\mu$. We do not know $\mu$, however. Whatever we choose, we also need to be able to say quantitatively how likely we believe it is that the mean $\pm$ the error limits will capture the true value of the variable.

Restated goal: to determine error limits
   on an estimated value that, with 95%     Answer = (value) ± (error limits)
confidence, captures the true value of $x$

$$(2.19)$$

The value of writing our goal this way is that, through sampling, we can address the restated goal.

As we discuss here, we can say a great deal about samples of a variable without knowing the details of the variable's underlying distribution. For the quantity we are measuring, we first agree to assume that its underlying pdf is a normal distribution of unknown mean $\mu$ and unknown standard deviation $\sigma$ (Equation 2.17).[8] Second, we take a sample of $n$ measurements and calculate $\bar{x}$ and $s$ for the sample set. Finally, we pose our questions. First, based on the sample set, what is a good estimate of the mean $\mu$ of the underlying distribution? Second, what ± error limits should we apply to the good estimate so that, with 95% confidence, the true value $\mu$ is captured in the range (value) ± (error limits)?

These questions deal with the sampling process. The sampling process introduces a new continuous stochastic variable, the sample mean $\bar{x}$. We saw in our commuting-time examples earlier in the chapter that when we draw a sample and calculate the sample mean $\bar{x}$, the observed mean varies from sample to sample. It is intuitive, perhaps, that the values of $\bar{x}$ will be in the neighborhood of the true value of $x$. This can be rigorously shown to be true [5, 38]. Formally, we say that when a random sample of size $n$ is drawn from a population of stochastic variable $x$, the expected value of the sample mean $\bar{x}$ is equal to $\mu$, the true value of the mean of the underlying distribution of $x$.[9] In terms of our goal, this result from statistics tells us that sample mean $\bar{x}$ is a good estimated value for $\mu$.

Expected value $E()$
of sample mean $\bar{x}$     $E(\bar{x}) = \mu$                     $(2.20)$
(any distribution)

Taking $\bar{x}$ as our estimate of a good value of $x$, we next ask about the probability that an individual observation of sample mean $\bar{x}$ will be close or not close to $\mu$, the mean of the underlying distribution of $x$. To answer this question, we start by quantifying "close" or "not close" to the true value by defining the *deviation* as the difference between the observed sample mean and the true value of the mean of $x$.

---

[8] We can relax this assumption later.
[9] See the literature discussion of the central limit theorem [15, 38].

$$\begin{array}{l} \text{Deviation between} \\ \text{the observed sample mean } \bar{x} \\ \text{and } \mu, \text{ the true mean} \\ \text{of the stochastic variable } x \end{array} \quad \text{deviation} \equiv (\bar{x} - \mu) \qquad (2.21)$$

The deviation defined in Equation 2.21 is also a continuous stochastic variable. Since the normal distribution is symmetric, the deviations are equally likely to be positive and negative, and overall the expected (mean) value of the deviation $(\bar{x} - \mu)$ is zero. As with all continuous stochastic variables, the probability of observing any specific value of the deviation is zero, but if we have the pdf for the deviation, we can calculate the probability that the deviation takes on values between two limits (using Equation 2.12). For example, for a measured average commuting time of $\bar{x} = 23.3$ min, if we knew the pdf $f(\bar{x} - \mu)$ of the deviation defined in Equation 2.21, we could answer the question, what is the probability that the deviation of our measured mean from the true mean, in either direction, is at most 5.0 min?

$$\begin{array}{l} \text{Maximum deviation} \end{array} \quad |(\bar{x} - \mu)_{max}| = 5.0 \text{ min} \qquad (2.22)$$

$$\begin{array}{l} \text{Probability that} \\ (\bar{x} - \mu) \text{ is in the} \\ \text{interval } [-5.0, 5.0] \end{array} \quad \Pr[-5.0 \leq (\bar{x} - \mu) \leq 5.0] \qquad (2.23)$$

$$\Pr[-5.0 \leq (\bar{x} - \mu) \leq 5.0] = \int_{-5.0}^{5.0} f(\bar{x}' - \mu) d(\bar{x}' - \mu) \qquad (2.24)$$

This question about the deviation is the same as asking, what is the probability that $\mu$, the true value of $x$, lies in the range $23.3 \pm 5.0$ min? By focusing on deviations, the error-limits problem now becomes the question of determining the pdf $f(\bar{x} - \mu)$ of the deviation between the sample mean $\bar{x}$ and the true mean $\mu$. This formulation has the advantage that to determine the pdf of the deviation we do not need to know $\mu$, the true mean value of $x$.

---

By focusing on the pdf of *deviation* $(\bar{x} - \mu)$, rather than of $\bar{x}$, we avoid having to know $\mu$, the true mean value of $x$.

---

To recap, in a previous section we established the pdf's role as the tool needed to calculate the probability that a stochastic variable takes on a value within a range (Equation 2.12). When the pdf of underlying experimental errors is not known, we customarily assume that the underlying distribution is a normal distribution (this is a good assumption for many experimental errors), and we sample the distribution (obtain $n$ replicates). We use the mean of the sample set $\bar{x}$ to determine a good value of $x$. To determine error ranges for our good value of $x$, we then seek the pdf of a new continuous stochastic variable, $(\bar{x} - \mu)$, the deviation of the sample mean from the true value of the mean. Concentrating on $(\bar{x} - \mu)$ means we avoid having to know the value of $\mu$.

The pdf we seek, $f(\bar{x} - \mu)$, the pdf of the deviation between a sample mean and the true mean of a normally distributed population of unknown standard deviation, has been determined [16, 34], and we present its mathematical form next (it is Student's $t$ distribution with $(n - 1)$ degrees of freedom, Equation 2.29). The derivation of $f(\bar{x} - \mu)$ is sufficiently complex that we do not present it here. It is useful, however, to sketch out the properties of the distribution by collecting our expectations for the distribution. This exercise helps us understand the answer from the literature, which we use to construct error limits.

**Characteristics of the pdf of the deviation of a sample mean from the true mean $(\bar{x} - \mu)$:**

1. For all sample sets, it seems reasonable that the most likely value of $\bar{x}$ is $\mu$ and thus that the most likely value of the deviation $(\bar{x} - \mu)$ is zero.
2. Since errors are random, positive and negative deviations are equally likely, and thus the probability density function of $(\bar{x} - \mu)$ is expected to be symmetric around zero.
3. We do not expect observed deviations $(\bar{x} - \mu)$ to be very large; thus the probability that $(\bar{x} - \mu)$ is large-positive or large-negative is very small.
4. If the standard deviation of a sample set $s$ is large, this suggests that random variations are large, and the probability density of the deviation $(\bar{x} - \mu)$ will be more spread out.
5. If the standard deviation of a sample set $s$ is small, this suggests that stochastic variations are small, and the probability density of the deviation $(\bar{x} - \mu)$ will be more tightly grouped near the maximum of the pdf, which is at $(\bar{x} - \mu) = 0$.
6. If the sample size $n$ is small, we know less about the variable $x$, and thus the probability density of the deviation $(\bar{x} - \mu)$ will be more spread out.
7. If the sample size $n$ is large, we know more about the variable $x$, and thus the probability density of the deviation $(\bar{x} - \mu)$ will be grouped closer to the maximum of the pdf, which is at $(\bar{x} - \mu) = 0$.

The probability density function of the deviation of the sample mean from the true mean has been worked out for the case when the underlying distribution is a normal distribution of unknown standard deviation [5, 16, 34, 38]. In agreement with our list of the characteristics of this pdf, the distribution depends on the sample standard deviation $s$ and the sample size $n$. To write the distribution compactly, it is expressed in terms of a dimensionless *scaled deviation*, $t$:

$$\text{Scaled deviation} \qquad \boxed{t \equiv \frac{(\bar{x} - \mu)}{s/\sqrt{n}}} \qquad (2.25)$$

where $s$ and $n$ are the sample standard deviation and sample size, respectively. The deviation $(\bar{x} - \mu)$ is scaled by the replicate standard error $s/\sqrt{n}$, producing the unitless stochastic variable $t$.

$$\text{Standard error of replicates} \qquad \boxed{e_{s,random} \equiv \frac{s}{\sqrt{n}}} \qquad (2.26)$$

The quantity $s/\sqrt{n}$ is also called the standard random error. For individual observations of the mean $\bar{x}$ of a sample set of size $n$, $t$ represents the number of standard errors the mean lies from the true value of the mean $\mu$ (Figure 2.8).

The quantity $s/\sqrt{n}$ in Equation 2.25 has meaning. It appears during consideration of the properties of the underlying distribution of $x$, the distribution we are sampling. When the standard deviation of the underlying normal distribution is known, its sampling distribution is also a normal distribution, and we can easily show (see Appendix E, Section E.4) that the standard deviation of the sample mean is $\sigma/\sqrt{n}$.

$$\begin{array}{c}\text{Normal} \\ \text{population} \\ \text{(known } \sigma)\end{array} \quad \begin{array}{c}\text{Standard} \\ \text{deviation} \\ \text{of the mean}\end{array} \quad = \frac{\sigma}{\sqrt{n}} \qquad (2.27)$$

When the standard deviation of the underlying distribution not known, the standard deviation of the mean is estimated by substituting $s$ for $\sigma$, where $s$ is the sample standard deviation.

$$\begin{array}{c}\text{Sampling normal} \\ \text{with unknown} \\ \text{standard deviation}\end{array} \quad \begin{array}{c}\text{Standard} \\ \text{deviation} \\ \text{of the mean} \\ \text{(estimate)}\end{array} \quad = \frac{s}{\sqrt{n}} \qquad (2.28)$$

The pdf of the sampling distribution of an underlying normal distribution of unknown standard deviation is not a normal distribution. The problem of determining the statistics of sampling an underlying normal distribution of unknown standard deviation was worked out in the late 1800s by William Sealy

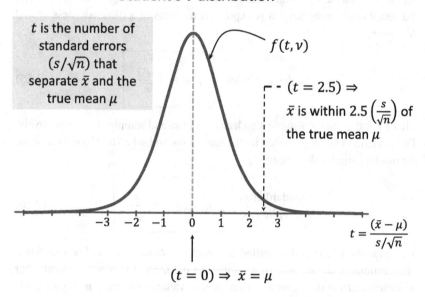

**Student's $t$ distribution**

$t$ is the number of standard errors $(s/\sqrt{n})$ that separate $\bar{x}$ and the true mean $\mu$

$f(t, \nu)$

$(t = 2.5) \Rightarrow$
$\bar{x}$ is within $2.5\left(\frac{s}{\sqrt{n}}\right)$ of the true mean $\mu$

$$t = \frac{(\bar{x} - \mu)}{s/\sqrt{n}}$$

$(t = 0) \Rightarrow \bar{x} = \mu$

Figure 2.8 The scaled variable $t$ indicates how many standard errors $s/\sqrt{n}$ separate the observed sample mean $\bar{x}$ and the true mean $\mu$. The Student's $t$ probability density distribution allows us to calculate the probability of observing values of $t$ within a chosen range.

Gosset, an analyst for the Guinness Brewery in Ireland, and the distribution is named for the pseudonym under which he published ("Student") [15, 16].

Student's $t$ distribution pdf of the scaled deviation $t$ for various degrees of freedom $\nu$

$$f(t, \nu) = \frac{\Gamma\left(\frac{\nu+1}{2}\right)}{\sqrt{\nu\pi}\,\Gamma\left(\frac{\nu}{2}\right)} \left(1 + \frac{t^2}{\nu}\right)^{-\left(\frac{\nu+1}{2}\right)}$$

(2.29)

The Student's $t$ distribution is the pdf of $t$, the scaled deviation of the sample mean from the true mean; it is written as a function of $t$ and $\nu$, where $\nu$ is called the degrees of freedom (Figure 2.9).[10] $\Gamma()$ is a standard mathematical function called the gamma function [54]. The Student's $t$ probability distribution function $f(t, \nu)$ is plotted in Figure 2.9 for various values of $\nu$, and we see that, as expected, it is symmetric, peaked at the center, and with very little density in its tails at low and high values of $t$. For the problem of sampling means

[10] Degrees of freedom are important when we are estimating parameters – it is important to avoid overspecifying a problem [5, 38].

## Student's *t* distribution

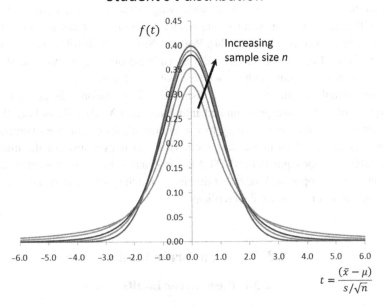

Figure 2.9 The Student's *t* distribution $f(t, \nu)$ describes how the value of the sample mean is distributed for repeated sampling of a normally distributed stochastic variable of unknown standard deviation $\sigma$. The Student's *t* probability distribution depends on the sample size *n*, and $f(t, \nu)$ gives the probability density as a function of the scaled deviation $t = \frac{(\bar{x}-\mu)}{s/\sqrt{n}}$. As *n* approaches infinity, the Student's *t* distribution approaches the standard normal distribution. Lines shown are for different sample sizes, expressed in terms of degrees of freedom $\nu = (n - 1)$, where the values of $\nu$ in the figure are $\nu = 1, 2, 5, 10$, and $\infty$.

for samples of size *n*, the applicable sampling distribution is the Student's *t* distribution with $\nu = (n - 1)$ degrees of freedom.

Recall our purpose: we seek to determine error limits on measurements of a stochastic variable *x*. The mean $\bar{x}$ of a set of replicates of the variable is a good estimate of the variable (Equation 2.20). To quantify variability in the sample set, we seek the sampling distribution of the mean of the sample set (Equation 2.29), which is based on the scaled deviation *t*. The use of deviation avoids the need to know the true mean $\mu$, which we do not know. The scaled deviation *t* (Equation 2.25) is scaled by an estimate of the standard deviation of the sample mean. This estimate is based on the more straightforward case when the standard deviation is known, the case when the underlying distribution is the normal distribution. Not knowing the standard deviation of the underlying distribution changes and complicates the pdf of the sampling distribution. When the standard deviation of the underlying distribution is not known, the pdf of the sampling distribution of the mean is the Student's $t$ distribution with $(n - 1)$ degrees of freedom [16, 34]. With the Student's $t$ distribution pdf and the methods of creating error limits discussed in Section 2.2, we are now able to determine error limits with known levels of confidence.

The formula for the Student's $t$ distribution in Equation 2.29 as well as integrals of $f(t, \nu)$ are programmed into Excel and MATLAB, making the Student's $t$ distribution very easy to use in error-limits calculations (several examples are provided in the next section). Excel is used most of the time; see Table D.1 for equivalent MATLAB commands). In the next section we discuss how to apply the Student's $t$ distribution to the problem of determining appropriate error limits for data replicates.

## 2.3  Replicate Error Limits

### 2.3.1  Basic Error Limits

This chapter is about quantifying replicate error, a type of *random* uncertainty present in our data. Replicate error shows up when we make repeated measurements of a stochastic variable – we do not get the same number every time. As discussed in the previous sections, random statistics teaches us that the average of repeated measurements is a good estimate of the true value of the variable. Statistics also guides us as to how to write error limits for this estimated value. The essential tool for writing limits due to random error is the Student's $t$ distribution, introduced in Section 2.2. This distribution allows us to quantify likely variability of the data based on the properties of a sample of the variable. In this section, we show how to use the Student's $t$ distribution to answer some very practical questions about uncertainty in experimental data.

We present two introductory examples, the first one addressing the question of a warrantied average value for a commodity and the second using the Student's $t$ distribution to assess different choices for the number of significant figures to report for an experimental result. These two initial examples lead to the definition of the most common form of the error limit, the 95% confidence interval, which we discuss in depth in Section 2.3.2.

**Example 2.4: Probability with continuous stochastic variables: guaranteeing the mean stick length.** *A shop sells sticks intended to be 6 cm long, and the vendor claims that the average stick length of his stock is between 5.5 cm and 6.5 cm. To assess this guarantee, we measure the lengths of 25 sticks and* *find that the lengths vary a bit. We calculate the mean length to be $\bar{x} = 6.0\,cm$, and the standard deviation of the sample set is $s = 0.70\,cm$. Based on these data, what is the probability that the true average stick length in the vendor's stock is between 5.5 and 6.5 cm, as claimed?*

**Solution:** This is a question about the mean of the distribution of stick lengths. The best estimate of average stick length is the sample mean, $\bar{x} = 6.0$ cm, which for our sample has just the value that the vendor hoped it would have. The lengths vary, however. Based on the sample of stick lengths obtained, how confident should the vendor be that the average stick length is between 5.5 and 6.5 cm? In terms of probability, based on the sample, what is the probability that the true population-average stick length is between 5.5 and 6.5 cm?

We can ask this question in terms of deviation: what is the probability that the maximum deviation of the sample mean $\bar{x}$ from the true mean of the population of stick lengths will be 0.5 cm?

$$\begin{array}{cc} \text{Maximum deviation} \\ \text{of the mean from the true} \end{array} \quad |(\bar{x} - \mu)_{max}| = 0.5 \text{ cm} \qquad (2.30)$$

The Student's $t$ distribution with $(n - 1)$ degrees of freedom expresses the probability of observing deviations of various magnitudes, if we have a sample from the population. The Student's $t$ distribution is the pdf for the stochastic variable $t$ (Equation 2.25). As we discussed in Section 2.2, the pdf of a continuous stochastic variable allows us to calculate the probability that the stochastic variable will take on a value within a range; thus, we are able to use the Student's $t$ distribution to calculate the probability that a value of $t$ lies within some range.

To translate our question about mean stick length into a question about values of $t$, we examine the definition of $t$:

$$\text{Scaled deviation } t \qquad t = \frac{(\bar{x} - \mu)}{s/\sqrt{n}} \qquad (2.31)$$

The quantities in the definition of $t$ refer to three properties of a sample of a stochastic variable (sample mean $\bar{x}$, sample standard deviation $s$, and number of observations in the sample $n$) and one property of the underlying population that has been sampled ($\mu$, the mean of the population). For the current discussion of stick lengths, we do not know the population mean $\mu$, but we know all three properties of a sample and we know the target maximum deviation $(\bar{x} - \mu)_{max}$. Thus, we can calculate a value of $t_{limit}$ associated with the target deviation. The question we are seeking to answer asks about the mean stick length deviation being no more than $\pm 0.5$ cm. Thus, for $t_{limit}$

calculated from the maximum deviation of 0.5 cm, the range of $t$ from $-t_{limit}$ to $+t_{limit}$ exactly expresses our mean-stick-length question: the probability that $t$ falls between $-t_{limit}$ and $+t_{limit}$ ($v = n - 1$) is the same probability that the population mean stick length $\mu$ will be between 5.5 cm and 6.5 cm, as claimed by the vendor.

Our first step, then, is to calculate $t_{limit}$, the scaled deviation that represents the maximum deviation of mean stick length.

$$t = \frac{(\bar{x} - \mu)}{s/\sqrt{n}} \tag{2.32}$$

Scaled
maximum
deviation
$$t_{limit} = \frac{(\text{maximum deviation})}{s/\sqrt{n}} \tag{2.33}$$

$$= \frac{(0.5 \text{ cm})}{0.70 \text{ cm}/\sqrt{25}}$$

$$= 3.57142 \quad (\text{unitless, extra digits shown})$$

The sampling distribution of the mean is the Student's $t$ distribution with $(n - 1)$ degrees of freedom. The probability that $-t_{limit} \le t \le t_{limit}$ is given by the integral in Equation 2.12, with the pdf function being the pdf of the Student's $t$ distribution and with the limits given by $\pm t_{limit}$ and for $v = (n - 1) = 24$ (Figure 2.10).

Probability that
scaled deviation $t$
is between $- t_{limit}$ and $t_{limit}$
$$\boxed{\Pr = \int_{-t_{limit}}^{+t_{limit}} f(t'; n - 1)dt'} \tag{2.34}$$

$$= \int_{-3.57142}^{3.57142} f(t'; 24)dt' \tag{2.35}$$

This integral over the pdf of the Student's $t$ distribution, and hence the probability we seek, is readily calculated in Excel, as we now discuss.

Carrying out integrals of the Student's $t$ probability density distribution is a very common calculation in statistics; Excel has built-in functions that evaluate integrals of $f(t, v)$ for the Student's $t$ distribution, and we discuss now how to use these to evaluate Equation 2.35. The Excel function T.DIST.2T($t_{limit}, n-1$) is called the two-tailed cumulative probability distribution of the Student's $t$ distribution. For the Student's $t$ distribution, the function T.DIST.2T($t_{limit}, n-1$) gives the area underneath the two probability-density tails that are outside the interval in which we are interested (Figure 2.10):

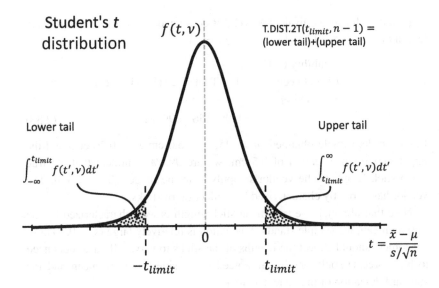

Figure 2.10 The integral under the Student's $t$ distribution pdf from $-\infty$ to $+\infty$ is 1; if we integrate from $-t_{limit}$ to $t_{limit}$ we leave behind two tails, each with the same amount of probability (since the distribution is symmetric). The area under the curve between $\pm t_{limit}$ is just the area in the two tails subtracted from the total area, which is 1.

$$\text{T.DIST.2T}(t_{limit}, n - 1) \equiv \int_{-\infty}^{-t_{limit}} f(t'; n - 1)dt' + \int_{t_{limit}}^{\infty} f(t'; n - 1)dt'$$

$$(2.36)$$

Thus, the probability we seek in Equation 2.35 is the difference between the total area under the curve, which is equal to 1, $(\int_{-\infty}^{\infty} f(t', n - 1)dt' = 1)$, and the value yielded by the Excel two-tailed function.

Probability that
scaled deviation $t$ is $\equiv \Pr[-t_{limit} \leq t \leq t_{limit}] = \int_{-t_{limit}}^{+t_{limit}} f(t'; n - 1)dt'$
between $- t_{limit}$ and $t_{limit}$

$$\boxed{\Pr[-t_{limit} \leq t \leq t_{limit}] = 1 - \text{T.DIST.2T}(t_{limit}, n - 1)} \qquad (2.37)$$

The availability of the function T.DIST.2T() in Excel makes calculating the probability we seek a matter of entering a simple formula into Excel (see

Appendix D for the equivalent MATLAB command.). For the sample of stick length $t_{limit} = 3.57142$ and $v = (n - 1) = 24$:

Probability that
$t$ is between    $= 1 - T.DIST.2T(3.57142, 24)$
$-t_{limit}$ and $t_{limit}$

$$= 0.998456 = 99.8\% \qquad (2.38)$$

Based on the sample obtained ($n = 25; \bar{x} = 6.0$ cm; $s = 0.70$ cm) and the target maximum deviation of 0.5 cm, we are 99.8% confident that the true mean stick length of the vendor's supply is in the range $6.0 \pm 0.5$ cm. The vendor has correctly characterized his collection of sticks.

Note that although the true mean stick length is well characterized by the sample mean, the sample standard deviation is relatively large ($s = 0.70$ cm). Customers should expect the lengths of the sticks to vary.[11] If customers need to have precise stick lengths, they need to consider both the mean and the standard deviation of the vendor's stock.

Example 2.5 applies the Student's $t$ distribution to the problem of determining the number of significant figures to associate with a measurement of density. Appendix C contains a list of Excel functions that are useful for error-limit calculations.

**Example 2.5: Play with Student's $t$: sig figs on measured fluid density.**
*A team of engineering students obtained 10 replicate determinations of the density of a 20 wt% aqueous sugar solution: $\rho_i (g/cm^3) = 1.0843$, 1.06837, 1.07047, 1.0635, 1.09398, 1.0879, 1.07873, 1.05692, 1.07584, 1.07587 (extra digits from the calculator are reported to avoid downstream round-off error). Given the variability of the measurements, how may significant figures should we report in our answer?*

**Solution:** The value of the mean density that we calculate from the students' data is $\bar{x} = \bar{\rho} = 1.075588$ g/cm$^3$, but the measurements ranged from 1.05692 to 1.09398 g/cm$^3$, and the standard deviation of the sample set is $s = 0.011297$ g/cm$^3$ (extra digits shown). When calculating the average density from the data, the computer gives 32 digits, but clearly we cannot justify that degree of precision in our reported answer, given the variability of the data.

---

[11] In fact, for this vendor's stock, the probability of actually having the length of a stick chosen at random fall between 5.5 and 6.5 cm is only 51%; see Problem 2.32. We know the mean of the stick population, with high confidence, to be between 5.5 and 6.5 cm, but the distribution of stick lengths leading to that well-characterized mean is broad.

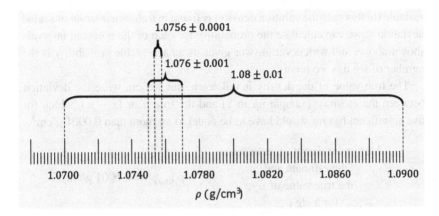

Figure 2.11 Error limits shown correspond to density reported to five sig figs (narrowest curly brackets), four sig figs, and three sig figs. When we specify an answer within the significant figures convention, it implies that the true value will be found in the range created by toggling the least certain digit by 1. When we specify fewer significant figures, we are presenting a broader range and indicating we are less certain of the true value.

Since the question concerns significant figures, can we perhaps apply the sig-figs rules from Appendix B? Unfortunately, these rules are only applicable when values of known precision are combined or manipulated. This is not our current circumstance; rather, the uncertainty in density is due to the sample-to-sample variability of the data (replicate variability). We must determine the uncertainty from the variability of the data and subsequently assign the correct number of significant figures.

We can address the question in this example by returning to the fundamental meaning of the sig-figs convention and thinking in terms of deviation $(\bar{x} - \mu)$. The sig figs idea is that the last digit retained is uncertain by plus or minus one digit. Another way of expressing the question of this example is as follows: what are the probabilities that the true value of the density is found in the following intervals (Figure 2.11):

$$\text{true value} \overset{?}{=} 1.0756 \pm 0.0001 \text{ g/cm}^3 \text{ (5 sig figs)}$$

$$\text{true value} \overset{?}{=} 1.076 \pm 0.001 \text{ g/cm}^3 \text{ (4 sig figs)}$$

$$\text{true value} \overset{?}{=} 1.08 \pm 0.01 \text{ g/cm}^3 \text{ (3 sig figs)}$$

The three $\pm$ values are three possible values of maximum deviation $(\bar{x} - \mu)_{max}$. The sampling distribution of the sample mean [the Student's $t$ distribution, $\nu = (n - 1)$] allows us to assess the probability that the true value of the variable (in this case the solution density) is found in a chosen interval of scaled deviation $t$; we can calculate the probability for each of the potential intervals shown above, and whichever answer gives us an acceptable probability is the number of sig figs we report.

The true value of the density is unknown, but we can write the deviation between the estimate (sample mean $\bar{x}$) and the true $\mu$ as $(\bar{x} - \mu)$, which for five significant figures would have to be equal to no more than 0.0001 g/cm$^3$.

$$\begin{array}{c} \text{Maximum deviation between} \\ \text{the estimate and} \\ \text{the true value of } \rho \\ \text{for 5 sig figs} \end{array} \qquad \left|(\bar{x} - \mu)_{max}\right| = 0.0001 \text{ g/cm}^3$$

The variable $t$ in the Student's $t$ distribution is the deviation $(\bar{x} - \mu)$ expressed in units of replicate standard error $s/\sqrt{n}$, which is based on sample properties (sample standard deviation $s$, number of samples $n$).

$$\text{Scaled deviation} \qquad t = \frac{(\bar{x} - \mu)}{s/\sqrt{n}}$$

We know the replicate standard error $s/\sqrt{n}$ from the dataset, and thus we can determine the value of the scaled deviation $t_{limit}$ such that the dimensionless interval between $-t_{limit}$ and $t_{limit}$ corresponds to a maximum deviation of 0.0001 g/cm$^3$.

$$\begin{array}{c} \text{Scaled} \\ \text{maximum} \\ \text{deviation} \end{array} \qquad t_{limit} = \frac{\text{max deviation}}{s/\sqrt{n}} = \frac{(\bar{x} - \mu)_{max}}{s/\sqrt{n}} \qquad (2.39)$$

$$t_{limit, 10^{-4}} = \frac{0.0001 \text{ g/cm}^3}{0.01129736 \text{ g/cm}^3/\sqrt{10}} = 0.02799 \text{ (unitless)}$$

The probability that, when we take a sample, the observed scaled deviation $t$ will be in the interval $-t_{limit} \le t \le t_{limit}$ is given by the area under the pdf of the Student's $t$ distribution with $v = (n - 1) = 9$ in the interval between $-t_{limit}$ and $t_{limit}$.

$$\text{Pr} = \int_{-t_{limit}}^{t_{limit}} f(t', v) dt' \qquad (2.40)$$

Using the same Excel function introduced in Example 2.4, and for the three sig-figs intervals under consideration [corresponding to maximum deviations $(\bar{x} - \mu)$ of $10^{-4}$, $10^{-3}$, and $10^{-2}$ g/cm$^3$], we obtain the following probabilities.

For five significant figures:

$$t_{limit, 10^{-4}} = \frac{(\bar{x} - \mu)_{max}}{s/\sqrt{n}}$$

$$= \frac{0.0001}{3.5725 \times 10^{-3}} = 0.02799$$

$$\int_{-0.02799}^{0.02799} f(t'; 9)dt' = 1 - \text{T.DIST.2T}(0.02799, 9) = \boxed{2\% \text{ (5 sig figs)}}$$

For four significant figures:

$$t_{limit, 10^{-3}} = \frac{0.001}{3.5725 \times 10^{-3}} = 0.2799$$

$$\int_{-0.2799}^{0.2799} f(t'; 9)dt' = 1 - \text{T.DIST.2T}(0.2799, 9) = \boxed{21\% \text{ (4 sig figs)}}$$

For three significant figures:

$$t_{limit, 10^{-2}} = \frac{0.01}{3.5725 \times 10^{-3}} = 2.799$$

$$\int_{-2.799}^{2.799} f(t'; 9)dt' = 1 - \text{T.DIST.2T}(2.799, 9) = \boxed{98\% \text{ (3 sig figs)}}$$

The calculation shows us that intervals associated with five significant figures ($\pm 10^{-4}$ g/cm$^3$, only 2% confidence of capturing the true value; review Figure 2.7) and four significant figures ($\pm 10^{-3}$ g/cm$^3$, just 21% confidence) are not justified; based on the variability of the data, we should report no more than three significant figures or expect a deviation of at least $\pm 10^{-2}$ g/cm$^3$ if we want to be reasonably sure that the reported interval includes the true value of the density, given the variability of the observations. Specifically, we can be 98% confident that the true value of the density is within the interval associated with reporting three significant figures. If we choose to report four significant figures, we are taking a substantial risk, as the sample statistics imply that we should only have 21% confidence that we will bracket the true value of the density with this narrower choice.

From Examples 2.4 and 2.5, we see that we can answer some interesting and practical questions with the Student's $t$ distribution and Excel. While sig-figs rules are helpful when manipulating quantities of known uncertainty, sampling gives us direct access to the variability of a quantity. The Excel (or MATLAB) functions make using the statistics fast and easy.

At the beginning of Section 2.2.3 we defined our restated goal as

Restated goal: to determine error limits
on an estimated value that, 95%         Answer = (value) ± (error limits)
of the time, captures the true value of $x$

$$(2.41)$$

We have made progress on this goal. Based on a sample $(n, \bar{x}, s)$, we now know how to calculate the probability of finding the true mean $\mu$ in a chosen interval. The steps are given here.

**Calculate the likelihood of finding the true mean $\mu$ in a chosen interval, based on a sample of size $n$ with mean $\bar{x}$ and standard deviation $s$:**

1. Choose the magnitude of the maximum deviation $|(\bar{x} - \mu)_{max}|$. This determines the chosen interval for the error limits.
2. Calculate the maximum scaled deviation $t_{limit}$ from $|(\bar{x} - \mu)_{max}|$, the definition of $t$, and the sample properties $n$ and $s$ (Equation 2.39, repeated here).

$$\text{Scaled maximum deviation} \quad t_{limit} = \frac{\text{max deviation}}{s/\sqrt{n}} = \frac{(\bar{x} - \mu)_{max}}{s/\sqrt{n}} \quad \text{(Equation 2.39)}$$

3. Use Equation 2.34 (repeated here) to calculate the probability that the observed deviation $|(\bar{x} - \mu)|$ is no larger than the chosen value of maximum deviation.

$$\text{Probability that } t \text{ is between } -t_{limit} \text{ and } t_{limit} = \int_{-t_{limit}}^{+t_{limit}} f(t'; n - 1)dt' \quad \text{(Equation 2.34)}$$

$$= 1 - \text{T.DIST.2T}(t_{limit}, n - 1)$$

4. Report the answer for the predicted mean as $\bar{x} \pm$ (max deviation) at a confidence of (result).

The definition of the scaled sampling deviation $t$ and the identification of the Student's $t$ distribution (with $\nu = n - 1$) as the appropriate probability density distribution for the sampling distribution of the mean, when the standard deviation is unknown, are the advances that make these error-limit and probability calculations possible.

Calculating the probability for a chosen maximum deviation clarifies error-limit and sig-figs choices, but choosing the maximum deviation (step 1) is not always the preferred way of addressing error questions. We would prefer to choose the confidence level with which we are comfortable and turn the problem around and calculate the limits $\pm t_{limit}$ that correspond to the chosen confidence level. We show how to do this in Section 2.3.2. The probability we choose is 95%, and the interval calculated, when expressed in terms of error limits on $\bar{x}$, is called the 95% confidence interval of the mean. In the next section we discuss the issue of how to back-calculate $t_{limit}$ from sample sets to obtain 95% confidence intervals.

We have a final comment on the validity of using the Student's $t$ distribution for estimating sampling properties of the mean. The use of the Student's $t$ distribution is based on the assumption that the underlying distribution of $x$ is the normal distribution with unknown standard deviation, but rigorous calculations show that even if the underlying distribution is not normal, if it is at least a centrally peaked distribution, we may continue to use the Student's $t$ distribution as the sampling distribution of the sample mean [5, 38].

## 2.3.2  Confidence Intervals of the Mean

The sig-figs convention, based on plus/minus "1" in the last digit, is a coarse expression of error limits that we have seen may not precisely reflect the likely uncertainty in measurements. In Example 2.5 the four significant figures choice for average density (that is, chosen maximum deviation of $\pm 10^{-3}$ g/cm$^3$) gave a too low amount of confidence at 21% (the true value is captured within this range only about 1 in 5 times that a sample is processed), but the three significant figures choice ($\pm 10^{-2}$ g/cm$^3$) forced us to the perhaps too conservative 98% confidence level. The jump from 21% confidence to 98% confidence was rather abrupt, and it was forced by thinking of error limits as having to be plus/minus "1" of a decimal place. It may make more sense to choose our confidence level and let the plus/minus increment be whatever corresponds to that confidence level. We explore this approach in the next example.

**Example 2.6: Error limits driven by confidence level: measured fluid density, revisited.** *A team of engineering students obtained 10 replicate determinations of the density of a 20 wt% aqueous sugar solution (see Example 2.5 for the data). With 95% confidence and based on their data, what is the value of the density along with appropriate error limits on the determined value? (Answer is a range.)*

**Solution:** The calculation of sample mean density $\bar{\rho}$ and sample standard deivation $s$ from the data proceeds as in Example 2.5, and the remaining problem is to construct the appropriate error limits such that the probability of capturing the true value of density is 95%. As with the calculations in Example 2.5, we arrive at a confidence value by integrating the probability density function of the sampling distribution between $-t_{limit}$ and $t_{limit}$.

$$\Pr[-t_{limit} \le t \le t_{limit}] = \int_{-t_{limit}}^{t_{limit}} f(t'; n-1)dt' \qquad (2.42)$$

where $f(t; n-1)$ is the pdf of the Student's $t$ distribution with $(n-1)$ degrees of freedom. Previously we chose the maximum deviation $|(\bar{x} - \mu)_{max}|$, determined $t_{limit}$ from the maximum deviation and properties of the sample, and calculated the probability of observing that deviation from the Student's $t$ distribution. For the current example, the desired confidence is given as 95%; $\Pr = 0.95$ is thus the value of the integral in Equation 2.42. What is not known in this case is the value of the scaled deviation $t_{limit}$ to be used in the limits of the integration so as to obtain that chosen value of the probability.

$$\Pr[? \le t \le ?] = 0.95 = \int_{?}^{?} f(t'; n-1)dt' \qquad (2.43)$$

To determine the 95% probability limits, we must back-calculate $t_{limit}$ from the value of the integral (that is, 0.95) so that Equation 2.43 holds. Once we know $t_{limit}$, we can write the range in which we expect to find the true value of the density as follows:

$$\begin{array}{c} \text{Maximum} \\ \text{scaled} \\ \text{deviation} \end{array} \quad t_{limit} = \frac{|(\bar{x} - \mu)|_{max}}{s/\sqrt{n}}$$

$$\begin{array}{c} \text{Maximum deviation} \\ \text{(error limits)} \end{array} \quad \pm(\bar{x} - \mu) = \pm t_{limit}\left(\frac{s}{\sqrt{n}}\right) \qquad (2.44)$$

Solving Equation 2.44 for $\mu$, we expect, with 95% confidence, the true mean $\mu$ will be in this interval:

$$\mu = \bar{x} \pm t_{limit}\frac{s}{\sqrt{n}} \qquad (2.45)$$

Excel performs the inversion we seek (Equation 2.43) with its function T.INV.2T($\alpha$, $n-1$) (see also Appendices C and D). The significance level $\alpha$ is defined as 1 minus the confidence level (probability) sought.

$$\text{Significance level} \quad \boxed{\alpha \equiv 1 - \text{Pr}} \quad (2.46)$$

For 95% confidence, $\alpha = 0.05$. The parameter $\nu = (n - 1)$ is the number of degrees of freedom of the sampling process. The general integral of interest in this type of problem is

Probability that
$t$ is between $\qquad \text{Pr}[-t_{limit} \le t \le t_{limit}]$
$-t_{limit}$ and $t_{limit}$

$$\boxed{\text{Pr} = (1 - \alpha) = \int_{-t_{limit}}^{t_{limit}} f(t', \nu) dt'} \quad (2.47)$$

The back-calculation in Equation 2.47 of $t_{limit}$ from known $\alpha$ and $\nu = (n - 1)$ is performed by Excel or MATLAB:

Calculate limits $\pm t_{limit}$
from Equation 2.47
for set significance $\alpha$ $\qquad \boxed{t_{limit} = t_{\frac{\alpha}{2}, \nu} = \text{T.INV.2T}(\alpha, \nu)} \quad (2.48)$
(Excel)

(MATLAB) $\qquad \boxed{t_{limit} = t_{\frac{\alpha}{2}, \nu} = -\text{tinv}\left(\frac{\alpha}{2}, \nu\right)} \quad (2.49)$

We explain the subscript "$\alpha/2$" nomenclature in the discussion that follows.

The value of $t_{limit}$ we seek is the value that will give an area of $(1 - \alpha)$ when $f(t', n - 1)$ is integrated between $-t_{limit}$ and $t_{limit}$ (Figure 2.12). There are two pdf "tails" containing excluded areas: one between $-\infty$ and $-t_{limit}$ and one between $t_{limit}$ and $\infty$ (compare with Equation 2.36). The total area in the two tails is $\alpha$. In the Excel function T.INV.2T() (2T="two tailed"), one variable we specify is the total amount of probability in the two tails ($\alpha$); the other variable is the number of degrees of freedom $\nu = (n - 1)$. The area under $f(t; \nu)$ below $-t_{limit}$ is $\alpha/2$; thus the convention is to write $t_{limit} = t_{\frac{\alpha}{2}, \nu}$.

We now apply this calculation to the current problem of fluid density error limits. For 95% confidence ($\alpha = 0.05$), with $\nu = (n - 1) = 9$ degrees of freedom, the $t_{limit}$ we obtain is:

$$t_{limit} = t_{0.025, 9} = \text{T.INV.2T}(0.05, 9)$$
$$= 2.262157$$

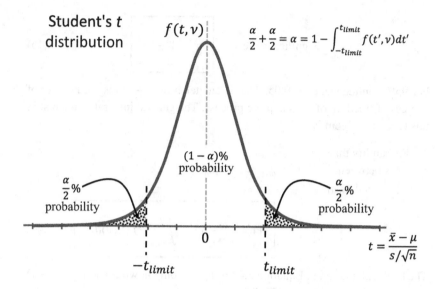

**Figure 2.12** For confidence intervals, we are interested in knowing how far out toward the tails we need to go to capture $(1 - \alpha)\%$ of the probability between the limits $-t_{limit}$ and $t_{limit}$. The small probability that resides in the two tails represents improbable observed values of the sample mean.

Once again the calculation is reduced to a simple function call in Excel or MATLAB. From $t_{limit}$ we now calculate the range in which we expect to find the true value of the density.

$$\text{Maximum scaled deviation} \qquad t_{limit} = \frac{(\bar{x} - \mu)_{max}}{s/\sqrt{n}}$$

$$\mu = \bar{x} \pm t_{limit} \frac{s}{\sqrt{n}}$$

$$= 1.075588 \pm 2.262157 \left( \frac{0.01129736}{\sqrt{10}} \right)$$

$$\begin{array}{c} \text{20 wt\% solution density} \\ \text{(95\% confidence)} \end{array} = \boxed{1.076 \pm 0.008 \text{ g/cm}^3} \qquad (2.50)$$

This interval is shown in Figure 2.13 along with the sig figs–based error limits from Example 2.5. The 95% confidence interval is more precise than the broad, three sig-figs error limits, while still corresponding to a very reasonable (and known) confidence level, 95%.

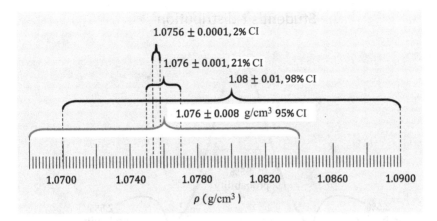

Figure 2.13 Confidence intervals from Examples 2.5 and 2.6. The confidence level varies depending on the error limits chosen. With 95% confidence intervals, we specify a confidence level and calculate the error limits.

Note that in writing our answer in Equation 2.50, we retain only one digit on error. Also, the error is in the third digit after the decimal, and this uncertainty determines how many digits we report for the density (we usually keep only one uncertain digit). For more on significant figures when writing error limits (including exceptions to the one-uncertain-digit rule), see Section 2.4.

The process followed in Example 2.6 is a statistical way of knowing the stochastic "give and take" amounts we mentioned earlier in the chapter. A common choice is to report the amount of "give and take" that will, with 95% confidence, make your estimate right: "right" means we are 95% confident that the calculated range captures the true value of the mean of the distribution, $\mu$. The range that, with 95% confidence, includes the true value of the mean is called a *95% confidence interval of the mean* (Figure 2.14).

$$\begin{array}{l} \text{95\% confidence interval} \\ \text{of the mean} \end{array} = \left( \begin{array}{c} \text{range that, with 95\% confidence,} \\ \text{contains } \mu, \text{ the true mean} \\ \text{of the underlying distribution} \end{array} \right) \quad (2.51)$$

The 95% confidence interval[12] of the mean is the usual way to determine error limits when only random errors are present. For a stochastic variable

___

[12] If we create a large number of 95% confidence intervals from different samples, 95% of them will contain the true mean $\mu$. The range for any *one sample*, however, either does or does not contain $\mu$. Thus, the probability that any one confidence interval contains the true mean $\mu$ is not 95%: It is either one or zero. Rather, we are 95% confident that the mean is in that interval.

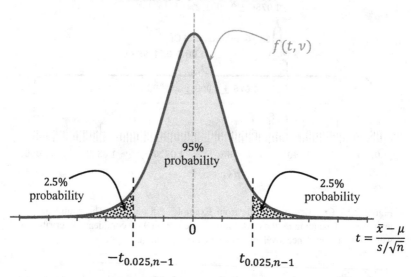

## Student's $t$ distribution

Figure 2.14 The Student's $t$ probability distribution, $f(t; \nu)$, is used to construct 95% confidence intervals on the mean. The central 95% region represents the most likely values we will observe for the sample mean in a sample of size $n$. The 5% of the probability that resides in the two tails represents improbable observed values of sample mean – improbable, but not impossible. They are observed about 5% of the time (they will be observed 1 in 20 times that a sample of size $n$ is tested).

$x$ sampled $n$ times with sample mean $\bar{x}$ and sample variance $s^2$, the 95% confidence interval of the mean is calculated as follows:

95% confidence interval
of the mean $\bar{x}$
(replicate error only)

$$\mu = \bar{x} \pm t_{0.025, n-1} e_{s, random} \qquad (2.52)$$

$$e_{s, random} \equiv \frac{s}{\sqrt{n}}$$

$$t_{0.025, n-1} = \text{T.INV.2T}(0.05, n-1) \qquad \text{Excel}$$

$$= -tinv(0.025, n-1) \qquad \text{MATLAB}$$

where $e_{s, random} = \frac{s}{\sqrt{n}}$ is the standard error on the mean of $x$ and $t_{0.025, n-1}$ is a value associated with the Student's $t$ distribution that ensures that 95% of the probability in the distribution is captured between the limits given in Equation 2.52. Three-digit values of $t_{0.025, n-1}$ are given in Table 2.3 as a function of sample size $n$; these values are calculated more precisely with

Table 2.3. *The Student's t distribution approximate values of $t_{0.025, n-1}$ for use in constructing 95% confidence intervals for samples of size n. The numbers in **bold** are equal to "**2**" to one digit. Accurate values of $|t_{0.025, n-1}|$ may be calculated with Excel's function call T.INV.2T(0.05, n − 1) or with −tinv(0.025, n − 1) with MATLAB.*

| $n$ | 2 | 3 | 4 | 5 | 6 | 7 | 8 | 9 | 10 | 20 | 50 | 100 | ∞ |
|---|---|---|---|---|---|---|---|---|---|---|---|---|---|
| $n-1$ | 1 | 2 | 3 | 4 | 5 | 6 | 7 | 8 | 9 | 19 | 49 | 99 | ∞ |
| $t_{0.025, n-1}$ | 12.71 | 4.30 | 3.18 | 2.78 | 2.57 | **2.45** | **2.36** | **2.31** | **2.26** | **2.09** | **2.01** | 1.98 | 1.96 |

Excel using the function call T.INV.2T(0.05, $n$ − 1).[13] Note that for large $n$, $t_{0.025, n-1}$ approaches a value of about 2.

For practice, we now apply the Student's $t$ distribution and Equation 2.52 to determine 95% confidence intervals on the commuting-time measurements in Examples 2.1 and 2.2.

**Example 2.7: Commuting time, revisited.** *Over the course of a year, Eun Young took 20 measurements of her commuting time under all kinds of conditions. Ten of her observations are shown in Table 2.1 (Example 2.1) and 10 other measurements taken using the same stopwatch and the same timing protocols are shown in Table 2.2 (Example 2.2). Calculate an estimate of Eun Young's time-to-commute first using the first dataset and then using all 20 data points. Express your answers for the estimates of commuting time with error limits consisting of 95% confidence intervals of the mean.*

**Solution:** Using the first set of $n = 10$ data points found in Table 2.1, we employ Equation 2.52 with $\bar{x} = 28$ min, $s = 11$ min, and $t_{0.025, 9} = 2.26$ [T.INV.2T(0.05, 9)]. The 95% confidence interval of the mean is:

$$
\begin{array}{c}
\text{95\% confidence interval} \\
\text{of the mean} \\
\text{Set 1; } n = 10
\end{array}
\qquad \mu = \bar{x} \pm t_{0.025, n-1} \frac{s}{\sqrt{n}}
$$

$$
= 28 \pm (2.26) \frac{11}{\sqrt{10}}
$$

$$
= 28 \pm 8 \text{ min}
$$

Using all $n = 20$ data points found in Tables 2.1 and 2.2, we calculate $\bar{x} = 30$ min, $s = 10$ min, and from T.INV.2T(0.05, 19) we find $t_{0.025, 19} = 2.09$. The 95% confidence interval of the mean for these data is as follows:

[13] In this formula, the $\alpha = 0.05$ corresponds to the 95% confidence level selected; for a 99% confidence level, replace 0.05 with 0.01.

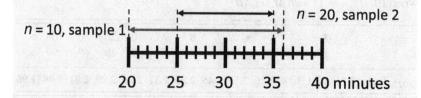

Figure 2.15 The first dataset with $n = 10$ produced a wider 95% confidence interval than the second dataset with $n = 20$. The two predictions overlap and thus are consistent in their predictions for the true value of the mean of the variable, $\mu$.

$$
\begin{aligned}
& \text{95\% confidence interval} \\
& \text{of the mean} \\
& \text{Combined set; } n = 20
\end{aligned}
\qquad \mu = \bar{x} \pm t_{0.025,\,n-1}\frac{s}{\sqrt{n}}
$$

$$
= 30 \pm (2.09)\frac{10}{\sqrt{20}}
$$

$$
= 30 \pm 5 \text{ min}
$$

Taking a sample size of 10 yielded an estimate of between 20 and 36 min for the mean commuting time (Figure 2.15, sample 1). Taking a sample of size 20 yielded an estimate that agrees that the true mean is between those two numbers, but that narrows the prediction to be between 20 and 30 min at the same level of confidence (Figure 2.15, sample 2). Because of the $\sqrt{n}$ in the denominator of Equation 2.52, increasing $n$ narrows the confidence interval and more precisely predicts the mean commuting time. Note that in neither case could we rely on sig-figs rules to express the appropriate precision. The true uncertainty is reflected in the variability of the data and had to be determined through sampling.

As Example 2.7 shows, when we have data replicates, it is straightforward to use software to calculate an expected value of the mean and a 95% confidence interval for that value. For predictions of the sample mean, obtaining more replicates narrows the range of the confidence interval (makes us more precise in our estimate of the true value of the mean at the same level of confidence).

$$\text{Predicted value} \atop \text{of the mean:}$$
$$\mu = \bar{x} \pm t_{0.025, n-1} \frac{s}{\sqrt{n}}$$
$$= \lim_{n \to \infty} \left( \bar{x} \pm t_{0.025, n-1} \frac{s}{\sqrt{n}} \right)$$
$$= \bar{x} \tag{2.53}$$

The 95% confidence interval quantifies the "give or take" that should be associated with an estimate of the mean of a stochastic quantity such as commuting time.

Note that the current discussion has focused on error limits on the mean. As $n$ increases to infinity, we become certain of the mean. One example we have discussed is the typical commuting time for Eun Young – the answer for typical commuting time is the mean commuting time and the error limits are the error limits on the mean. A different but related question is, what duration do we expect for Eun Young's commute *tomorrow*? This is a question about a "next" value for a variable. The answer to this question is also the mean of the dataset, but the variability is different – the variability expected in tomorrow's commuting time is much larger than the variability expected for the mean. More replicates refine our estimate of the mean, but larger samples do not make Eun Young's commute less variable from day to day. To address a "next" value of a variable, we use a prediction interval rather than a confidence interval; see Section 2.3.3 and Example 2.15.

In Example 2.8 we carry out another 95% confidence interval calculation for laboratory data.

**Example 2.8: Density from laboratory replicates.** *Ten student groups measure the density of Blue Fluid 175 following the same technique: using an analytical balance they weigh full and empty $10.00 \pm 0.04$ ml pycnometers.*[14] *Their raw results for density are given in Table 2.4. Note that they used ten different pycnometers. With 95% confidence, what is the density of the Blue Fluid 175 implied by these data? Assume only random errors are present.*

**Solution:** We use Excel to calculate the sample mean and standard deviation of the data in Table 2.4: the mean of the dataset is $\bar{\rho} = 1.73439$ g/cm$^3$ and the standard deviation is $s = 0.00485$ g/cm$^3$ (excess digits have been retained to avoid round-off error in downstream calculations). According to our understanding of the distribution of random events, with 95% confidence, the true value of a variable $x$ measured $n$ times is within the range $\bar{x} \pm t_{0.025, n-1} \frac{s}{\sqrt{n}}$

---

[14] For more on pycnometers, see Figure 3.13 and Examples 3.8 and 4.4.

2 Quick Start

Table 2.4. *Raw data replicates of the*
*room-temperature density of Blue Fluid*
*175 obtained by ten student groups.*

| Index | Density, $g/cm^3$ |
|-------|-------------------|
| 1 | 1.7375 |
| 2 | 1.7272 |
| 3 | 1.7374 |
| 4 | 1.7351 |
| 5 | 1.73012 |
| 6 | 1.7377 |
| 7 | 1.7398 |
| 8 | 1.72599 |
| 9 | 1.7354 |
| 10 | 1.7377 |

(Equation 2.52); the quantity $t_{0.025,n-1}$ is given in Table 2.3 (or calculated from the Student's $t$ distribution with Excel or MATLAB) as a function of $n$. For the density data in Table 2.4, we calculate:

$$n = 10$$

$$t_{0.025,9} = 2.262157 \quad [\text{using Excel: T.INV.2T(0.05,9)}]$$

$$\text{density} = \bar{x} \pm t_{0.025,9} \frac{s}{\sqrt{n}}$$

$$= 1.734391 \pm (2.262157) \left( \frac{0.004853}{\sqrt{10}} \right)$$

$$\boxed{\rho_{BF175} = \quad 1.734 \pm 0.003 \text{ g/cm}^3 \text{ (95\% CI)}}$$

From the data collected, the expected value of the density of the Blue Fluid 175 is $1.734 \pm 0.003$ g/cm$^3$ with 95% confidence. We are 95% confident that the true value of the density of the fluid is within this interval. Note that the calculated 95% confidence interval is $\pm 3$ in the last digit ($\pm 0.003$ g/cm$^3$). We cannot express this precise level of uncertainty with significant figures alone. Without the error limits, reporting $\rho = 1.734$ g/cm$^3$ implies by sig figs that the uncertainty is $\pm 0.001$ ("1" in the last digit), which is overly optimistic.

If we choose to report the uncertainty as $\pm 0.001$ g/cm$^3$, we can calculate the confidence we should have in such an answer, as we see in the next example.

**Example 2.9: Density from laboratory replicates: tempted by four sig figs.**
*In Example 2.8 we calculated the 95% confidence interval associated with some measurement replicates of density. If we instead take a shortcut and guess that we may report the final answer to 4 sig figs as $\rho = 1.734$ g/cm$^3$, what is the confidence level associated with this answer?*

**Solution:** Error limits of $\pm 0.001$ g/cm$^3$ imply a maximum deviation from the true value of $|(\bar{x} - \mu)_{max}| = 0.001$ g/cm$^3$. To calculate the probability of such a maximum deviation, we integrate the pdf of the Student's $t$ distribution between limits of $\pm t_{limit}$, where $t_{limit}$ is calculated from the maximum deviation (see Example 2.5).

$$\begin{aligned}
\text{Maximum scaled deviation} \quad t_{limit} &= \frac{|(\bar{x} - \mu)_{max}|}{s/\sqrt{n}} \\[2mm]
&= \frac{(0.001 \text{ g/cm}^3)}{0.004853 \text{ g/cm}^3/\sqrt{10}} \\[2mm]
&= 0.651612953 \\[2mm]
\text{Pr} &= \int_{-t_{limit}}^{t_{limit}} f(t'; n-1)dt' \\[2mm]
&= 1 - \text{T.DIST.2T}(t_{limit}, n-1) \\[2mm]
&= 1 - \text{T.DIST.2T}(0.6516129, 9) \\[2mm]
&= 47\%
\end{aligned}$$

Our confidence is less than 50% that the true value of the density is found in the narrow interval $\rho = 1.734 \pm 0.001$ g/cm$^3$.

Determining uncertainty of replicates is essential in scientific and engineering practice, so it is worthwhile to become familiar with the basic features of the Student's $t$ distribution. Looking at the values in Table 2.3, we see that for two replicates ($n = 2, \nu = 1$) and considering only one digit of precision, we must bracket almost 13 standard errors of the mean to be 95% sure that we have captured the true mean ($t_{0.025,1} \approx 13$, Figure 2.16). Increasing from $n = 2$ replicates to $n = 3$ replicates, however, reduces the number of standard errors needed for 95% confidence to about $t_{0.025,2} \approx 4$, which is a big improvement. Going to $n = 4$ narrows the confidence interval still further to $t_{0.025,3} \approx 3$ standard errors, which is also approximately the value of $t_{0.025,n-1}$ for $n = 5$ and 6 (to one digit). For $n \geq 7$, the value of $t_{0.025,n-1}$ is approximately 2. From

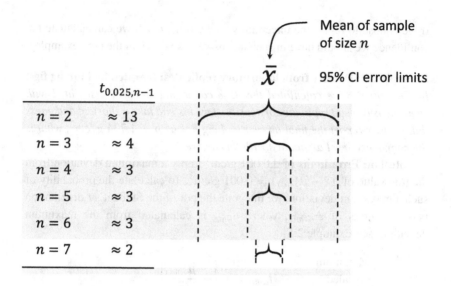

Figure 2.16 The number of standard errors to use when constructing replicate error limits varies strongly as $n$ changes from 2 to 3; the number continues to decrease until approximately $n \approx 7$, when it plateaus at about 2.

the values of $t_{0.025, n-1}$, we can understand why scientists and engineers often take at *least* data triplicates to determine an estimate of a stochastic quantity: we get a large gain in precision when we increase the number of measurements from $n = 2$ to $n = 3$, and less substantial increases when we increase $n$ further, particularly after $n = 7$.

The 95% confidence interval for replicates, given in Equation 2.52, is a widely adopted standard for expressing experimental results. In Appendix A we provide a replicate error worksheet as a reminder of the process used to obtain error limits on measured quantities subject to random error. For $n > 7$, about plus or minus two standard errors ($e_{s, random} = s\sqrt{n}$) corresponds to the 95% confidence interval.

**Random error**

95% CI of mean         $n$ replicates,    $\mu = \bar{x} \pm t_{0.025, n-1} e_{s, random}$    (2.54)
$(e_{s, random} = s\sqrt{n})$

$$n > 7 \text{ replicates, } \mu \approx \bar{x} \pm 2 e_{s, random}$$

In subsequent chapters we discuss ways to incorporate nonrandom errors into our error limits, and additional error worksheets are discussed there. When expressing 95% confidence intervals with nonrandom errors, we retain the structure of the 95% confidence interval as approximately $\pm 2e_s$ (two standard errors; see Appendix E, Equation E.5), but we use a different standard error that corresponds to the independent combination of replicate, reading, and calibration errors.

Standard errors combine in quadrature

$$e_{s,cmbd}^2 = e_{s,random}^2 + e_{s,reading}^2 + e_{s,cal}^2 \qquad (2.55)$$

The task of incorporating nonrandom errors into error limits requires us to determine an appropriate standard error $e_s$ for the nonrandom contributions (see Section 1.4).

**Nonrandom error**
95% CI of estimate
($e_{s,cmbd}$ = combination of random, reading, and calibration error)

$$\text{estimate} \approx \bar{x} \pm 2e_{s,cmbd} \qquad (2.56)$$

We pursue these calculations in Chapters 3 and 4.

The examples that follow provide some practice with 95% confidence intervals of the mean and with using the Student's $t$ values from Table 2.3 or from Excel or MATLAB.

**Example 2.10: Power of replicates: density from a pycnometer.** *Chris and Pat are asked to measure the density of a solution using pycnometers[15] and an analytical balance. They are new to using pycnometers, and they each take a measurement; Chris obtains $\rho_1 = 1.723\,g/cm^3$ and Pat obtains $\rho_2 = 1.701\,g/cm^3$. Chris averages the two results and reports the average along with the replicate error limits. Calculate Chris's answer and error limits ($n = 2$). Pat takes a third measurement and obtains $\rho_3 = 1.687\,g/cm^3$. Pat decides to average all three data points (Pat's two and Chris's measurement). Calculate Pat's answer and error limits, assuming that all three measurements are equally valid and affected only by random error. Comment on the effect of taking the third measurement versus reporting the average of a duplicate only.*

**Solution:** The 95% confidence interval of the mean of data replicates is

95% CI
of the mean:
$$\bar{x} \pm t_{0.025,n-1} \frac{s}{\sqrt{n}}$$

For Chris's two data points (Chris's measurement and Pat's first data point), the mean is $\bar{\rho} = 1.7120$ g/cm$^3$ and the standard deviation is $s = 0.0156$ g/cm$^3$, and thus Chris obtains:

[15] For more on pycnometers, see Figure 3.13 and Examples 3.8 and 4.4.

Chris's answer:                 $n = 2$

$$t_{0.025,1} = 12.706$$

$$\text{density} = 1.7120 \pm (12.706)\left(\frac{0.0156}{\sqrt{2}}\right)$$

$$= 1.71 \pm 0.14 \text{ g/cm}^3$$

Note that we chose to provide two uncertain digits in our answer since the leading error digit is "1" (see Section 2.4).

For Pat's answer, we use all three data points, calculating: $\bar{\rho} = 1.70367$ g/cm$^3$ and $s = 0.0181$ g/cm$^3$. Pat obtains:

Pat's answer:                   $n = 3$

$$t_{0.025,2} = 4.303$$

$$\text{density} = 1.70367 \pm (4.303)\left(\frac{0.0181}{\sqrt{3}}\right)$$

$$= 1.70 \pm 0.05 \text{ g/cm}^3$$

Because the leading error digit is 5, we provide only one digit on error.

The two answers are compared in Figure 2.17. Chris and Pat obtain similar estimates of the density, but the error limits on Chris's number are quite a bit wider than those on Pat's, since Pat used a data triplicate versus Chris's duplicate. There is a significant increase in precision when we are able to

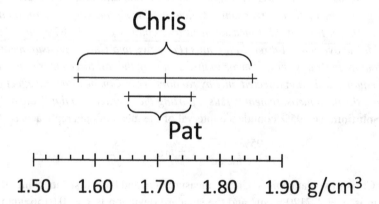

Figure 2.17 Chris and Pat obtain similar estimates of a sample's density (in g/cm$^3$), but Chris's answer is significantly less precise (wider 95% confidence interval).

obtain a data triplicate ($t_{0.025, 2} \approx 4.3$ for $n = 3$; see Table 2.3) compared to only having two replicates ($t_{0.025, 1} \approx 12.7$ for $n = 2$).

Replication provides the opportunity to obtain precise values of measured quantities. Sometimes, however, there is no opportunity to replicate, as in the next example.

**Example 2.11: Estimate density uncertainty without replicates.** *We seek to determine the density of a sample of Blue Fluid 175, but there is only a small amount of solution available, and we can make the measurement only once. The technique used is mass by difference with a $10.00 \pm 0.04$ ml pycnometer and an analytic balance (the balance is accurate to $\pm 10^{-4} g$). What are the appropriate error limits on this single measurement?*

**Solution:** We have discussed thus far how to obtain error limits when replicate measurements are available. In the current problem, we do not have any replicates to evaluate, so the techniques of this chapter do not allow us to assess the error limits.

The scenario in this problem is quite common, and we address this question further in Chapter 5. We can estimate error limits of a single data point calculation with a technique called *error propagation*. Error propagation is based on appropriately combining the estimated reading and calibration error of the devices that produce the numbers employed in the calculation. In the current problem, these would be the reading and calibration error of the analytical balance and the calibration error of the pycnometer. For now, we cannot proceed with this problem; we revisit this estimate in Examples 5.1 and 5.4.

Thus far, we have explored the appropriate methods for determining error limits on measurements when random error is present: when random errors are present, we take data replicates and construct 95% confidence intervals around the mean of the sample set. We have also indicated that when more than random error is present, we must estimate reading and calibration error and combine these with replicate error to obtain the correct combined error. In Chapter 3 we turn to determining reading error.

The remaining sections of this chapter address several topics related to replicate measurements as well as some topics related to measurement error in general. First, we discuss *prediction intervals*, which may be used to address problems such as estimating uncertainty in future ("next") data points. Next, we discuss a technique for turning potential systematic errors into more easily handled random errors. Finally, we formally present our convention for reporting significant figures on error. Together these topics complete our quick start on replicate error and set the stage for the discussion of systematic errors, which begins in Chapter 3.

### 2.3.3 Prediction Intervals for the Next Value of $x$

In the previous section on confidence intervals for the mean, we saw how to estimate a value for a replicated quantity and how to determine its uncertainty. The uncertainty of the sample mean is proportional to the standard error of replicates, $e_s = s/\sqrt{n}$, which gets smaller as the sample size $n$ gets larger. For large sample sizes, $\lim_{n \to \infty} s/\sqrt{n} = 0$ and the 95% confidence interval error limits will be very tight, and the true value of the mean of the underlying distribution of $x$ will be known with a high degree of precision. Note that *it is the true, mean value of the stochastic variable $x$*, calculated from the mean of samples, that is determined within tighter and tighter error limits.[16] Confidence intervals for the mean do not tell us about error limits for individual measurements. We explore this topic in the next example.

**Example 2.12: Using statistics to identify density outliers.** *In Example 2.8, we calculated the value of the density of Blue Fluid 175 to be $\rho_{BF175} = 1.734 \pm 0.003$ g/cm$^3$. The density was determined by calculating the mean of 10 measurements, and the uncertainty in the result was calculated from the 95% confidence interval of the mean. Looking back at the data used to calculate the mean (Table 2.4, plotted in Figure 2.18), we see that several of the original data points are outside of the 95% confidence interval of the mean. Does this indicate that some of the data points are outliers and should be discarded?*

**Solution:** No, that is not correct. We have calculated the 95% confidence interval *of the mean*. It indicates that we are 95% confident that the true mean value of the Blue Fluid 175 density measurement (the mean of the underlying distribution of density measurements) is in the interval $1.734 \pm 0.003$ g/cm$^3$. The confidence interval of the mean gets more and more narrow when we use larger and larger sample sizes; in other words, the more data we obtain, the more precisely we can state the value of the mean (the confidence interval becomes small).

$$\mu = \bar{x} \pm t_{0.025, n-1} \frac{s}{\sqrt{n}} \qquad (2.57)$$

$$\text{For } n \text{ large:} \quad \mu = \lim_{n \to \infty} \left( \bar{x} \pm t_{0.025, n-1} \frac{s}{\sqrt{n}} \right)$$

$$= \bar{x}$$

As $n$ increases, eventually nearly *all* measured data points will lie outside the 95% confidence interval of the mean.

---

[16] This assumes only random error influences the measurement; see Chapters 3 and 4.

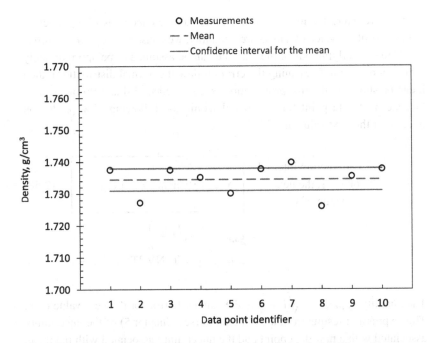

Figure 2.18 Several of the original data points from Example 2.8 fall outside the 95% confidence interval of the mean. The confidence interval of the mean only relates to how well we know the mean; it does not help us evaluate or understand the scatter of the individual observations of the variable.

When we look at the raw data (Figure 2.18), they are scattered, and their scatter reflects the random effects that influenced the values obtained in the individual, noisy measurements. The next (eleventh) measurement of density will also be subject to these random effects, and we expect subsequent measurements to be similarly scattered. The 95% confidence interval *of the mean* does not tell us about the magnitude of these random effects on individual measurements.

> The 95% confidence interval *of the mean* does not tell us about the magnitude of random effects on individual measurements.

The idea that individual data points may be suspect – that is, might be outliers (due to a blunder or other disqualifying circumstances) – can be evaluated with a different tool, the prediction interval. We discuss this approach next.

If we are interested in expressing an interval that encompasses a prediction of the next likely value of a measured quantity, we construct a *prediction interval*. If the underlying pdf of the variable can be assumed to be approximately normally distributed (meaning the errors follow the normal distribution), then it can be shown with error propagation (see reference [38] and Problem 2.41) that the next data point for $x$ will fall within the following 95% prediction interval of the next value of $x$ [5, 38]:

95% prediction interval
for $x_{n+1}$, the next
value of $x$

$$x_{n+1} = \bar{x} \pm t_{0.025, n-1} \, e_{s, next} \qquad (2.58)$$

$$e_{s, next} \equiv s\sqrt{1 + \frac{1}{n}}$$

$$t_{0.025, n-1} = \text{T.INV.2T}(0.05, n - 1)$$

The quantity $e_{s, next} = s\sqrt{1 + \frac{1}{n}}$ is the standard error for the next value of $x$. This expression results from a combination (see Chapter 5) of the uncertainty associated with a new data point and the uncertainty associated with the mean $\bar{x}$. Note that this interval shrinks a bit with increasing $n$ due to changes in $t_{0.025, n-1}$ and $n$ (Table 2.3), but for $n \geq 7$, the prediction interval plateaus at approximately $\pm 2s$.

We can practice creating prediction intervals by calculating a prediction interval for the Blue Fluid 175 density data from Example 2.12.

**Example 2.13: Prediction interval on students' measurements of fluid density.** *Based on the student data on density of Blue Fluid 175 given in Table 2.4, what value do we expect a new student group to obtain for $\rho_{BF175}$, given that they follow the same experimental procedure? The answer is a range.*

**Solution:** We assume that the new student group is of comparable ability and attentiveness to procedure as the groups whose data supplied the results in Table 2.4. Thus, the next value obtained will be subject to the same amount of random scatter as shown in the previous data. We can find the answer to the question by calculating the 95% prediction interval of the next value of $\rho_{BF175}$.

We construct the 95% prediction interval using Equation 2.58 and the data in Table 2.4. To avoid round-off error, extra digits are shown here for use in the intermediate calculations (see Section 2.4). The final answer is given with the appropriate number of significant figures.

$$\begin{bmatrix} 95\% \text{ PI} \\ \text{of next} \\ \rho_{BF175} \end{bmatrix} = \bar{x} \pm t_{0.025,n-1}s\sqrt{1 + \frac{1}{n}}$$

$$\bar{\rho} = 1.734391 \text{ g/cm}^3$$
$$n = 10$$
$$s = 0.004853 \text{ g/cm}^3$$
$$t_{0.025,9} = 2.262157 \quad [\text{T.INV.2T}(0.05,9)]$$

$$\begin{bmatrix} 95\% \text{ PI} \\ \text{of next} \\ \rho_{BF175} \end{bmatrix} = 1.734391 \pm (2.262157)(0.004853)(1.0488)$$

$$= \boxed{1.734 \pm 0.012 \text{ g/cm}^3} \qquad \text{(two uncertain digits)}$$

Our result[17] indicates that, with 95% confidence, a student group will obtain a value between 1.723 and 1.746 g/cm$^3$.

This prediction interval is wider than the confidence interval of the mean (Figure 2.19). Several individual data points fall outside the 95% confidence interval of the mean, but all of the data points are within the 95% prediction interval for the next value of $\rho_{BF175}$. None of the data points is particularly unusual, as judged at the 95% confidence level.

An important use for the prediction interval for the next value of $x$ is to determine if new data should be obtained or existing data discarded due to a probable mistake of some sort. The idea is that some experimental outcomes are very unlikely, and if an unlikely value is observed, it is plausible that the data point is a mistake or blunder and should not be used. Example 2.14 considers such an application.

**Example 2.14: Evaluating the quality of a new density data point.** *One day in lab, ten student groups measured the density of a sample of Blue Fluid 175, obtaining the results shown in Table 2.4. An eleventh student group also measured the density of the same fluid the next day, following the same procedure. The result obtained by Group 11 was $\rho_{BF175} = 1.755$ g/cm$^3$, which was higher than any value obtained by the first ten groups. Should we suspect that there is some problem with Group 11's result, or is the result within the range we would expect for this measurement protocol?*

---

[17] Two digits are used in expressing the error limits in this case because the leading error digit is "1"; see Section 2.4.

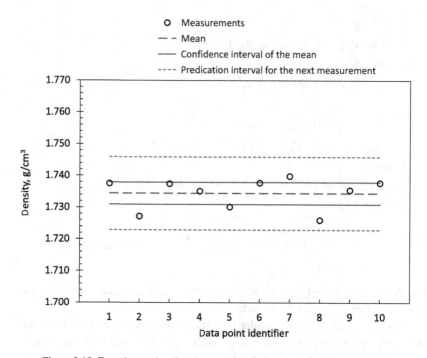

Figure 2.19  For $n$ larger than 3, it is unsurprising that several of the original data points fall outside the 95% confidence interval of the mean, since the confidence interval is constructed with $\pm t_{0.025, n-1} s/\sqrt{n}$ and $n$ is in the denominator. To address the question of how representative any given data point is of the entire set, we construct 95% prediction intervals of the next value of $x$. All the data points in the example fall within the prediction interval; by definition, if the effects are all random, 95% of the data will fall within the prediction interval.

**Solution:** We address this question by seeing whether the new result lies within the 95% prediction interval implied by the original data. In Example 2.13 we calculated that prediction interval:

$$\text{Based on Table 2.4 data } (n = 10) \begin{bmatrix} 95\% \text{ PI} \\ \text{of next} \\ \rho_{BF175} \end{bmatrix} = 1.734 \pm 0.012 \text{ g/cm}^3$$

This result tells us that with 95% confidence, the next measured value of $\rho_{BF175}$ will lie in the following interval:

$$1.722 \text{ g/cm}^3 \le \rho_{BF175} \le 1.746 \text{ g/cm}^3 \tag{2.59}$$

The value measured by Group 11 was $\bar{\rho} = 1.755$ g/cm$^3$, which is not within this interval. Thus, we conclude that their result is suspect, since with

95% confidence, individual results are expected to be within the prediction interval calculated. Perhaps there was a mistake in the execution of the procedure.

We could also ask a related question: what is the probability that a measurement at least as extreme as this measurement occurs? If the resulting probability is very small, that could be reason for rejecting the point. The posed question may be answered by finding the total probability within the region of the Student's $t$ distribution pdf tail that just includes this rare point; see Problem 2.29.

In Example 2.14 we did not conclude that the examined value is definitely in error; rather, we concluded that the value is suspect. We are using a 95% prediction interval, which means that, on average, one time in 20 we expect a next data point to lie outside of the interval. When working with stochastic variables, we can never be 100% certain that an unlikely outcome is not occurring. What we can do, however, is arrive at a prediction and deliver that prediction along with the level of our confidence, based on the samples examined.

We can use the prediction interval to determine the range of commuting times that Eun Young should expect on future trips, based on her past data (as discussed in Examples 2.1, 2.2, and 2.7). This is a classic question about a next value of a stochastic variable.

**Example 2.15: Commuting time, once again revisited.** *Over the course of a year, Eun Young took 20 measurements of her commuting time under all kinds of conditions, using the same stopwatch and the same timing protocols; the data are in Tables 2.1 and 2.2. What is Eun Young's most likely commuting time tomorrow? Provide limits that indicate the range of values of commuting time that Eun Young may experience.*

**Solution:** The most likely amount of time that Eun Young's commute will take is the mean of the 20 observations, with 95% confidence, which we determined in Example 2.7 to be $30 \pm 5$ min. The error limits on the mean indicate our confidence (at the 95% level) in the value of the mean.

To determine the range of values of commuting time that Eun Young may experience tomorrow ("next"), we need the 95% prediction interval of the next value of the variable. We calculate this with Equation 2.58 with $\alpha = 0.05$.

$$\begin{matrix} \text{95\% PI} \\ \text{of next} \\ \text{value of } x \end{matrix} \qquad x_i = \bar{x} \pm t_{0.025, n-1}\, s\sqrt{1 + \frac{1}{n}} \qquad (2.60)$$

For the 20 data points on Eun Young's commuting time, the 95% prediction interval of the mean is:

$$x_i = \bar{x} \pm t_{0.025,n-1}\, s\sqrt{1+\frac{1}{n}}$$

$$x = 29.9 \pm (2.093024)(9.877673)\sqrt{1+\frac{1}{20}}$$

$$= 29.9 \pm 21.18476$$

$$= 30 \pm 22 \text{ min}$$

Thus, Eun Young should expect a commuting time of up to 52 min (95% confidence level).

Replicate sampling, as we see from the discussion here, is a powerful way to estimate the true value of a quantity that we are able to sample. The only limitation on the power of replication is that *replicate error only accounts for random effects*. If nonrandom effects cause measurements to be different from the true value, then no amount of replication will allow us to find the true value of the variable or the proper range of the confidence or prediction intervals.

> If *nonrandom* effects cause measurements to be different
> from the true value, then no amount of replication
> will allow us to find the true value of the variable
> or the proper range of the confidence or prediction intervals.

To see what this dilemma looks like, consider the next example.

**Example 2.16: Repeated temperature observations with a digital indicator.** *Ian uses a digital temperature indicator equipped with a thermocouple to take the temperature at the surface of a reactor. He records the temperature reading every 5 min for 20 min and the values recorded are 185.1, 185.1, 185.1, 185.1, and 185.1 (all in °C). Assuming there are only random errors in these measurements, what is the temperature at the surface of the reactor? Include the appropriate error limits.*

**Solution:** Assuming that there are only random errors, we calculate the value of the surface temperature as the average of the five values, and we use the 95% confidence interval of the mean in Equation 2.52 for the uncertainty.

$$n = 5$$
$$\bar{T} = 185.1°C$$
$$s = 0.0°C$$
$$t_{0.025,4} = 2.78$$

$$\text{surface temperature} = \bar{T} \pm t_{0.025,4} \left( \frac{s}{\sqrt{n}} \right)$$

$$= \boxed{185.1000 \pm 0.0000°C}$$

According to this calculation, since we obtained five identical measurements, we know the surface temperature of the reactor *exactly*.

Of course, you should be skeptical of this conclusion and already wondering what went wrong with this logic. If that is so, you are correct: something has gone wrong. The sensor used in this analysis gave the same value every time a measurement was taken, and we (correctly) calculated a standard deviation of zero. The conclusion we can draw from the zero standard deviation is that the sensor is very consistent and there is no detectable random error.

It would be wrong to assume that we now know the temperature exactly, however. For one thing, since the indicator only gives temperature to the tenths digit, we cannot tell the difference between 181.12°C and 181.13°C with our indicator. We discuss this sort of error, called reading error, in Chapter 3 (see Problem 3.3). In addition, we have not explored the actual *accuracy* of the temperature indicator. Does the indicator give the actual temperature or does it read a bit high or low? Many home ovens can run high or low due to limitations in sensor accuracy, and good cooks know to adjust for this tendency. Matching a sensor's reading with the true value of the quantity of interest is called *calibration*. Uncertainty due to limitations of calibration is discussed in Chapter 4 (see Problem 4.23 for more on thermocouple accuracy). Both reading error and calibration error are systematic errors, and repeating the measurement will not make these systematic errors visible. We must track down systematic errors by other methods.

In this chapter we have discussed how to account for random errors in measurement. There are also nonrandom effects, and we discuss these beginning in the next chapter. Two common sources of nonrandom error are considered in this book: reading error (Chapter 3) and calibration error (Chapter 4). These nonrandom effects often dominate the uncertainty in experimental results and should not be neglected without investigation. Both reading error and calibration error are systematic errors, and repeating the measurement will not make them visible. Instead, we must track down systematic errors individually.

In the next section we introduce randomization, an experimental technique that helps us ferret out some types of nonrandom error. In Section 2.4 we close with a discussion of the significant figures appropriate for error limits.

### 2.3.4 Essential Practice: Randomizing Data Acquisition

The methods described in this book allow us to quantify uncertainty. When random errors are present, the values obtained for a measured quantity bounce around, even when all the known factors affecting the quantity are held constant. This makes random errors easy to detect when a measurement is repeated or replicated. As discussed earlier in this chapter, we can quantify random errors by analyzing replicate measurements: once we obtain replicate values, we apply the statistics of the sampling of stochastic variables to report a good value for the property (the mean of replicates) and the error limits (95% confidence interval of the mean).

Soon we will consider systematic contributions to uncertainty due to reading error and calibration errors. These are errors that we know are present. Because we recognize the presence of reading and calibration errors, we can explore the sources of the errors, reason about how these errors affect samples, and construct standard errors for each of these sources (we do this in Chapters 3 and 4). Since random, reading, and calibration errors are independent effects, they simultaneously act on measurements, and thus they add up in ways we can determine (they add like variances – that is, in quadrature. This topic was introduced in Chapter 1, and details are discussed in Section 3.3).

Somewhat more difficult to deal with are systematic errors that we do *not* know are present. Unrecognized systematic errors do not show up as scatter in replicates because these errors correlate, or act systematically; thus they can skew our data without us ever knowing they were present. To see this dilemma in action, we discuss a specific example.

**Example 2.17: Calibrating a flow meter.** *A team is assigned to calibrate a rotameter flow meter. The rotameter is installed in a water flow loop (see Figure 4.4; water temperature $= 19.1°C$) constructed of copper tubing, with the flow driven by a pump. To calibrate the rotameter, the flow rate is varied and independently measured by the "pail-and-scale" method: the water flow is collected in a "pail" (or tank, or other reservoir), and the mass of the water collected over a measured time interval is recorded. The mass measurement,*

Table 2.5. *Data from calibrating a rotameter by the pail-and-scale method with two team members systematically dividing up data acquisition duties. Extra digits provided; see also Table 6.9.*

| Index | Mass flow rate (kg/s) | Rotameter reading ($R\%$) |
|---|---|---|
| 1 | 0.04315 | 14.67 |
| 2 | 0.05130 | 17.33 |
| 3 | 0.05515 | 18.50 |
| 4 | 0.06512 | 21.50 |
| 5 | 0.07429 | 24.33 |
| 6 | 0.08545 | 27.80 |
| 7 | 0.09585 | 31.00 |
| 8 | 0.10262 | 33.00 |
| 9 | 0.11449 | 36.67 |
| 10 | 0.13201 | 42.00 |
| 11 | 0.14325 | 45.50 |
| 12 | 0.15100 | 51.00 |
| 13 | 0.16882 | 56.33 |
| 14 | 0.18326 | 60.67 |
| 15 | 0.19643 | 65.00 |
| 16 | 0.20447 | 67.00 |
| 17 | 0.21361 | 69.67 |
| 18 | 0.22880 | 74.67 |
| 19 | 0.23625 | 77.00 |
| 20 | 0.24323 | 79.00 |
| 21 | 0.26620 | 85.50 |
| 22 | 0.27831 | 90.67 |

along with the collection time, allows the team to calculate the observed mass flow rate for each reading on the rotameter (reading $R\%$ =% full scale).

The team divides up the calibration flow range, with one team member taking the low flow rates and the other taking the high flow rates. The data are shown in Table 2.5. What is the calibration curve for the rotameter? What is the uncertainty associated with the calibration curve?

**Solution:** This text presents the error-analysis tools needed to create a calibration curve for data such as that in Table 2.5: in Chapter 4 we introduce calibration and calibration error, and in Chapter 6, which builds on Chapter 5, we show how to produce an ordinary least-squares fit of a model to data. We discuss the current problem out of sequence to illustrate some

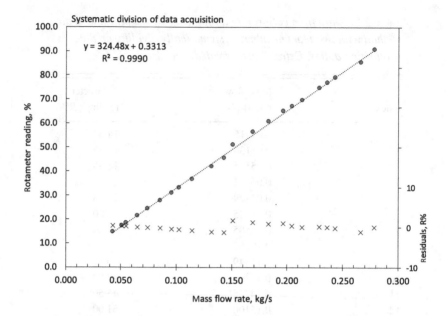

Figure 2.20 The rotameter calibration data are plotted for the case in which the work was systematically divided between the two experimenters, with one providing the low-flow-rate data and the other providing the high-flow-rate data. Also shown are the residuals at each value of $x$; this is the difference between the data and the fit. We observe a pattern in the residuals, a hint that a systemic error may be affecting the data.

common experimental-design and systematic errors that occur. We introduce the software tools we need as we go along. In the chapters that follow we offer a more thorough discussion of the model-fitting tools used here.

Systematic errors can creep into our data in ways that are difficult to anticipate. In the problem statement for this example, we learned a little bit about the experimental design for the rotameter calibration – in particular, the calibration team shared data acquisition duties, with one team member acquiring the low-flow-rate data points and the other team member acquiring the high-flow-rate ones. This arrangement seems harmless; both team members are presumably equally qualified to take the data. Splitting up the data acquisition in this way spreads out the work in an equitable way.

The data in Table 2.5 are plotted in Figure 2.20, and they look fine. Included in the figure is an ordinary least-squares curve fit, and the equation for the fit and its $R^2$ value (coefficient of determination; see Equation 6.17) are shown. This curve fit is obtained in Excel by right-clicking on the data in the plot and

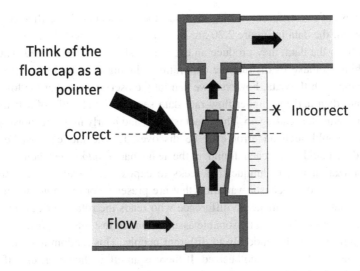

Figure 2.21 The rotameter is designed to be read from the position "pointed to" by the angled portion of the float. Sometimes new users think the reading comes from the location of the top of the float.

choosing to add a "Trendline"; in the Trendline dialog box we check the boxes that instruct Excel to display both the equation and the $R^2$ on the plot. The data follow a straight line, and $R^2$ is almost 1, indicating that the linear model is a good choice, as it is capturing the variation in the data. (We will have more to say on $R^2$ in Example 2.18 and in Chapter 6.)

The data look good, and the linear fit appears to be excellent. Unfortunately, there is a hidden systematic error in the data. The slope obtained by the trendline fit is 324 $R\%/(\text{kg/s})$; when the systematic error is found and eliminated, the correct slope obtained is 306 $R\%/(\text{kg/s})$. (A complete dataset without the systematic error is given in Table 6.9.) The data have an unrecognized systematic error.

We return to this problem in Example 2.18.

What hidden issue has affected the rotameter calibration? And how could we have prevented the problem? It turns out that the hidden systematic error in the data is the result of a misunderstanding in how to read the rotameter (Figure 2.21). The float in the rotameter has an angled indicator that "points" to a position on a scale; the reading that is expected by the manufacturer is the number that is pointed to by the indicator. Investigation revealed that one team member thought that the reading was determined by the position of the *top* of the float rather than by the position of the angled indicator.

The use of two different reading methods does not seem to have affected the data much: the data in Figure 2.20 are smooth and linear, and when a straight line is fit to the data, they produce an excellent value of $R^2 = 0.9990$. There is nothing to make us distrust the calibration. Having been told that there is a problem with the data, however, we can look closer and see that the lower-flow-rate data and the higher-flow-rate data seem to be slightly offset; they meet at rotameter reading 45. The effect is small and nearly invisible, however.

How would such an error ever be discovered? The hidden mistake in these data could have been found if the team had *randomized* their trials. Randomization is a technique that seeks to expose *each* experimental trial to the sources of stochastic variation that are present. For example, we may believe that it would make no difference who reads the rotameter during the calibration work. This is a reasonable assumption. To test this supposition, we can assign data trials randomly to different people. This random assignment may have no effect, as hypothesized. If there is an effect, however, the effect will appear randomly.

Why, we might ask, would we want to introduce variability? Why not just use one person so that the data are consistent? The reason to randomize, for example, the identity of the data-taker, is to guard against it *mattering* who took the data. If only one person takes the data, and if that person makes an error consistently, we will not be able to detect the mistake. Randomizing the data taker tests the hypothesis that it does not matter who takes the data. When we design a process to depend critically on one single aspect, such as entrusting all the data-taking to one individual's expertise or taking data in a systematic way from low to high values, we are exposing our experiments to the risk that all the data may contain hidden systematic errors. On a grand scale, when we scientists and engineers publish our results and invite others to attempt to reproduce the results, we are asserting that any competent investigator will obtain the same results as we obtained. If we have done our experiments correctly, and if our colleagues do the same, we will all get the same results. For our results to be of the highest quality, we should use randomization wherever possible to double-check that no systematic effects have been unintentionally introduced.

Our next question might be, why will randomizing help us find systematic errors? And, will we get any useful data out of the more scattered results we obtain when we randomize? If we randomize our experimental design and we were right that the change does not make a difference, we have shown that the change did not make a difference. This allows us to simplify future experiments with confidence, as we know definitively that the variable we randomized does not affect the outcome.

If we randomize an aspect of our experimental design and there *is* an effect, the effect will become visible, as it will introduce additional scatter in the data. The presence of scatter is a message to the experimenter that there are random effects in the experiment as currently designed and executed. If the scatter is not too large and can be tolerated, nothing additional needs to be done. If the scatter is large and cannot be tolerated, its presence becomes a reason to reevaluate the experimental design to find the source of the scatter. The magnitude of the scatter becomes a detector of systematic error that needs to be addressed. Unrecognized systematic errors are a serious threat to the quality of the conclusions we may make with our data; knowing that there is an unrecognized effect – one that shows up as extra scatter when the process is randomized – is a welcome first step to tracking down a potentially serious problem with our studies.

We can see the randomization effect on the rotameter calibration in Example 2.17 if we start over with a different dataset in which each of the researchers again obtained half the data points, but in this case, the flow rates were assigned randomly.

**Example 2.18: Calibrating a flow meter, revisited.** *A team is assigned to calibrate a rotameter flow meter. The rotameter is installed in a water flow loop (see Figure 4.4; water temperature = 19.1°C) constructed of copper tubing, with the flow driven by a pump. To calibrate the rotameter, the flow rate is varied and independently measured by the "pail-and-scale" method.*

*The team divides up the calibration flow range, with flow rates assigned randomly to the two team members. The data are shown in Table 2.6. What is the calibration curve for the rotameter? What is the uncertainty associated with the calibration curve?*

**Solution:** The data in Table 2.6 are plotted in Figure 2.22.[18] These data look more scattered than the original plot in Figure 2.20. An ordinary least-squares fit to a line has been obtained with a slope of 309 $R\%/$(kg/s) and yields an $R^2$ of 0.9954.

The randomization of the data acquisition led to data with more scatter. The scatter in the data is a message: something is affecting the data. The message is received when the team reflects on what they see, holds a team meeting to review the procedure, and discovers and corrects the problem.

---

[18] These data were obtained by the author with each rotameter reading recorded in both the correct way and the incorrect way for each mass flow rate measurement. Thus, we could produce this second version of a plot for the same set of experiments. All the datasets referenced in Examples 2.17 and 2.18 (and Problems 6.7, 6.10, and 6.17) are also given in Table 6.9 (limited number of digits provided).

Table 2.6. *Data from calibrating a rotameter by the pail-and-scale method; data acquisition duties were divided randomly. Extra digits provided; see also Table 6.9.*

| Index | Mass flow rate (kg/s) | Rotameter reading ($R\%$) |
|-------|-----------------------|---------------------------|
| 1     | 0.04315               | 17.67                     |
| 2     | 0.05130               | 17.33                     |
| 3     | 0.05515               | 18.50                     |
| 4     | 0.06512               | 24.50                     |
| 5     | 0.07429               | 27.00                     |
| 6     | 0.08545               | 27.80                     |
| 7     | 0.09585               | 31.00                     |
| 8     | 0.10262               | 33.00                     |
| 9     | 0.11449               | 39.00                     |
| 10    | 0.13201               | 45.00                     |
| 11    | 0.14325               | 45.50                     |
| 12    | 0.15100               | 48.33                     |
| 13    | 0.16882               | 53.80                     |
| 14    | 0.18326               | 60.67                     |
| 15    | 0.19643               | 62.33                     |
| 16    | 0.20447               | 62.50                     |
| 17    | 0.21361               | 67.00                     |
| 18    | 0.22880               | 74.67                     |
| 19    | 0.23625               | 77.00                     |
| 20    | 0.24323               | 79.00                     |
| 21    | 0.26620               | 83.00                     |
| 22    | 0.27831               | 90.67                     |

For the rotameter data, plotting the points with different symbols according to who took the data made it clear that there was a systematic effect tied to operator identity. The take-away from this example is that we need to be on guard when we plan our experiments – systematic errors can sneak in through seemingly innocuous choices. Another take-away is that scatter in data can be a good thing if it shows that something unrecognized is affecting the data systematically.

This example also shows that values of $R^2$ that approach 1 *do not guarantee that data are of high quality and free from systematic errors.* As we discuss in Chapter 6, $R^2$ reflects the degree to which a chosen model (in our case, a straight line with nonzero slope) represents the data, compared to the

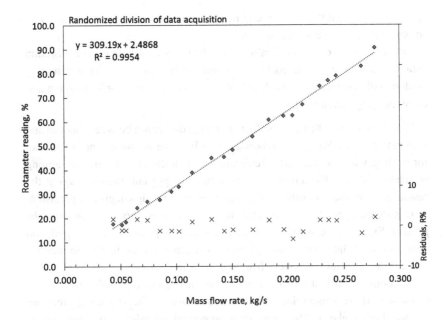

Figure 2.22 The rotameter calibration data are plotted for the task randomly divided between two experimenters. The scatter from the model line is greater than in Figure 2.20; this is reflected in the value of the standard deviation of $y$ at a value of $x$, $s_{y,x}$ as discussed in the text. The scatter barely affects $R^2$. The residuals in this dataset (data − model) have a more random pattern compared to those in Figure 2.20.

assumption that the data are constant (flat line). In an ordinary least-squares fit, the statistic that reflects the scatter of data with respect to the model line is $s_{y,x}$, the standard deviation of $y$ at a given $x$ (see Chapter 6, Equation 6.25). For the data discussed in this example, the values of $s_{y,x}$ are

$$\begin{array}{ll} \text{Shared data, assigned systematically} & s_{y,x} = 0.783 \ R\% \qquad (2.61) \\ \text{(Example 2.17)} & \end{array}$$

$$\begin{array}{ll} \text{Shared data, assigned randomly:} & s_{y,x} = 1.614 \ R\% \qquad (2.62) \\ \text{(Example 2.18)} & \end{array}$$

The statistic $s_{y,x}$ correctly reflects the qualities of the fits. When the data are all taken correctly, $s_{y,x}$ is small and the data points are all close to the model line; $s_{y,x}$ for correctly taken data is 0.456 $R\%$ (see Table 6.9 and Problem 6.7). In contrast, when the results randomly include data that are read incorrectly, $s_{y,x}$ is large. The systematic case for Example 2.17 has an intermediate value for

$s_{y,x}$, indicating that the error in reading the rotameter is hidden in the relatively smooth data (with the wrong slope).

Obtaining an accurate calibration curve begins with taking good calibration data. Good data are obtained by following the best practices for minimizing random and systematic errors. Randomization is a key tool for identifying and eliminating systematic errors.

The discussion in this section focuses on the difference between random and systematic errors. We prefer random errors because we have a powerful tool for dealing with random error: replication. When the only differences among replicate trials are the amounts of random error present, then averaging the results gives us the true value of the measurement to a high degree of precision. A requirement for this to be valid, however, is that *only random error be present*. Systematic error will not disappear when replicates are averaged, and thus no matter the number of replicates, it remains our obligation to identify and eliminate systematic errors from our measurements.

Randomization is a tool for making systematic errors visible, as we have discussed. If we randomize our data acquisition, switching things that we do not think make a difference, then unrecognized effects can show up as additional "random" error. It may seem undesirable to design our protocols to amplify random error, but that is an inappropriate conclusion. Randomization is an important part of experimental troubleshooting since it is better to discover and correct systematic effects than to leave them in place, hidden, and to believe, incorrectly, that our results are high quality.

Randomization works against the *selective* introduction of errors into a process. If some aspect of a process introduces errors, these errors are easier to identify and fix if they affect all the data points rather than just some points. If a subset of trials are isolated from a source of error, that error source becomes systematic and possibly invisible. Replicates that result from data acquisition that is free from arbitrary systematic effects are called true replicates. In the next example we explore the kind of questions we can ask to ensure that our experiments produce true replicates.

**Example 2.19: True replicates of viscosity with Cannon–Fenske viscometers.** *Students are asked to measure the kinematic viscosity of a 30wt% aqueous sugar solution using Cannon–Fenske ordinary viscometers (Figure 2.23). The viscometers' procedure calls for loading an amount of solution (following a standard process), thermal-soaking the loaded viscometer in a temperature-controlled bath, and measuring the fluid's efflux time by drawing the fluid into the top reservoir and timing the travel of the fluid meniscus between timing marks. The viscometers are pre-calibrated by the manufacturer,*

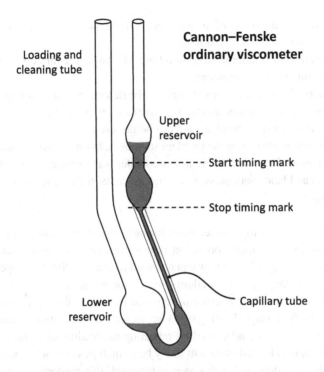

**Cannon–Fenske**
**ordinary viscometer**

Loading and
cleaning tube

Upper
reservoir

Start timing mark

Stop timing mark

Lower
reservoir

Capillary tube

Figure 2.23 The Cannon–Fenske ordinary viscometer uses gravity to drive a fluid through a capillary tube. The viscosity comes from measuring the efflux time $\Delta t_{eff}$ for the fluid meniscus to pass from the upper timing mark to the lower timing mark. The calibration coefficient $\tilde{\alpha}$ is supplied by the manufacturer. The viscosity $\tilde{\mu}$ is equal to $\tilde{\alpha} \rho \Delta t_{eff}$, where $\rho$ is the fluid density. For more on Cannon–Fenske viscometers, see Example 5.8.

*and kinematic viscosity $\tilde{\mu}/\rho$ is obtained by multiplying the efflux time by a manufacturer-supplied calibration constant. How can the student teams produce valid replicates of kinematic viscosity with these viscometers?*

**Solution:** This is a question about experimental design. To address this question, we remind ourselves that a true replicate is exposed to all the elements that introduce error into the process. To plan for true replicates, then, we must reflect on all the elements that might potentially introduce error into the process.

We identify the following issues:

1. Loading the standardized volume may be subject to uncertainty in volume.
2. The temperature bath must accurately maintain the instrument at the desired temperature.

3. The timing of the travel of the meniscus will be impacted by the operator's reaction time with the stopwatch.
4. The viscometer's calibration constant must be known accurately at the temperature of the experiment.
5. The particular viscometer used must not be defective; that is, it must perform the way the manufacturer warranties it to perform.
6. The viscometer must be clean and dry before use.
7. We did not mention it in the problem statement, but it is also important that when the liquid is flowing the viscometer is maintained in a vertical position and held motionless, since gravity drives the flow in these instruments.

We have many issues to consider when formulating our experimental plan.

Some of the viscometer operation issues presented can be addressed by replication. For example, the flow can be timed repeatedly. Since the operator's response time may lead to random positive or negative deviations in the measured efflux time, the average of repeated timings will give a good value of efflux time for a single loading of a viscometer. Likewise, the uncertainty in sample volume may be addressed by repeating the loading of samples. Each time the sample is loaded there will likely be a small positive or negative error in the volume added, and if this step is repeated, this random effect can be averaged out. The possibility of a defective or dirty viscometer can be explored by filling three or more different viscometers and averaging results across these different, but presumably equivalent, devices. This kind of repetition would also address a calibration problem associated with a single viscometer.

The issue of the water temperature cannot be addressed by replication (unless multiple baths are feasible), but must instead be addressed by calibration. The temperature of the bath must be measured and controlled with devices of established accuracy to eliminate the impact of a temperature offset.

The issue of the vertical placement of the viscometer may be addressed by using specially designed holders that ensure reproducible vertical placement. If these are not available, replication with different, careful operators will allow this source of uncertainty to be distributed across the replicates and, if the effect is random, it will average out.

Following a discussion of these issues, the class agreed that their true replicates would be obtained as follows:

1. A well-calibrated temperature indicator would be used to determine the bath temperature. All groups would use the same, carefully calibrated bath, which would be designed to hold the viscometers vertically.

2. To reduce the impact of timing issues on efflux time, each group would draw up and time the flowing solution three times and use the average of the three timings to produce a single value of efflux time, which would be converted to a single value of kinematic viscosity.
3. Three groups would measure mean efflux time using three different viscometers and following the standard procedure, resulting in three replicates of viscosity.
4. The three independent measurements of kinematic viscosity (by the three groups) would be considered true replicates and would be averaged to yield the final kinematic viscosity value and its error limits (replicate error only).

To see this protocol at work, see Example 5.8, which considers the data analysis of Cannon–Fenske measurements on an aqueous sugar solution.

The thinking process used in Example 2.19 is a general solution and is recommended when a high degree of accuracy is called for: possible sources of both random and systematic error are identified; procedures are chosen to reduce or at least randomize the errors; and finally, true replicates are taken and averaged. Reflecting alone does not guarantee that we will think of all the sources of error in our measurements, but certainly it is an essential step toward ensuring better data acquisition and error reduction.

The last section of this chapter addresses the convention for significant figures on error limits.

## 2.4  Significant Figures on Error

When we determine a number from a measurement of some sort, we do not know that number with absolute certainty. In the previous sections we saw that for data subjected to random error only, we can take multiple measurements, average the results, and express the expected value of the quantity we are measuring as the calculated average along with an appropriate 95% confidence interval of the mean (Equation 2.52).

When presenting the result of such an exercise, we are faced with the choice of how many digits to show for both the expected value, $\bar{x}$, and for the error limits, $\pm t_{0.025,n-1} e_s$. The accepted practice for writing these results is to follow the significant-figures convention – that is, retain all certain digits and one uncertain digit. Since the error limits indicate the amount that the measurement may vary, we can adopt the following rule for error:

> **Rule 1: Sig Figs on Error**
> Report only one digit on error.
> (one uncertain digit)

Thus, a density 95% confidence interval limit of, for example, $2e_s = \pm 0.0323112$ g/cm$^3$ for a mean of $\bar{x} = 1.2549921$ g/cm$^3$ would be expressed as

$$\rho = 1.25 \pm 0.03 \text{ g/cm}^3 \quad (95\% \text{ CI, one uncertain digit})$$

Note that the error limits make the digit 5 on the density uncertain, and therefore to follow the significant-figures convention, we round our results and report only up to that one, uncertain, decimal place.

Although Rule 1 says we keep only one digit on error, we do make an exception to the one-digit rule in some cases, as we see in Rule 2.

> **Rule 2: Sig Figs on Error**
> We may report two digits on error
> if the error digit is 1 or 2.
> (two uncertain digits)

If the uncertainty for density had been $2e_s = 0.0123112$ g/cm$^3$ (error digit is 1 – that is $\pm 0.01$ g/cm$^3$) on an expected value of 1.2549921 g/cm$^3$, the result would be expressed with an additional uncertain error digit:

$$\rho = 1.255 \pm 0.012 \text{ g/cm}^3 \quad (95\% \text{ CI, two uncertain digits}) \quad (2.63)$$

The reasoning behind this second rule is that when the error digit is 1 or 2, the next digit to the right has a large effect when rounding, and keeping the extra digit will allow the reader to calculate the 95% confidence interval with less round-off effect.[19] The user of the number in Equation 2.63 must remember, however, to correctly interpret the density to have only three significant figures, even though four digits are shown (two digits are uncertain). The presence of the error limit with two digits shown makes the two-digit uncertainty clear.

It bears repeating that it is good practice to tell the reader what system you are using to express your uncertainty. We use 95% confidence limits or about two standard errors. Others may use 68% (approximately one standard error) or 99% (approximately three standard errors; see Problem 2.37). If the author fails to indicate which meaning is intended, there is no sure way of knowing which standard is being employed.

---

[19] When the digit you are rounding is 5, there is no good justification to choose to round up or to choose to round down. The best we can do is to randomize this choice. This can be achieved by targeting obtaining an even number after rounding. This practice, over many roundings, is unbiased, whereas always rounding the digit 5 up (or down) is systematic and can introduce bias.

Discussing the sig-figs rules brings up another related issue, that of round-off error. When we round off or truncate numbers in a calculation, we introduce calculation errors. Calculations done by computers may be the result of thousands or millions of individual mathematical operations. Round-off errors are undesirable, and for this reason, within the internal functioning of calculators and computers, those devices retain many digits (at least 32) so that the round-off errors affect only digits that are well away from the digits that we are going to retain.

In our own calculations, we should follow the same practices. If we are doing follow-up calculations with a value that we calculated or measured, in the intermediate calculations we should use all the digits we have. This requires us to record extra digits from intermediate calculations and to use the most precise values of constants obtained elsewhere. In performing 95% confidence and prediction interval calculations, when a value of $t_{0.025, n-1}$ is needed from the Student's $t$ distribution, we should use the most accurate available value, by employing, for example, Excel's T.INV.2T() function call rather than using the truncated numbers from Table 2.3. Shortcuts (using truncated or rounded values) may be employed for estimates, but high-precision numbers are best to use for important calculations. Keeping extra digits is comparable to what a computer or calculator does internally: it keeps all the digits it has. It is only at the last step, when we report to others our final answer of a quantity, that we must report the appropriate number of significant figures. Rounding off intermediate calculations can severely degrade the precision of a final calculation and should be avoided.

## 2.5 Summary

In this chapter, we present the basics of using statistics to quantify the effects of random errors on measurements. The methods discussed are summarized here; the discussion of the impact of reading errors begins in Chapter 3, followed by calibration error in Chapter 4 (both are systematic errors).

### Summary of Properties of a Sample of $n$ Observations of Stochastic Variable $x$, Subject Only to Random Error

- A quantity subject to random variation is called a stochastic variable. Values obtained from experimental measurements are continuous stochastic variables.
- Stochastic variables are represented by an expected value and error limits on the expected value.

- When a stochastic variable is sampled, the mean of a sample set (mean $\bar{x}$; size $n$; sample standard deviation $s$) is an unbiased estimate of the expected value of $x$. This is a formal statement of the correctness of using averages of true replicates to estimate the value of a stochastic variable, $x$.
- With 95% confidence, the true value of $x$ (the mean of the underlying distribution of the stochastic variable) is within the range $\approx \bar{x} \pm 2s \left( \frac{1}{\sqrt{n}} \right)$ (for $n \geq 7$) or the range $\bar{x} \pm t_{0.025,n-1} \left( \frac{s}{\sqrt{n}} \right)$ (for all $n$, but especially for $n < 7$).
- With 95% confidence, the next value of $x$, if the measurement were repeated, is within the range $\approx \bar{x} \pm 2s \sqrt{\left( 1 + \frac{1}{n} \right)}$ (for $n \geq 7$) or the range $\bar{x} \pm t_{0.025,n-1} s \sqrt{\left( 1 + \frac{1}{n} \right)}$ (for all $n$, but especially for $n < 7$) as established by a previous sample set of size $n$, mean $\bar{x}$, and sample standard deviation $s$.
- The accepted convention is to use one digit (one uncertain digit) on error (except if the error digit is 1 or 2, in which case use two uncertain digits on error). Report the value $x$ to no more than the number of decimal places in the error.
- Do not round off digits in intermediate calculations; carry several extra digits and round only the final answer to the appropriate number of significant figures.
- It is recommended to round in an unbiased way. When the digit you are rounding is 5, there is no good justification to choose to round up or to choose to round down. The best we can do is to randomize this choice. This can be achieved by seeking to obtain an even number after rounding. This practice, over many roundings, is unbiased, whereas always rounding the digit 5 up (or down) is systematic and can introduce bias.
- **Bonus advice:** In engineering practice, we usually have no more than two or three significant figures, and we can even expect to have only one significant figure in some cases. Only with extreme care can we get four significant figures. If you have not rigorously designed and executed your measurement with the aim of eliminating error, you have no more than three sig figs, and quite likely you have two or one sig figs. Our advice: **avoid reporting four or more significant figures in an engineering report**. Two significant figures is the most likely precision in engineering work; only if you can justify it should you use three significant figures or more.

## 2.6 Problems

1. Which of the following are stochastic variables and which are not? What could be the source of the stochastic variations?

   (a) Weight of a cup of sugar
   (b) The number of days in a week
   (c) The temperature at noon in Manila, Philippines
   (d) The number of counties in the state of Michigan
   (e) The number typical of counties in a U.S. state

2. In Example 2.1 we calculate Eun Young's mean commuting time to be 28 min, but in Example 2.2 we calculate her commuting time to be 32 min. Which mean is correct? Explain your answer.

3. In Example 2.1 we calculate Eun Young's commuting time to be 28 min. Looking at the data used to calculate this mean time, how much would you expect her actual commuting time to vary? We are not asking for a mathematical calculation in this question; rather, using your intuition, what would you expect the commuting time to be, most of the time? (Once you have made your estimate, see Example 2.15 for the mathematical answer.)

4. How do we determine probabilities from probability density functions (pdf) for stochastic variables? In other words, for the stochastic variable commuting time (see Examples 2.1 and 2.2), if we knew its pdf, how would we calculate, for example, the probability of the commute taking between 50 and 55 min?

5. In Example 2.3 we presented the pdf for the duration of George's daily work commute (Equation 2.13). What is the probability that George's commuting time is 35 min or longer?

6. For the pdf provided in Example 2.3 (Equation 2.13), what is the probability that it takes George between 20 and 25 min to commute?

7. From the pdf provided in Example 2.3 (Equation 2.13), what is George's mean commuting time?

8. Reproduce the plot in Figure 2.5, which shows the pdf of George's commuting time, using mathematical software. Describe in a few sentences the implications of the shape of the pdf.

9. Reproduce the plot in Figure 2.6, which shows the pdf of the normal distribution, using mathematical software. What is the normal distribution? Describe it in a few sentences.

10. Reproduce the plot in Figure 2.9, which shows the pdf of the Student's $t$
    distribution, using mathematical software. What is the Student's $t$
    distribution? Describe it in a few sentences.
11. A sample set of packaged snack food items is sent to the lab. Each item is
    weighed. For a sample for which $n = 25$, $\bar{x} = 456.323$ g, and
    $s = 6.4352$ g (extra digits supplied), calculate the error limits on the
    mean mass and assign the correct number of significant figures. Discuss
    your observations.
12. A sample of 16 maple tree leaves is collected; we measure the mass of
    each leaf. For the leaf mass data shown in Table 2.7, calculate the sample
    mean $\bar{x}$, standard deviation $s$, and standard error from replicates $s/\sqrt{n}$.
    What is the mean maple leaf mass in the sample? Include the appropriate
    error limits.
13. A sample of 16 maple tree leaves is collected; we measure the length
    from the leaf stem to the tip of the leaf. For the leaf length data shown in
    Table 2.7, calculate the sample mean $\bar{x}$, standard deviation $s$, and

Table 2.7. *Sample of maple tree*
*leaves masses and lengths.*

| $i$ | Mass (g) | Length (cm) |
|---|---|---|
| 1 | 0.93 | 9.5 |
| 2 | 1.38 | 11.7 |
| 3 | 1.43 | 10.9 |
| 4 | 1.41 | 10.4 |
| 5 | 0.78 | 8.4 |
| 6 | 1.07 | 10.0 |
| 7 | 2.17 | 11.5 |
| 8 | 1.43 | 12.0 |
| 9 | 1.34 | 11.5 |
| 10 | 0.92 | 8.8 |
| 11 | 0.73 | 7.8 |
| 12 | 0.85 | 8.5 |
| 13 | 1.49 | 11.5 |
| 14 | 1.11 | 9.5 |
| 15 | 1.29 | 9.8 |
| 16 | 0.59 | 8.4 |

Table 2.8. *Sample of lilac bush leaves masses and lengths.*

| $i$ | Mass (g) | Length (cm) |
|-----|----------|-------------|
| 1   | 0.65     | 5.9         |
| 2   | 0.37     | 4.7         |
| 3   | 0.66     | 5.9         |
| 4   | 0.41     | 4.7         |
| 5   | 0.99     | 6.4         |
| 6   | 0.74     | 5.7         |
| 7   | 1.01     | 6.0         |
| 8   | 0.64     | 5.3         |
| 9   | 0.42     | 4.2         |
| 10  | 0.5      | 5.0         |
| 11  | 0.53     | 4.4         |
| 12  | 0.6      | 5.2         |
| 13  | 0.47     | 4.5         |
| 14  | 0.44     | 4.2         |
| 15  | 0.56     | 4.9         |
| 16  | 1.06     | 6.6         |
| 17  | 0.73     | 5.5         |

standard error from replicates $s/\sqrt{n}$. What is the mean maple leaf length in the sample? Include the appropriate error limits.

14. A sample of 17 leaves from a lilac bush is collected; we measure the mass of each leaf. For the leaf mass data supplied in Table 2.8, calculate the sample mean $\bar{x}$, standard deviation $s$, and standard error from replicates $s/\sqrt{n}$. What is the mean lilac leaf mass in the sample? Include the appropriate error limits.

15. A sample of 17 leaves from a lilac bush is collected; we measure the length across the broadest part of the leaf. For the leaf length data supplied in Table 2.8, calculate the sample mean $\bar{x}$, standard deviation $s$, and standard error from replicates $s/\sqrt{n}$. What is the mean lilac leaf length in the sample? Include the appropriate error limits.

16. A sample of 15 leaves from a flowering crab tree is collected; we measure the mass of each leaf. For the leaf mass data supplied in Table 2.9, calculate the sample mean $\bar{x}$, standard deviation $s$, and standard error

Table 2.9. *Sample of flowering crab tree leaves masses and lengths.*

| i | Mass (g) | Length (cm) |
|---|---|---|
| 1 | 0.56 | 5.5 |
| 2 | 0.37 | 4.5 |
| 3 | 0.32 | 4.3 |
| 4 | 0.36 | 5.0 |
| 5 | 0.47 | 4.3 |
| 6 | 0.61 | 5.6 |
| 7 | 0.43 | 4.8 |
| 8 | 0.36 | 4.1 |
| 9 | 0.49 | 4.7 |
| 10 | 0.45 | 4.7 |
| 11 | 0.27 | 3.5 |
| 12 | 0.61 | 4.9 |
| 13 | 0.27 | 4.7 |
| 14 | 0.59 | 5.2 |
| 15 | 0.50 | 5.2 |

from replicates $s/\sqrt{n}$. What is the mean flowering crab leaf mass in the sample? Include the appropriate error limits.

17. A sample of 15 leaves from a flowering crab tree is collected; we measure the length across the longest part of the leaf, from stem to tip. For the leaf length data supplied in Table 2.9, calculate the sample mean $\bar{x}$, standard deviation $s$, and standard error from replicates $s/\sqrt{n}$. What is the mean flowering crab leaf length in the sample? Include the appropriate error limits.

18. The process for manufacturing plastic 16-oz drinking cups produces seemingly identical cups. We weigh 19 cups to see how much their masses vary. For the data shown in Table 2.10, calculate the sample mean $\bar{x}$, standard deviation $s$, and standard error from replicates $s/\sqrt{n}$. What is the mean cup mass in the sample? Include the appropriate error limits.

19. For the Cannon–Fenske viscometer efflux time replicates in Table 5.3, calculate the sample mean $\bar{x}$, standard deviation $s$, and standard error from replicates $s/\sqrt{n}$. What is the 95% confidence interval of efflux time for each of the three viscometers? See Table 5.4 for some of the answers to this problem.

Table 2.10. *Sample of masses of plastic cups.*

| $i$ | Mass (g) |
|---|---|
| 1 | 8.47 |
| 2 | 8.48 |
| 3 | 8.53 |
| 4 | 8.45 |
| 5 | 8.44 |
| 6 | 8.46 |
| 7 | 8.49 |
| 8 | 8.52 |
| 9 | 8.48 |
| 10 | 8.51 |
| 11 | 8.42 |
| 12 | 8.45 |
| 13 | 8.47 |
| 14 | 8.49 |
| 15 | 8.44 |
| 16 | 8.47 |
| 17 | 8.45 |
| 18 | 8.48 |
| 19 | 8.47 |

20. For the three viscosity replicates in Table 5.4, calculate the sample mean $\bar{x}$, standard deviation $s$, and standard error from replicates $s/\sqrt{n}$. What is the 95% confidence interval of solution viscosity?

21. For the maple-leaf mass data given in Table 2.7, what are the sample mean and sample standard deviation? Calculate the level of confidence associated with reporting two, three, or four significant figures on the mean mass. Express your answer for the mean with the appropriate number of significant figures.

22. For the plastic cup mass data given in Table 2.10, what are the sample mean and sample standard deviation? Calculate the level of confidence associated with reporting two, three, or four significant figures on the mean mass. Express your answer for the mean with the appropriate number of significant figures.

23. For the Blue Fluid 175 density data given in Example 2.8, what are the sample mean and sample standard deviation? Calculate the level of confidence associated with reporting two, three, or four significant figures on the mean density. Express your answer for the mean with the appropriate number of significant figures.

24. In Example 2.4 we reported on measurements of length for a sample of 25 sticks from a shop. If the mean stick length is reported with three significant figures, what is the level of confidence we are asserting? Repeat for two and one sig figs. How many sig figs will you recommend reporting?

25. Sometimes lab workers take a shortcut and assume an error limit such as 1% error as a rule of thumb. For the plastic cup mass data in Table 2.10, what are the $\pm 1\%$ error limits on the mean? What level of confidence is associated with these error limits? Comment on your answer.

26. For the sample of packaged food weights described in Problem 2.11, what is the probability that the true mean is within $\pm 1\%$ of the measured mean? Comment on your answer.

27. Based on the sample of plastic cup data given in Table 2.10, what is the probability that a cup (a "next" cup) will weigh less than 8.42 g?

28. The density of an aqueous sugar solution was measured ten times as reported in Example 2.5. What are the mean and standard deviation of the density data? What is the standard deviation of the mean? An eleventh data point is taken and the result is 1.1003 $g/cm^3$ (extra digits supplied). How many standard deviations (of the mean) is this result from the mean? In your own words, what is the significance of this positioning?

29. In Example 2.14 we identified a new data point on density $(\rho_{BF,n+1} = 1.755$ $g/cm^3)$ as being outside the expected 95% prediction interval of the dataset, implying that there may have been a problem with the measurement. If we broadened our prediction interval to, for example, 96% confidence, perhaps the data point would be included. What is the smallest prediction interval (the smallest percent confidence) that we could choose to use to make the new obtained value in that example consistent with the other members of the dataset? Comment on your findings.

30. Using the "worst-case method" discussed in the text, estimate the duration of Eun Young's commute, along with error limits. Perform your calculation for both datasets provided (Examples 2.1 and 2.2). Calculate also the 95% confidence interval and the 95% prediction interval for the

combined dataset. Comment on your answers. What do you tell Eun Young about the probable duration of her commute tomorrow?

31. For Eun Young's commute, as discussed in Examples 2.1 and 2.2, what is the probability that tomorrow the commute will take more than 40 min? Use the combined dataset ($n = 20$) for your calculation.

32. For the stick vendor's data from Example 2.4, what is the 95% prediction interval for the next value of stick length? What is the probability that a stick chosen at random from the vendor's collection is between 5.5 and 6.5 cm? *Hint: we know the next value in terms of deviation. What is the probability?*

33. For the maple leaf data in Table 2.7, calculate the 95% confidence interval of the mean and the 95% prediction interval of the next value of leaf mass. Create a plot like Figure 2.19 showing the data and the intervals. In your own words, what do these two limits represent?

34. For the lilac leaf data in Table 2.8, calculate the 95% confidence interval of the mean and the 95% prediction interval of the next value of leaf length. Create a plot like Figure 2.19 showing the data and the intervals. In your own words, what do these two limits represent?

35. For the flowering crab leaf data in Table 2.9, calculate the 95% confidence interval of the mean and the 95% prediction interval of the next value of both leaf mass and length. Create a plot like Figure 2.19 showing the data and the intervals. In your own words, what do these two limits represent?

36. For the plastic cup data in Table 2.10, calculate the 95% confidence interval of the mean and the 95% prediction interval of the next value of cup mass. Create a plot like Figure 2.19 showing the data and the intervals. In your own words, what do these two limits represent?

37. For data that follow the normal distribution (pdf given in Equation 2.17), calculate the confidence level associated with error limits of $\mu \pm \sigma$, $\mu \pm 2\sigma$, and $\mu \pm 3\sigma$.

38. For all probability density distributions, the integral of the pdf from $-\infty$ to $\infty$ is 1. Verify that this is the case for the normal distribution (Equation 2.17).

39. Do these two numbers agree: $62 \pm 14°C$; $58 \pm 2°C$? Justify your answer with a graphic.

40. An $(x, y)$ dataset is given here: (20, 16.234); (30, 16.252); (40, 16.271); (50, 16.211); (60, 16.201); (70, 16.292); (80, 16.235). Plot $y$ versus $x$. Does $y$ depend on $x$ or is $y$ independent of $x$ within a reasonable amount of uncertainty in $y$? Justify your answer.

41. Equation 2.58 gives the 95% prediction interval for the next value of $x$:

95% prediction interval
for $x_i$, the next          $x_i = \bar{x} \pm t_{0.025,\,n-1}\, s\sqrt{1 + \dfrac{1}{n}}$   (Equation 2.58)
value of $x$:

Using the error-propagation techniques of Chapter 5, show that Equation 2.58 holds.

42. Based on the sample of plastic cup data given in Table 2.10, what is the probability that a cup will weigh more than 8.55 g?

# 3

# Reading Error

We begin this chapter by presenting a situation that is influenced by *reading error*.

**Example 3.1: Temperature indicators of differing resolutions.** *Two accurate temperature indicators simultaneously monitor the temperature in a room over the course of 24 hours. Indicator 1 shows temperature values out to the tenths digit, while Indicator 2 does not give the tenths digit. Data for one 24-hour period from the two indicators are given in Table 3.1. For each dataset, calculate the average measured temperature and error limits on the average temperature (95% confidence interval of the mean; replicate error only). Do the predicted mean temperatures agree?*

**Solution:** We learned in Chapter 2 how to evaluate the random error in data using statistical tools, and now we are asked to apply these techniques to two sets of temperature data. Assuming there are only random errors present in the datasets, we determine the error limits as follows: we calculate the mean $\bar{T}$ and standard deviation $s$ of each dataset ($n$ observations) and construct 95% confidence intervals of the mean for each (Equation 2.54).

$$\begin{array}{l} \text{24-hour average} \\ \text{room temperature} \end{array} = \bar{T} \pm t_{0.025, n-1} \frac{s}{\sqrt{n}} \qquad (3.1)$$

$$t_{0.025, n-1} = \text{T.INV.2T}(0.05, n-1)$$

For the temperature data provided in Table 3.1, these calculations are summarized in Table 3.2, and the two confidence intervals are compared in Figure 3.1 in the form of error bars on the mean temperatures. (For instructions on how to add custom error bars in Excel, see [42].)

Looking at the data in Table 3.1, the data from the two indicators are identical except for the lack of precision on the second indicator. The mean temperatures are different, however (Table 3.2), and the replicate

Table 3.1. *Temperature versus elapsed time from two indicators for use in Example 3.1.*

| Elapsed time (hr) | Indicator 1 (°C) | Indicator 2 (°C) | Elapsed time (hr) | Indicator 1 (°C) | Indicator 2 (°C) |
|---|---|---|---|---|---|
| 0 | 24.4 | 24 | 13 | 24.4 | 24 |
| 1 | 24.4 | 24 | 14 | 24.4 | 24 |
| 2 | 24.4 | 24 | 15 | 24.4 | 24 |
| 3 | 24.4 | 24 | 16 | 24.4 | 24 |
| 4 | 24.4 | 24 | 17 | 24.4 | 24 |
| 5 | 24.4 | 24 | 18 | 24.4 | 24 |
| 6 | 24.4 | 24 | 19 | 24.4 | 24 |
| 7 | 24.4 | 24 | 20 | 24.4 | 24 |
| 8 | 24.4 | 24 | 21 | 25.0 | 25 |
| 9 | 24.4 | 24 | 22 | 25.2 | 25 |
| 10 | 24.4 | 24 | 23 | 25.4 | 25 |
| 11 | 24.4 | 24 | 24 | 25.4 | 25 |
| 12 | 24.4 | 24 | | | |

Table 3.2. *Summary of the replicate error calculations of Example 3.1. To avoid round-off error, extra digits beyond those that are significant are retained in these intermediate calculations.*

| Quantity | Indicator 1 | Indicator 2 |
|---|---|---|
| mean, $\bar{T}$ | 24.536°C | 24.160°C |
| $s$ | 0.325°C | 0.374°C |
| $n$ | 25 | 25 |
| $t_{0.025,n-1}$ | 2.064 | 2.064 |
| $e_{s,random} = \frac{s}{\sqrt{n}}$ | 0.0650°C | 0.0748°C |
| $\pm t_{0.025,n-1}e_s$ | ±0.134°C | ±0.154°C |
| 95% CI upper limit | 24.7°C | 24.3°C |
| 95% CI lower limit | 24.4°C | 24.0°C |

95% confidence intervals calculated are not large enough to allow overlap (Figure 3.1), implying that the two measurements do not agree.

The lack of resolution in the second indicator makes the data less precise. In addition, when we rely solely on replicate error, we appear to underestimate the error in this case, since the confidence intervals do not overlap, even

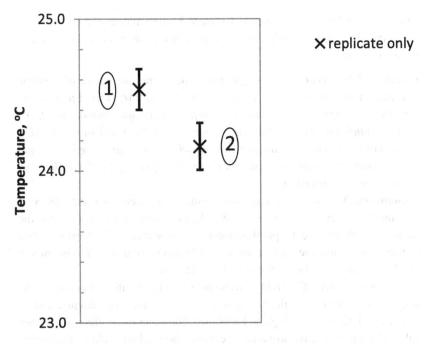

Figure 3.1 In Example 3.1 we compare the means of two datasets taken with two accurate temperature indicators in the same room; the confidence intervals do not overlap. The two indicators have different levels of precision. The error bars indicate 95% confidence intervals based on replicate error only.

though they represent the same experiment and both indicators are known to be accurate. This is a dilemma. It appears we have overlooked a source of uncertainty that is larger than the random error, which is the only error thus far taken into account.

The data for this example were carefully selected to illustrate a problem that can occur as a result of the finite precision offered by laboratory instruments. Looking at Table 3.1, we see that the round-off of the tenths digit for the data shown is fairly large and is repeated for nearly all of the points. The net effect is that, in the truncated data provided by Indicator 2, the mean temperature is underestimated. The data are very consistent, however, so the standard deviation of the dataset is not too large. Thus the overall uncertainty for Indicator 2 is also underestimated.

The problem in Example 3.1 is an aspect of reading error. As we will see in this chapter, when we properly account for reading error, we do not get a better value, but we do get a more realistic estimate of the error limits in situations like those encountered in Example 3.1 (the dilemma in Example 3.1 is resolved in Example 3.10). Another aspect of reading error is showcased in the next example.

**Example 3.2: Error on a single fluctuating measurement of electric current.** *A digital multimeter (Fluke Model 179, DMM) is used to measure the electric current in an apparatus, and for a particular measurement, the signal displayed by the DMM fluctuates between 14.5 mA and 14.8 mA. What value should we report for the current? We wish to use our value in a formal report. What uncertainty limits should we use for reporting the single observed value of the electric current?*

**Solution:** We have a single measurement of electric current. Since it fluctuates between 14.5 mA and 14.8 mA, we might choose to estimate the value as 14.65 mA, the midpoint between the two extremes. Perhaps we could further estimate that the error limits are $\pm 0.15$ mA since that is the amount that toggles our answer between the two observed values.

The problem with this train of thought is that the limited precision on the display means that when the DMM reads 14.5 mA, the true current could be as low as 14.45 mA or as high as 14.54 mA. The same issue affects the higher value. If we postulate the worst-case scenario, the reading could be fluctuating between 14.45 mA and 14.84 mA, which implies a wider spread than our original estimate.

It is desirable to develop a method that allows us to deal with this reading issue appropriately and systematically. Further, the reading uncertainty is in addition to both random effects that may be present in the signal being measured as well as any calibration effects. Thus, we need a standardized way to express the reading error so that we can combine it with replicate variations and with uncertainties from other sources. We also need a way to assess the confidence we have in the error limits developed. We revisit the fluctuating-current scenario in Example 3.6.

Examples 3.1 and 3.2 point out some uncertainties that can be generated by the finite precision of devices. These effects are accounted for through reading error, written as $e_R$, or expressed as standard reading error $e_{s,reading} = e_R/\sqrt{3}$. In the sections that follow, we justify these quantities and learn to use them.

## 3.1 Sources of Reading Error

When we take a measurement using an instrument, we may *read* a value in various ways. For example, we may read from a digital display, or from a pointer indicating a value on a scale, or from a fluid level compared to a scale (Figure 3.2). In each of these cases there is a limit to how precisely we can obtain the reading (limited resolution). Another reading limitation occurs because no device is infinitely sensitive; rather, some signal threshold must be met before the device responds with a change in reading. Fluctuations may also degrade the precision of a reading. Each time we estimate the reading error, we must consider all possible reading-error sources before settling on our estimate of the effect.

Our presentation focuses on three common sources of reading error: sensitivity issues, limited scale subdivisions, and fluctuations (Figure 3.3). We discuss each of these in turn, using a reading error worksheet (Appendix A and Figure 3.4) as our guide to making estimates and proceeding systematically. We first discuss how to obtain a good estimate of reading error $e_R$ for each source. The overall $e_R$ determined is the largest of the three contributions:

Reading error
(nonstandard): $\quad e_R = \max(\text{sensitivity, resolution, fluctuations}) \quad$ (3.2)

Subsequently, we discuss the need to convert $e_R$ into standard reading error $e_{s,reading} = e_R/\sqrt{3}$ (Section 3.2), a quantity that philosophically matches the other errors we define in this book – standard replicate error (Chapter 2) and standard calibration error (Chapter 4) – and which we use when

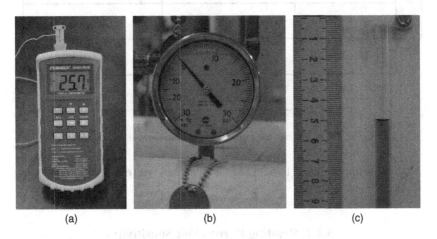

       (a)              (b)              (c)

Figure 3.2 Instrument readings are taken from a digital display, perhaps, or from a pointer indicating a value on a scale, or from a fluid level compared to a scale. (a) A digital temperature indicator using a thermocouple; (b) a Bourdon pressure gauge, and (c) fluid in a manometer tube.

**Sources of Reading Error, $e_R$**

1. Limits of instrument ***sensitivity***
2. Limits of the degree of ***subdivision*** of the scale or display (resolution)
3. Instrument reading ***fluctuations***

Figure 3.3 We identify three sources of reading error $e_R$: sensitivity, limited scale subdivisions, and signal fluctuations.

| | | | | |
|---|---|---|---|---|
| | **Device name:** | | | |
| | Measured Quantity: (give symbol) | | | |
| | Representative value: | (include units) | | Quantity, or **N**ot **A**pplicable |
| | **issue** | **contribution to error** | | |
| Reading error, $e_R$: | *Sensitivity* (from manufacturer or rule of thumb) | How much signal does it take to cause the reading to change? | 1 | |
| Reading error, $e_R$: | *Resolution:* limitation on marked scale or digital readout | Half smallest division or decimal place | 2 | |
| Reading error, $e_R$: | *Fluctuations* with time of observation | (max-min)/2 | 3 | |
| Reading error, $e_R$: | | Maximum of 1, 2, & 3: | $e_R =$ | |
| | Standard reading error: | $e_{s,reading} = e_R/\sqrt{3}$ | $e_s =$ | (units) |
| | | 95% Confidence Interval based on reading error: | $\pm 2e_s =$ | (units) |

Figure 3.4 An excerpt of the reading error worksheet in Appendix A.

combining uncertainties (when adding quantities we must add "apples to apples").

### 3.1.1 Reading Error from Sensitivity

The first and trickiest consideration when determining reading error is determining the instrument's sensitivity. Sensitivity reflects that devices are limited in their ability to respond to small signals.

### 3.1.1.1 Sensitivity Defined

Devices that provide measurements function through clever exploitations of material behavior and physics. For example, a Bourdon gauge (Figure 3.2b) indicates pressure without using electricity or external pneumatic power. This device is composed of a flattened, flexible metal tube that has been curved into a "C" shape. When water or air under pressure enters the tube, the fluid pushes and tries to straighten out the "C," just like your breath unrolls a paper party horn. To turn this effect into an indicator, the motion of the straightening of the "C" is mechanically connected to rotate a pointer around a scale. The flattened tube is carefully constructed so that the angular deflection of the pointer is proportional to a precise pressure increment; the calibrated apparatus is a compact and effective pressure gauge (see YouTube for a look at the insides of a Bourdon gauge).

The scale on a Bourdon gauge indicates the range of pressures that it is capable of measuring. In addition to being limited to the values on its scale, the gauge is limited by its inherent sensitivity to changes in pressure. If the gauge is reading a mid-range pressure – for example, 13.2 psig – the dial will likely move if the pressure increases by 1 psig. Will it move for an increase of 0.1 psig? Will it move for 0.01 psig? The least amount of change that may be "sensed" or reflected by a change in the instrument's display is its *sensitivity*.

To test for sensitivity, we need to apply a very small, known change in whatever the sensor is measuring and to record the amount of signal needed to cause the reading to change. For an analytical balance, for example, we could use standard weights to test how much mass must be applied to have the reading change. The smallest detectable increment in mass is an estimate of the sensitivity for this instrument.

Experimental investigations of sensitivity can be time-consuming since we must also consider that the sensitivity may vary over the measurement range of the device. Manufacturers report sensitivity for their devices, and if we have the manufacturer's specification sheets (or can find them on the Internet), we can look up those values. If the device sensitivity is not readily found in the literature, we must determine sensitivity experimentally or we may choose to estimate it, at least on a preliminary basis. We recommend making sensitivity estimates using some combination of conventional wisdom, experience, and worst-case guesswork.[1] Figure 3.5 contains some suggested rules of thumb

---

[1] Reading and calibration errors are categorized in the field of measurement science (metrology) as "Type B" uncertainties, meaning they are arrived at by logic, reflection, and experience rather than by a statistical process. "Type A" uncertainties are determined through statistics. For more on this topic, see Appendix E.

**Rules of Thumb for Sensitivity**

> <u>Optimistic</u>: Device can sense the smallest change shown on its display.
>
> **Choose: "1" of the last digit**                    display = 74.3°C
>                                                        sensitivity = 0.1°C

> <u>Pessimistic</u>: The last display digit is presumed to be a second uncertain digit included to minimize round-off. Device senses changes only when the input registers *more* than ±1 in the second-to-last digit of the display.
>
> **Choose: 15 × "1" of the last digit**            display = 74.3°C
>                                                    sensitivity = 1.5°C

Figure 3.5  Suggested rules of thumb to use for error quantities. See Example 3.3 for a discussion of the logic behind these limits.

for estimating sensitivity; the logic behind these suggestions is explored in Example 3.3.

**Example 3.3: Rules of thumb for sensitivity (temperature).** *A temperature indicator shows values of temperature including the tenths digit; for example, it may show 74.3°C. What would be reasonable estimates for the sensitivity for this device?*

**Solution:** We do not have very much information about the temperature indicator, but we can explore possible assumptions to make, setting lower and upper bounds on what the sensitivity might be.

*Optimistic assumption.* The manufacturer included the tenths digit of temperature in °C on the display; this may mean that the device is sensitive to increments of ±1 in the smallest digit displayed. Thus, for a reading of 74.3°C, an optimist would assume that this device could sense the difference between 74.3°C and 74.4°C. The sensitivity is estimated to be 0.1°C (1 times the last displayed digit). If the device is more sensitive than this, the sensitivity would be invisible; thus ±1 of the last digit is an upper bound on sensitivity.

*Pessimistic assumption.* Although the manufacturer included the tenths digit in the device display, this may have been because the manufacturer chose to include *two* uncertain digits. "Why would they do that?" you may reasonably ask. It may be that the true sensitivity limit is slightly higher than ±1 in the second-to-last digit. In our temperature example, if the sensitivity were ±1.5 times the second-to-last digit, this would make the sensitivity ±1.5°C, and we would only sense the difference between 74.3°C and 75.8°C with our device.

For such a sensitivity, retaining the second uncertain digit allows us to take the larger uncertainty into account without losing additional precision to round-off error. The sensitivity in this case is estimated to be 1.5°C (15 times "1" of the last displayed digit).

Assuming that manufacturers would not under-sell their devices (that is, sell them as less sensitive than they actually are), the optimistic estimate of sensitivity just described is the most optimistic we could be about sensitivity. The pessimistic assumption, however, is not the most pessimistic assumption we could make. Any fractional sensitivity might cause a manufacturer to report a second uncertain digit; for example, the sensitivity could be ±2.5°C or even lower. Working with the device should give insight as to its performance, and a user should always be monitoring performance to understand the quality of the measurements obtained. The most accurate way to determine sensitivity is to measure it or obtain the manufacturer's reported sensitivity.

In this sensitivity discussion, we have suggested that the user make estimates of error quantities rather than seek rigorous values. We pause here to consider the ins and outs of *estimating* quantities, especially estimating contributions to uncertainty.

### 3.1.1.2 An Aside on Making Estimates

Estimating does not come naturally to students in the physical sciences. In quantitative fields of study, we are most comfortable with determining the "right" answers to questions. In practical applications of science and engineering, however, we may not wish to spend the time or other resources needed to track down some details of a calculation, particularly in the early stages of design and planning. Instead, we rely on an "engineering estimate" and push through with a calculation, returning later to validate our assumptions (and estimates) only if the project proceeds or other circumstances demand it. Knowing how to make good estimates – in our case of uncertainty – is a valuable quantitative skill that keeps our work moving forward. Equally important is developing the discipline of keeping track of our assumptions and estimates so that we never forget the foundations on which our decisions rest. Some of our error estimates never need to be improved upon because they are swamped by other, larger contributions to the uncertainty. Other estimates, however, end up being critical to a project. If that turns out to be the case, those estimates must be revisited, validated, or revised through rigorous measurements or calculations.

We use our best judgment when we are led to make estimates, of course, but it is likely that we will get the estimates wrong from time to time. This is less perilous than it may seem: remember that we are going to combine many

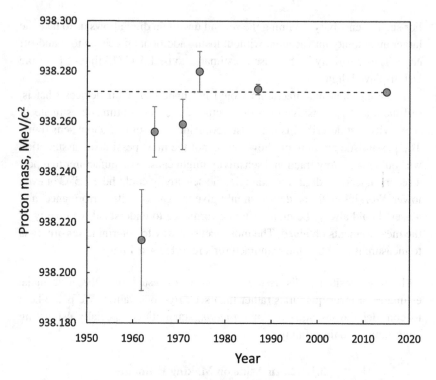

Figure 3.6 The world average value of the proton mass as a function of date of reporting; error bars are those supplied by the original researchers. Data replotted from Lyons [29] with the original source cited as the Particle Data Group. It is not clear what level of confidence was asserted.

of our error estimates. Consider the three reading error sources listed in Figure 3.3, only the largest of which is retained when reporting overall reading error. In Chapter 5 we explain error propagation, which formally combines errors. Often only the most dominant error contribution survives all this combining, so we should not be too nervous about any individual estimate we make. If we are too casual in our estimates, however, we may end up neglecting the dominant error or identifying a spurious dominant error, misleading our decision making.

Estimating error is tricky, and even the experts do not get it right every time. Consider the data in Figure 3.6, which represent the experimentally determined mass of a proton as published between 1960 and 2015 (this comparison was constructed by Lyons [29]). The mass of a proton has not changed since time immemorial, yet there is a clear variation in the reported values. In addition, although uncertainty is acknowledged in Figure 3.6 through the presence of error ranges around the best estimates (shown in the form of error bars on each

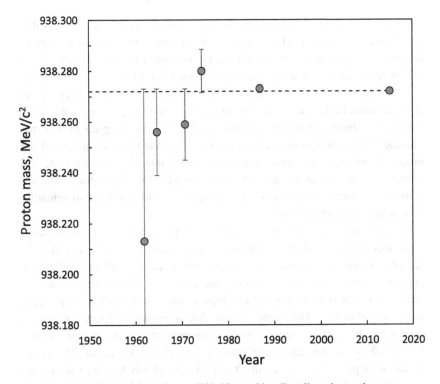

Figure 3.7 The data from Lyons [29]. New, arbitrarily adjusted error bars are included; these indicate how large the error bars would need to be to allow all the data to be consistent with the most precisely known number.

point), the error bars do not overlap. If the error ranges had been correctly estimated each year, they would have looked a bit more like the adjusted error bars shown in Figure 3.7, which overlap and capture the value that ultimately was determined to be the true value of the proton mass. The measurements depicted on these graphs are very good measurements that were made by highly competent researchers – their results are reported with six significant digits – and yet, for the first two decades of research, the uncertainty was significantly underestimated.

What caused the underestimation? We do not know for sure, but we infer that, although the researchers did their best to identify sources of uncertainty in their methods, they must have missed or underestimated some error sources when they calculated their error limits. The predicted values became closer and closer to the true value over time, demonstrating that investigators were able to hypothesize refinements to the measurement methods, permitting research teams to improve the measurements over time, and, happily, the correct value was ultimately determined to a high degree of precision. The usual way to improve an experiment is to identify the largest source of error and attack and eliminate the cause in a subsequent experimental iteration.

The lesson derived from this discussion is not that it is hopeless to estimate error. Rather, we can draw comfort from knowing that even the finest researchers struggle with error estimates, and thus our struggles have a fine precedent. We can also state with confidence that the ultimate success of the project to measure the mass of a proton was guided by reflection on, and elimination of, sources of error in measurements. Thus, the process we are learning is exactly the process we need to master to reach sound scientific conclusions from observations.

How does this discussion of estimates integrate with our error method? First, this discussion dictates that we should proceed in a way that frankly records all estimates, so that each estimate can be reevaluated later in the data-analysis process. Second, as we combine errors, we should test worst-case scenarios to see if making more pessimistic error estimates would, perhaps, make no difference. Doing this would determine whether the outcome is sensitive to this error estimate. Third, in downstream calculations, we should keep track of the impact of the errors on the calculated result. The error-propagation process discussed in Chapter 5 very clearly tells us which parts of a calculation are sensitive to which error, and which parts are not at all sensitive. If we use software to carry out error-propagation calculations, we can test various, revised (more conservative) error estimates with almost no effort, allowing us to zero in on the aspects of our experiments that are the most critical to improving the quality of the final results (for an exercise of this sort, see Example 5.7). Making better decisions from data is our primary goal.

With this new appreciation for the role of making estimates in error analysis, we continue with our task of determining reading error.

### 3.1.1.3 Determining Sensitivity

The discussion in Section 3.1.1.2 sought to establish the appropriateness, usefulness, and validity of making thoughtful approximations. Returning now to the determination of reading error from sensitivity, we offer three approaches to obtaining a value for sensitivity: (1) make an estimate; (2) obtain a value from the manufacturer of a device; and (3) measure it. These are listed from easy to more difficult, but the same order is also from least reliable to most reliable.

**Determining Sensitivity**

1. We could **estimate** it. We would use our experience to decide whether to assume a conservative (that is, pessimistic) or optimistic rule-of-thumb number as an estimate (see Figure 3.5 and Example 3.4). If we suspect that sensitivity is not the dominant contribution to reading error, we could even ignore this contribution (indicate "not applicable" or "NA" in the worksheet).
2. If making such an estimate is not deemed sufficiently reliable, sensitivity may be obtained from the **instrument specifications** ("specs") provided by the manufacturer. The specifications for an instrument can usually be found on the Internet by searching for the instrument model number. One difficulty in reading the manufacturer's specs is that we may encounter different and contradictory vocabulary describing error specifications; the definitions used by the manufacturer must be examined carefully and interpreted correctly (see Example 3.9). In addition, the manufacturer's values will be for the pristine, as-manufactured instrument, which may not reflect the condition of the actual instrument in use.
3. Finally, if sensitivity is judged to be a critical issue, it may be necessary to plan experiments to **determine sensitivity in situ** (as installed and used in the laboratory). This would be the "gold standard" method of determining sensitivity, but it has a high cost in terms of time and effort. To test the sensitivity, the laboratory must have the capability to apply very small known inputs to the instrument over its entire range; it can be expensive to buy the necessary standards and to develop the expertise to carry out a test of sensitivity.

It is left to our judgment as to which method to follow to determine sensitivity. In all cases, it is best practice to note throughout the process (on the supplied worksheet, for example; see Figure 3.4), which assumptions are being made. In the event that revisions or refinements to any values are considered, we need to know what assumptions were made the first time around.

## 3.1.2 Reading Error from Scale Markings

A second, straightforward source of reading error is the finite limit in the degree of subdivision of the scale of an instrument's display (Figure 3.8). This is called the display's resolution. When reading a ruler, for example, if the smallest subdivision is a millimeter, we can perhaps roughly guess, in some circumstances, a reading down to about a third or half a millimeter, but we

Figure 3.8 When we take a reading from an instrument, the precision of the device is limited by the degree of subdivision of its scale. On the left we show a rotameter, a kind of flow meter (compare with the sketch in Figure 2.21). The reading is 42% of full scale and the scale is marked in increments of 2%. On the right is a pneumatic control valve. The position of the valve is marked like an inch ruler, in increments of 1/8 with "full open" at the top and "closed" near the bottom. The reading shown is 5/8 open.

cannot be more precise than that. In Figure 3.8, we see examples of analog scales that present this subdivision type of reading error, which in digital displays takes the form of a finite number of digits displayed (to see an example of a digital scale, see the temperature indicator in Figure 3.2). The subdivision reading-error contribution is easy to determine: we estimate the resolution as half the smallest subdivision on the scale or as half the smallest digit on a digital readout.

$$
\text{Subdivision reading-error contribution} \qquad \boxed{ e_{R,\,est,\,res} = \left( \frac{\text{smallest division}}{2} \right) } \qquad (3.3)
$$

In Figure 3.8, for example, the subdivision reading error for the rotameter is $\frac{1}{2}$ (2%); for the pneumatic valve it is $\frac{1}{2} \left( \frac{1}{8} \text{open} \right)$; and for the temperature indicator in Figure 3.2 it is $\frac{1}{2}$ (0.1°C). In the reading error worksheet, the resolution appears in line 2 (see Figure 3.4 and Example 3.4).

### 3.1.3 Reading Error from Fluctuations

The third origin of reading error is fluctuation (if present) of a digital or analog signal. In a digital display, the last digit may flicker between two or more numbers. On an analog display, the pointer may chatter over a range of scale divisions. Fluctuations in a reading can represent noise in the signal or perhaps noise in the electronics of the device. Fluctuations worsen if the signal is amplified: when viewed on a coarse scale, a signal may not be fluctuating, but after amplification the fluctuations can appear. One approach when fluctuations appear could be to avoid or limit amplification; if we did this, however, the reading error from limited subdivision would increase; thus, the effect is transferred rather than going away.

Reading error from fluctuations is straightforward to determine (line 3 in Figure 3.4). There may be no fluctuation, in which case this contribution is zero or "not applicable." If the pointer or digit is fluctuating, we follow this protocol: estimate the reading error as half the range of the fluctuation.

Fluctuation reading-error contribution (estimated)
$$e_{R,est,fluc} = \left( \frac{(max - min)}{2} \right) \tag{3.4}$$

A pressure gauge with a pointer chattering, for example, between 1 and 3 psi would have a reading error $e_{R,est,fluc} = \frac{1}{2}(3 \text{ psi} - 1 \text{ psi}) = 1$ psi.

All three contributions to reading error are included in the reading error worksheet in Appendix A, part of which is shown in Figure 3.4. The overall $e_R$ determined is the largest of the three contributions:

Reading error (nonstandard)
$$e_R = max(\text{sensitivity, resolution, fluctuations}) \tag{3.5}$$

For practice with the reading error worksheet, Example 3.4 shows how one could determine reading error on a digital temperature indicator recording room temperature.

**Example 3.4: Reading error of a temperature indicator.** *Figure 3.2 shows a digital temperature indicator reading 25.7°C. Determine the standard reading error for this device. Calculate also the error limits (95% level of confidence) based on reading error only ($\pm 2e_{s,reading}$).* [2]

---

[2] When determining the 95% confidence interval error limits with data replicates (large $n$, a Type A uncertainty), the error limits are $\pm t_{0.025,n-1}e_s \approx \pm 2e_s$. We can use this result to propose that $\pm 2e_s$ will also be appropriate error limits for not strictly statistical, or Type B, uncertainties (see Appendix E). The use of $\pm 2e_s$ for error limits in general is intuitive and approximately correct, although it is not universal [21]. The "2" in this approximation is called the *coverage factor*. We discuss this issue and some of the limitations in Section 3.3 and Appendix E.

**Solution:** Using the reading error worksheet, the first step is to estimate the sensitivity. Using the rule of thumb from Figure 3.5, the optimistic estimate would be 0.1°C, since the indicator shows one decimal place. The pessimistic estimate would be 15 times that or 1.5°C (see Problem 3.2). We are new to this indicator, so we choose a number in the middle, neither optimistic nor pessimistic, and we write 0.7°C. This choice is arbitrary, but we hope it is reasonable.

The second question is to write the resolution, which is half the smallest decimal place or 0.05°C for this indicator. Resolution is usually this straightforward.

The last question is to note any fluctuations. We write "NA" or "not applicable" since we do not have any information about fluctuations.

The maximum of these three estimates is our sensitivity estimate at 0.7°C. It is a bit concerning that the number we estimated in a somewhat casual way is the dominant error. That does happen – our advice is to note on the worksheet that this is the case and proceed. If, after combining this error with replicate error (if available) or with calibration error (see Chapter 4), this is still the dominant error, then we should consider gathering more information and revisiting the estimate of sensitivity.

Reading error
(nonstandard)
$$e_R = \max(\text{sensitivity}, \text{resolution}, \text{fluctuations})$$

$$= \max(0.7, 0.05, \text{NA}) = 0.7°C \quad \text{(sensitivity estimate)}$$

We obtain the standard reading error for the temperature indicator in a straightforward application of a formula – we justify the standard reading error formula in Section 3.2.

Standard
reading error
(see Section 3.2)
$$e_{s,reading} = \frac{e_R}{\sqrt{3}} = \frac{0.7}{\sqrt{3}} = 0.404°C \text{ (extra digits provided)}$$

Error limits
95% level of confidence $= \pm 2e_s = \pm 0.8°C$
(see Section 3.3)

$$\boxed{T = 25.7 \pm 0.8°C}$$

Reading error only,
sensitivity dominant,
95% level of confidence

Note the language choice of *95% level of confidence* rather than *95% confidence interval*. With Type B errors, which come from estimates, we use the level-of-confidence terminology, indicating the less statistically linked nature of Type B errors compared to the more statistically rigorous Type A errors (see Appendix E).

Example 3.4 shows two important aspects of the consideration of reading error: (1) likely sources are considered and determined or estimated, and (2) the reasoning used is recorded, in case the estimates and other judgments need to be revisited. The next step is to prepare the reading error to be combined with other sources of uncertainty. This is done by creating the standard reading error.

## 3.2 Standard Reading Error

The reading error $e_R$ that we determined in the last section is a measure of the uncertainty due to instrument reading. We identify $e_R$ as the largest of three effects: limited sensitivity, limited extent of scale subdivisions, and the presence of fluctuations. The quantity (reading) $\pm e_R$ defines a range that could reasonably contain the true value of the measurement.

The quantity $e_R$ is acceptable as a measure of reading error, but it is an arbitrary measure. In addition, we encounter a difficulty when we consider our ultimate goal, which is to combine this error with the replicate error discussed in Chapter 2 and with the calibration error of Chapter 4. To combine independent errors, the measures we combine must each represent the same kind of error measure. There is no one correct way to combine *arbitrarily* identified error measurements, but if we use error measures based on a common principle, there can be a consistent way to combine them. The accepted approach to this issue is to define a standardized error measure, the standard error $e_s$, defined for all types of error as the standard deviation of the error's *sampling distribution* [5, 38]. We discussed sampling distribution in Chapter 2 with respect to Gosset's work establishing the Student's $t$ distribution as the sampling distribution of the mean (when sampling an underlying normal distribution of unknown mean and variance). The sampling distribution is the probability density distribution that indicates, through the integral in Equation 2.12, the probability that a single observation of a stochastic variable will fall in a particular range $[a, b]$.

Definition of $f$:
probability from
probability density function (pdf)
(continuous stochastic variable)

$$\Pr[a \leq x \leq b] \equiv \int_a^b f(x')dx'$$

(Equation 2.12)

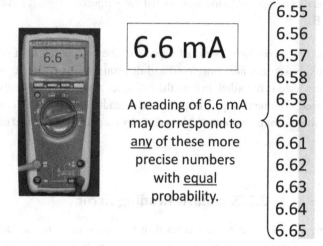

Figure 3.9 If a device such as a digital multimeter (shown) reads 6.6 mA, for example, the true value may be any value between 6.55 and 6.65 mA with equal probability. Knowing this we can deduce the pdf of the sampling distribution of the deviation of the reading from the true value of the measurement.

We turn now to the task of determining the sampling distribution of reading error.

Reading error is caused by limitations in reading values provided by a gauge or a display. These values are not normally distributed (the probability distribution is not a normal distribution; see Equation 2.17), and the sampling distribution of reading error is not the Student's $t$ distribution. We can determine the sampling distribution of instrument reading by reflecting on the sampling process of taking a reading.

Consider the experiment of using the digital multimeter (DMM) shown in Figure 3.9 to measure an electric current in milliamps (mA). The multimeter shown gives only one decimal place for electric current. The true, 2-decimal-place-precise value of current, however, assuming that the multimeter is perfectly accurate, could be as small as 6.55 mA, which would be rounded up by the multimeter to 6.6 mA, or it could be as large as 6.65 mA, which would be rounded down to the same value. All the values between these two limits are equally likely to have given rise to the observed reading. If the reading is 6.6 mA (and the digital multimeter is accurate), there is no chance (zero

**Rectangular probability distribution of half-width $e_R$:**

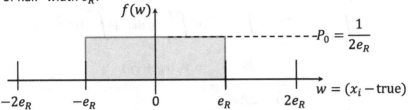

Figure 3.10 The deviation $w$ is a continuous stochastic variable characterized by the rectangular probability density function $f(w)$ shown.

probability) that the true reading is 6.50 or 6.70 mA. There are, however, an infinite number of more precise values that could be the true electric current: all two-decimal-place numbers between 6.55 and 6.65 mA are equally likely to be the true value of the electric current when the digital multimeter reads 6.6 mA.

We see from this discussion that there is a difference between the reading, $x_i = 6.6$ mA in the example discussed, and the true value, perhaps 6.562 mA. The deviation $w_i$ between the reading and the true value of the measured quantity is a continuous stochastic variable related to reading error. The standard reading error is the standard deviation of the sampling distribution of $w_i = (x_i - \text{true})$, where $x_i$ is the reading and "true" is the actual, infinitely precise value of the measurement. The shape of the sampling probability density distribution is rectangular with equal probability for all values between $-e_R$ and $e_R$ ($e_R = 0.05$ mA in the example) and zero probability for values outside this range (Figure 3.10). The half-width of the distribution is $e_R$, which is our reading error. A rough expression of the probability density function is therefore given by

$$
\begin{array}{l}
\text{Rough expression of pdf} \\
\text{of sampling distribution} \\
\text{of reading deviation}
\end{array}
\qquad
f(w) = \left\{
\begin{array}{ll}
0 & w < -e_R \\
P_0 & -e_R \leq w < e_R \\
0 & w \geq e_R
\end{array}
\right.
\qquad (3.6)
$$

where $P_0$ is the unknown probability density associated with observing individual values of $w_i$ between $-e_R$ and $e_R$. We can calculate $P_0$ by recognizing that the area under the probability density function must be 1.

$$\int_{-\infty}^{\infty} f(w')dw' = 1$$

$$\int_{-\infty}^{\infty} f(w')dw' = \int_{-\infty}^{-e_R} 0\,dw' \int_{-e_R}^{e_R} P_0\,dw' + \int_{e_R}^{\infty} 0\,dw'$$

$$= P_0 x' \Big|_{-e_R}^{e_R} = P_0\,(e_R + e_R)$$

$$1 = 2P_0 e_R$$

$$\boxed{P_0 = \frac{1}{2e_R}} \tag{3.7}$$

The normalized probability density distribution of $f(w)$, therefore, is

Probability density function for
reading error deviation
(rectangular distribution)
$$\boxed{f(w) = \begin{cases} 0 & w < -e_R \\ \frac{1}{2e_R} & -e_R \leq w < e_R \\ 0 & w \geq e_R \end{cases}}$$

$$\tag{3.8}$$

This distribution has a zero mean and a standard deviation proportional to $e_R$. In Example 3.5, we calculate $\bar{w}$ and the standard deviation $\sigma$ of the pdf for $w$.

**Example 3.5: Mean and standard deviation of the sampling distribution of reading error deviation.** *What are the mean and standard deviation of the rectangular probability density distribution given in Equation 3.8?*

**Solution:** We begin with the sampling distribution of reading error, which is a normalized rectangular distribution of half-width $e_R$.

$$f(w) = \begin{cases} 0 & w < -e_R \\ \frac{1}{2e_R} & -e_R \leq w < e_R \\ 0 & w \geq e_R \end{cases} \qquad \text{(Equation 3.8, revisited)}$$

For a continuous stochastic variable, the mean of a variable characterized by probability density distribution $f(w)$ is calculated as follows ([21]; Appendix E):

$$\text{Mean of a continuous stochastic variable} \qquad \bar{w} = \int_{-\infty}^{\infty} f(w')w'dw' \qquad (3.9)$$

$$= \int_{-e_R}^{e_R} \frac{1}{2e_R} w' dw'$$

$$= \frac{1}{2e_R} \frac{w'^2}{2}\Big|_{-e_R}^{e_R}$$

$$= \frac{1}{4e_R}\left(e_R^2 - e_R^2\right) = 0$$

For a continuous stochastic variable characterized by probability distribution $f(w)$, the variance is calculated as [38]:

$$\text{Variance of a continuous stochastic variable} \qquad \sigma^2 = \int_{-\infty}^{\infty} \left(w' - \bar{w}\right)^2 f(w')dw' \qquad (3.10)$$

where $\bar{w}$ is the mean from Equation 3.9. For the distribution under consideration $\bar{w} = 0$, and thus we obtain

$$\sigma^2 = \int_{-e_R}^{e_R} \frac{1}{2e_R} w'^2 dw'$$

$$= \frac{1}{2e_R} \frac{w'^3}{3}\Big|_{-e_R}^{e_R}$$

$$= \frac{1}{6e_R}\left(e_R^3 + e_R^3\right) = \frac{1}{3}e_R^2$$

$$\text{Variance of a rectangular distribution of half-width } e_R \qquad \sigma^2 = \frac{e_R^2}{3} \qquad (3.11)$$

The rectangular distribution in Equation 3.8 is the sampling distribution of reading error. The standard error is defined as the standard deviation of the sampling distribution; therefore, the standard reading error is

$$\text{Standard reading error (rectangular distribution)} \qquad e_{s,reading} = \frac{e_R}{\sqrt{3}} \qquad (3.12)$$

With this result, we now have a standardized quantity by which we can report reading error. The significance of this is that we used the same mathematics here as we used for standard replicate error: standard error is standard deviation of the sampling distribution of the quantity. Thus, the standard reading error in Equation 3.12 may appropriately be combined with standard replicate error and with other similarly standardized errors to determine combined standard uncertainty (we combine standard errors by following how variances of independent stochastic variables combine, in quadrature [21]). We discuss the statistical reasoning behind combining standard errors in the next section.

We can practice with our new reading-error tools by revisiting Example 3.2. In that example we were presented with a device with a fluctuating signal. Fluctuations are part of reading error, and with the framework just developed, we are now better able to address the issues raised by fluctuations.

**Example 3.6: Error on a single fluctuating measurement of electric current, revisited.** *A digital multimeter (Fluke Model 179, DMM) is used to measure the electric current in an apparatus, and for a particular measurement the signal displayed by the DMM fluctuates between 14.5 mA and 14.8 mA. What value should we report for the electric current? We wish to use our value in a formal report. What uncertainty limits should we use for reporting the single observed value of the electric current?*

**Solution:** We do not have replicates, and therefore replicate error is impossible to assess. We do have enough information to assess reading error, however.

To determine $e_R$ for this device in this installation, we use the reading error worksheet (Figure 3.11 and Appendix A), which guides us in our decision making. We neglect sensitivity – that is, enter NA ("not applicable") or "neglect" – since we suspect the fluctuations dominate the reading error. For resolution, half the smallest digit is 0.05 mA. The fluctuation contribution is

$$\text{Fluctuation contribution} \quad \frac{(\text{max} - \text{min})}{2} = \frac{14.8 - 14.5}{2} = 0.15 \, \text{mA}$$

which is larger than the resolution contribution. Following the worksheet we calculate that

$$\text{Reading error} \quad e_R = \text{max}(\text{sensitivity, resolution, fluctuations})$$
$$= \text{max}(\text{NA}, 0.05, 0.15) = 0.15 \, \text{mA} \quad (3.13)$$

$$\begin{array}{l} \text{Standard} \\ \text{reading error} \end{array} \quad e_{s,reading} = \frac{e_R}{\sqrt{3}}$$
$$= \frac{0.15}{\sqrt{3}} = 0.0866 \, \text{mA} \quad (\text{extra digits shown})$$

The error limits at the 95% level of confidence based on reading error alone are $\pm 2e_{s,reading} = 0.173$ mA. Our final answer for the reading is

| Device name: | Digital Multimeter measuring current | | |
|---|---|---|---|
| Measured Quantity: (give symbol) | Current, I | | |
| Representative value: | (include units) 14.5-14.8 mA | | Quantity, or Not Applicable |
| issue | contribution to error | | |
| Sensitivity (from manufacturer or rule of thumb) | How much signal does it take to cause the reading to change? | 1 | neglect |
| Resolution: limitation on marked scale or digital readout | Half smallest division or decimal place | 2 | 0.05 mA |
| Fluctuations with time of observation | (max-min)/2 | 3 | 0.15 mA |
| | Maximum of 1, 2, & 3: | $e_R$ = | 0.15 mA |
| Standard reading error: | $e_{s,reading} = e_R/\sqrt{3}$ | $e_s$ = | (units) 0.0866 mA |
| 95% Confidence Interval based on reading error: | | $\pm 2e_s$ = | (units) 0.173 mA ≈ 0.17 mA |

The left side of the table is labeled vertically: Reading error, $e_R$.

Figure 3.11 The reading error worksheet guides us through the steps to determine the reading error for this problem. In addition, it allows us to document the assumptions we make.

| Single, fluctuating, electric-current measurement | current = 14.65 ± 0.17 mA | Reading error only, fluctuation-dominant, 95% level of confidence |
|---|---|---|

We have not yet considered calibration error (Chapter 4), but reading error estimates have already given us a lower bound on the error for this device. We consider calibration error for the digital multimeter in Example 4.7.

## 3.3 Combining Standard Errors

Our overall goal in this book is to determine appropriate error limits for experimentally determined quantities. We discussed that random errors may affect the value we measure; we have developed a standard replicate error to quantify this effect.

$$\text{Standard random error } e_{s,random} = \frac{s}{\sqrt{n}} \tag{3.14}$$

where $n$ and $s$ are, respectively, the number of measurements and the standard deviation of a sample set. As we also noted, reading a quantity from a meter, scale, or other display introduces a second type of error, reading error, and we developed a standard reading error to express this contribution.

$$\text{Standard reading error } e_{s,reading} = \frac{e_R}{\sqrt{3}} \tag{3.15}$$

where $e_R$ is the maximum of reading errors due to limited sensitivity, limited resolution, and fluctuations of signal. Each of these standard errors is the standard deviation of the sampling distribution for each type of error.

The two errors we have discussed (replicate and reading) both independently influence the outcome of an experiment; thus, we must combine them to get their cumulative effect on what we observe. It is not appropriate to simply add them together, however, as we now discuss.

Measurements are continuous stochastic variables. A stochastic variable is a variable that can take on a variety of possible values. When we make an individual measurement, the signal we obtain is the sum of the true value of the quantity that interests us, plus a stochastic contribution from random error, plus stochastic contributions from systematic errors such as reading and calibration error.

$$\begin{array}{l} \text{observed value} \\ \text{(stochastic)} \end{array} = \text{true value} + \begin{array}{l} \text{random error} \\ \text{(stochastic)} \end{array} + \begin{array}{l} \text{systematic errors} \\ \text{(stochastic)} \end{array}$$
$$\tag{3.16}$$

The "true value" in Equation 3.16 is a number that does not have a stochastic character, but the other quantities are all stochastic variables. We thus see that questions about combining errors are questions about how stochastic variables (and their statistics $\bar{x}$ and $\sigma^2$) combine.

The answer to how stochastic variables combine is a well-known result ([5, 21, 38] and Chapter 5). If $x$ and $y$ are independent stochastic variables with variances $\sigma_x^2$ and $\sigma_y^2$, respectively, and $z$ is their sum with variance $\sigma_z^2$, then $z$ is also a stochastic variable, and the three variances are related as follows [5, 38]:

$$z = x + y \tag{3.17}$$
$$\sigma_z^2 = \sigma_x^2 + \sigma_y^2 \tag{3.18}$$

This result tells us how to report a combined standard uncertainty for the case where we have determined the replicate and reading error: we should add the errors as variances – that is, in quadrature. Because we standardized error using a definition related to variance ($e_s$ = standard deviation of a sampling distribution = square root of the variance of the sampling distribution), the combined standard uncertainty $e_{s,cmbd}$ becomes

$$e_{s,cmbd}^2 = e_{s,random}^2 + e_{s,reading}^2 \qquad (3.19)$$

The quadrature rule extends to the sum of more than two independent stochastic variables as well.[3] For the simple additive combination of $m$ stochastic variables $y_j$:

$$z = y_1 + y_2 + \cdots + y_j + \cdots + y_m \qquad (3.20)$$
$$\sigma_z^2 = \sigma_1^2 + \sigma_2^2 + \cdots + \sigma_j^2 + \cdots + \sigma_m^2 \qquad (3.21)$$

Thus, once we have determined the standard calibration error in Chapter 4, we can add this independent error contribution to the combined standard uncertainty.

Standard errors
add in
quadrature:
$$\boxed{e_{s,cmbd}^2 = e_{s,random}^2 + e_{s,reading}^2 + e_{s,cal}^2} \qquad (3.22)$$

Combined
standard
error:
$$e_{s,cmbd} = \sqrt{e_{s,random}^2 + e_{s,reading}^2 + e_{s,cal}^2} \qquad (3.23)$$

The simplicity and correctness of this result is due to our careful standardization of the errors as standard deviations of the sampling distributions.

We offer a final word on error limits. In this chapter we have constructed error limits for $e_{s,reading}$ by somewhat casually doubling the error, making an analogy with how replicate error $e_s = s/\sqrt{n}$ works out for high degrees of freedom $\nu = n - 1$ (see the footnote in Example 3.4).

Replicate
error limits
at high $\nu$
$$\lim_{\nu \to \infty} \left( \pm t_{0.025,n-1} e_s \right) \approx \pm 2 e_s \qquad (3.24)$$

A shift in how we develop error limits is necessary because we arrived at $e_{s,reading}$ by means that involved judgment and estimates rather than pure statistics. Without the firm statistical foundation, we cannot determine 95% confidence intervals for quantities. Standard errors found through judgment and estimates and similar means are called Type B" (Appendix E, Section E.3). Both reading and calibration error are Type B errors.

---

[3] We have indicated that the errors are *independent* stochastic variables. For the quadrature rule to be true, a lesser criterion is actually required, that of being *uncorrelated*. See Chapter 5 and the literature for more on this point [5, 38].

$$\boxed{\begin{array}{c} \text{Error limits} \\ \text{for Type B–influenced} \quad \approx \pm 2e_s \\ \text{standard errors} \end{array}} \qquad (3.25)$$

The combined standard uncertainty $e_{s,cmbd}$ will usually include both Type A (statistics-based) and Type B standard errors; thus we will use the $\pm 2e_s$ error limits for combined standard error as well. Combined standard error describes error on single measurements when replicate, reading, and calibration errors are combined, as well as error on calculated quantities in which the combined error is the result of some formula or mathematical manipulation (Chapter 5).

We further assert that $\pm 2e_s$ error limits have a level of confidence of 95%. We provide justification for this choice in Appendix E. The use of the phrase "with 95% level of confidence" to describe Type B–influenced error limits rather than calling the error limits "95% confidence intervals" (reserved for Type A errors) is designed to mark the results as resulting from Type B–influenced error estimates. We recommend this language when reporting results [21].

## 3.4  Working with Reading Error

To practice working with reading error, we present four examples. The first two examples concern reading error on volumetric devices, while the third explores sensitivity; we also return to the temperature data from Example 3.1, which turns out to be a situation with error dominated by limited resolution. With the new tool of standard reading error, we are able to re-estimate the error limits for the two indicators in Example 3.1 and resolve the issue of having the two indicators agree (within uncertainty).

**Example 3.7: Reading error of a graduated cylinder.** *What is the standard reading error for fluid volume measured with the graduated cylinder shown in Figure 3.12?*

**Solution:** The graduated cylinder shown has scale markings every 2 ml. Consulting the reading error worksheet, we first consider sensitivity (how much signal does it take to cause the reading to change?). Looking at the cylinder markings, we read the volume as 204.5 ml – we judge we could read differing levels of approximately one quarter of the marked interval; that is, we could *sense* the difference between 204.0, 204.5, 205.0, etc. Thus, the sensitivity is 0.5 ml.

Figure 3.12 Graduated cylinders are versatile vessels that can be used to measure out variable volumes of liquids. Volume is read from matching the fluid meniscus level to a scale. When reading such a level, our eye should be level with the fluid level to avoid parallax error [30].

Next is resolution, defined as half the smallest division (resolution = 1 ml). Finally, we consider fluctuations (not applicable). We conclude that

$$\text{Reading error} \quad e_R = \max(\text{sensitivity, resolution, fluctuations})$$
$$= \max(0.5, 1, \text{NA}) = 1 \text{ ml} \quad (3.26)$$

Standard reading error $\quad e_{s,reading} = \dfrac{e_R}{\sqrt{3}} = \dfrac{1 \text{ ml}}{\sqrt{3}} = 0.577 \text{ ml}$

$$\boxed{e_{s,reading} = 0.6 \text{ ml}}$$

Considering only reading error, the volume measurement shown would be written as (answer) $\pm 2e_{s,reading}$.

Figure 3.13 Pycnometers are carefully manufactured to hold precise volumes of liquid; they are used to measure the densities of fluids. There is no reading error associated with pycnometers.

| Single, graduated-cylinder volume measurement | volume $= 204.5 \pm 1.2$ ml | Reading error only, resolution-limited, 95% level of confidence |

**Example 3.8: Reading error of a pycnometer.** *A pycnometer (Figure 3.13) is a piece of specialized glassware that is manufactured to have a precise, known volume. The cap of the pycnometer has a capillary through its center. The vessel is full to its calibrated volume when fluid completely fills that capillary. What is the standard reading error for the pycnometer shown in Figure 3.13?*

**Solution:** Consulting the reading error worksheet, we first consider sensitivity. Sensitivity is not applicable in the pycnometer because there is no value to "sense" – the device is designed to isolate one particular volume. Next, we consider half the smallest division: there are no divisions and therefore this is also not applicable. Finally, fluctuations: also not applicable due to the absence of a reading level to interpret. None of the contributors to reading error apply to the pycnometer.

Reading error $\quad e_R =$ max(sensitivity, resolution, fluctuations)

$\qquad\qquad\qquad\quad =$ max(NA, NA, NA) = NA

| Pycnometer reading error | $=$ | NA (not applicable) | (3.27) |

Reading error is not applicable to pycnometers. As we will see in Chapter 4, it is solely calibration error that determines the uncertainty when we use fixed-volume devices such as pycnometers and volumetric flasks (Example 4.4).

**Example 3.9: Reading error for electrical current using a digital multimeter (non-fluctuating).** *In Example 3.6 we considered how to present a measurement of electric current for a digital multimeter when fluctuations are present. For the digital multimeter shown in Figure 3.9, what is the reading error when measuring a non-fluctuating DC current signal in the range 4.0– 20.0 mA?*

**Solution:** To determine the reading error, we follow the reading error worksheet. There are three issues to consider: sensitivity, resolution, and fluctuations. It has been specified that there are no fluctuations in the reading we encounter here (previously, the fluctuations dominated), so we consider sensitivity and resolution only. Also, we previously estimated that sensitivity would be small in a digital multimeter, but resolution is also small for this measurement. When fluctuations dominated, we could be comfortable with our estimates. Without fluctuations, it would be prudent to check up on sensitivity for this question.

The difference between sensitivity and resolution can be confusing. By our definitions: *sensitivity* is the amount of signal needed to cause the reading to change; *resolution* is half the smallest division or decimal place displayed by the instrument. To determine sensitivity reliably, we must turn to the manufacturer of the device; thus we look up device sensitivity on the manufacturer's website.

The device in Figure 3.9 is a Fluke model 179 digital multimeter. When searching for instrument specifications, we sometimes find a link to specs or an indication that we can download a datasheet for the device. On the datasheet for the Fluke model 179 digital multimeter, we find that for DC current, one of the columns gives "resolution," which for our model is 0.01 mA. To check what Fluke means by resolution (making this check is essential, as we shall soon see), we search for the definition on the manufacturer's website, and we find it in its training library. Fluke defines resolution as "The smallest increment an instrument can detect and display" [11]. We see that we were justified in checking Fluke's definition of resolution, because, in fact, the company defines "resolution" differently than we do.

In this book, *the smallest increment an instrument can detect* is what we call *sensitivity*. For the word *resolution*, we use the definition "half the smallest increment an instrument can display." For the Fluke 179, therefore (using our definitions):

$$\text{Sensitivity} = 0.01 \text{ mA}$$

$$\text{Resolution} = \frac{1}{2}(0.1 \text{ mA}) = 0.05 \text{ mA}$$

Note that we are considering the case when the device is set to measure milliamps and is displaying down to the tenths digit.

Taking the largest of the three issues related to reading error, we find

Reading error $\quad e_R = \max(\text{sensitivity, resolution, fluctuations})$

$$= \max(0.01, 0.05, \text{NA}) = 0.05 \text{ mA} \qquad (3.28)$$

The standard reading error is

Standard
reading error $\quad e_{s,reading} = \dfrac{e_R}{\sqrt{3}} = \dfrac{0.05}{\sqrt{3}} \text{ mA}$

$$\boxed{e_{s,reading} = 0.02887 \text{ mA}}$$
Reading error only,
resolution-dominated,
extra digits displayed
to avoid round-off error

Note that the reading error for electrical current measurement with a Fluke 179 digital multimeter is very small; this is a highly precise device. In Chapter 4 (Example 4.7), we investigate the calibration error for the Fluke 179 to see the impact made by that type of uncertainty.

In the final example of the chapter, we assess the reading errors of the temperature indicators of Example 3.1.

**Example 3.10: Temperature indicators of differing resolution, revised.** *Two accurate temperature indicators (Example 3.1) simultaneously monitor the temperature in a room over the course of 24 hours. Indicator 1 shows temperature values out to the tenths digit, while Indicator 2 does not give the tenths digit. Data for one 24-hour period from the two indicators are given in Table 3.1. For each dataset, calculate the average measured temperature and error limits on the average temperature (95% level of confidence), considering both replicate error and reading error). Do the predicted mean temperatures agree?*

**Solution:** We began this problem in Example 3.1 and evaluated the standard replicate error of the data provided. To determine the standard reading error of

**Digital Temperature Indicator (low resolution)**

| Device name: | Digital Temperature Indicator (low resolution) | | |
|---|---|---|---|
| Measured Quantity (give symbol) | Temperature, $T$ | | |
| | (include units) | | |
| Representative value: | 25°C | | Quantity, or Not Applicable |
| Issue | contribution to error | | |
| Sensitivity (from manufacturer or rule of thumb) | How much signal does it take to cause the reading to change? | 1 | —— |
| Resolution: limitation on marked scale or digital readout | Half smallest division or decimal place | 2 | 0.5°C |
| Fluctuations with time of observation | (max−min)/2 | 3 | N/A |
| | Maximum of 1, 2, & 3: | $e_R$ = | 0.5°C |
| Standard reading error: | $e_{s,reading} = e_R/\sqrt{3}$ | $e_s$ = | 0.289°C (units) |
| | 95% Confidence Interval based on reading error: | $\pm 2e_s$ = | 0.6°C (units) |

(Reading error, $e_R$)

**Digital Temperature Indicator (high resolution)**

| Device name: | Digital Temperature Indicator (high resolution) | | |
|---|---|---|---|
| Measured Quantity (give symbol) | Temperature, $T$ | | |
| | (include units) | | |
| Representative value: | 25.0°C | | Quantity, or Not Applicable |
| Issue | contribution to error | | |
| Sensitivity (from manufacturer or rule of thumb) | How much signal does it take to cause the reading to change? | 1 | —— |
| Resolution: limitation on marked scale or digital readout | Half smallest division or decimal place | 2 | 0.05°C |
| Fluctuations with time of observation | (max−min)/2 | 3 | N/A |
| | Maximum of 1, 2, & 3: | $e_R$ = | 0.05°C |
| Standard reading error: | $e_{s,reading} = e_R/\sqrt{3}$ | $e_s$ = | 0.0289°C (units) |
| | 95% Confidence Interval based on reading error: | $\pm 2e_s$ = | 0.06°C (units) |

(Reading error, $e_R$)

Figure 3.14 For Example 3.10 we use the reading error worksheet to organize our estimates of reading error for two digital temperature indicators.

the devices, we use the reading error worksheet to obtain $e_R$; the standard reading error is obtained from $e_R/\sqrt{3}$. Finally, the replicate and reading independent error contributions are combined in quadrature.

The three parts of reading error are limits on sensitivity, limits on subdivision, and fluctuations. The sensitivity of the temperature indicator is not known, so for a first pass we will neglect it (assume the indicator has been designed to be sensitive to small changes and therefore sensitivity is less of an issue than other contributions to reading error). There is no fluctuation in the reading, so the fluctuation contribution is also zero. The subdivision issue is nonzero and is different in the two indicators. For the first indicator, the smallest digit is the tenths digit; $e_R$ is thus 0.05°C. For the second indicator, the smallest digit is the ones digit; $e_R$ in this case is 10 times larger, $e_R = 0.5°C$.

| Reading error | $e_R = \max(\text{sensitivity}, \text{resolution}, \text{fluctuations})$ | |
|---|---|---|
| Reading error, indicator 1 | $e_R = \max(\text{NA}, 0.05, \text{NA}) = 0.05°\text{C}$ | (3.29) |
| Reading error, indicator 2 | $e_R = \max(\text{NA}, 0.5, \text{NA}) = 0.5°\text{C}$ | (3.30) |

The standard reading errors of the two indicators are subsequently calculated per the reading error worksheet (Figure 3.14).

The final step is to add the replicate (Example 3.1, Table 3.2) and reading standard errors as variances (in quadrature) to obtain $e_{s,cmbd}$ (Equation 3.19). Extra digits are shown in the intermediate calculations presented here.

Indicator 1

$$e_{s,random} = 0.0650°C$$

$$e_{s,reading} = \frac{0.05°C}{\sqrt{3}} = 0.0289°C \quad \text{(resolution dominant)}$$

$$e_{s,cmbd,1}^2 = e_{s,random}^2 + e_{s,reading}^2$$

$$= 0.0051(°C)^2$$

$$e_{s,cmbd,1} = 0.0711°C \tag{3.31}$$

$$2e_{s,1} = 0.14°C \text{ (2 uncertain digits reported on error)}$$

Indicator 2

$$e_{s,random} = 0.0748°C$$

$$e_{s,reading} = \frac{0.5°C}{\sqrt{3}} = 0.2887°C \quad \text{(resolution dominant)}$$

$$e_{s,cmbd,2}^2 = e_{s,random}^2 + e_{s,reading}^2$$

$$= 0.0889(°C)^2$$

$$e_{s,cmbd,2} = 0.2982°C \tag{3.32}$$

$$2e_{s,2} = 0.6°C \text{ (1 uncertain digit on error)}$$

Following Equation 3.25, the error limits at the 95% level of confidence are equal to $\pm 2e_{s,cmbd}$ for each indicator (reading and replicate error considered; added in quadrature).

The new error estimates are reflected in the error bars in Figure 3.15, and the contrast with the purely replicate error bars from Example 3.1 is immediately apparent. Adding in the reading error did not change the error limits much for Indicator 1, but for low-precision Indicator 2 the uncertainty has become quite a bit larger. As we might expect, an indicator that is less precise has more reading uncertainty. By evaluating the standard reading error and adding it in quadrature to the standard replicate error, we now have error bars that better represent the differing precisions of the two indicators.

Well-determined error limits should encompass the true value of the measured quantity with 95% level of confidence; for two accurate indicators of different precision, we expect the error bars of the less precise indicator to completely encompass the error bars from the more precise indicator. With the inclusion of reading error, this is now the case for the example under consideration. Thus we conclude that the two indicators agree within the known uncertainties.

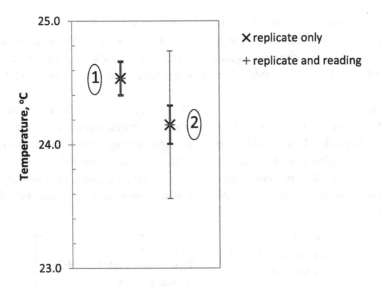

Figure 3.15 When reading error is taken into account along with replicate error, the mean temperature estimates of Example 3.1 agree. The overall error bars, which include both replicate and reading error, correctly reflect the greater uncertainty in the data from the less precise meter. Both sets of error bars overlap and are therefore consistent.

## 3.5 Summary

In this chapter we presented reading error $e_R$ and standard reading error $e_{s,reading} = e_R/\sqrt{3}$. Reading error $e_R$ is the largest of three contributions: errors due to limited sensitivity, limited extent of scale subdivisions, and the presence (or not) of fluctuations.

| Reading error | $e_R = \max(\text{sensitivity}, \text{resolution}, \text{fluctuations})$ |
|---|---|

| Standard reading error | $e_{s,reading} = \dfrac{e_R}{\sqrt{3}}$ |
|---|---|

In Section 3.1.1.2, we also discussed how the definition of standard error allows us to combine independent errors: we do so by adding them as variances of stochastic variables (that is, in quadrature). Errors are standardized by expressing them as the standard deviation of the error's sampling distribution.

We also presented a discussion of estimating error – we may quite reasonably choose to estimate an error contribution that we do not know. We should always record which quantities are estimates, however, in case the estimate dominates the final error determined. When error estimates dominate, it is prudent to revisit them.

In Section 3.3, we discussed Type B–influenced error limits that provide a 95% level of confidence. These limits, also called *expanded uncertainty* (see Appendix E and [21]), are used when reading error and calibration error are included in determining error limits. Reading and calibration errors are determined from nonstatistical methods, and as a result the pathway to appropriate error limits with these standard errors is less direct than for replicate errors.

$$\begin{array}{c}\text{Expanded} \\ \text{uncertainty at} \\ \text{a level of} \\ \text{confidence of 95\%}\end{array} \quad \boxed{\begin{array}{c}\text{Error limits} \\ \text{for Type B–influenced} \\ \text{standard errors}\end{array} = \pm 2e_s} \quad (3.33)$$

In the next chapter we discuss a third significant error source, calibration error. We anticipate that once a standard calibration error is determined, we can include that standard error in the combined standard uncertainty, in quadrature, to obtain better error limits. This will be the case.

## 3.6 Problems

1. In your own words, explain how reading error is standardized. Why is it necessary to standardize error?
2. A temperature indicator is manufactured and found to have a sensitivity of 1.6°C, meaning the device does not register a change in temperature has occurred unless the temperature has changed by at least 1.6°C. The manufacturer is choosing between attaching a display that shows a tenths-place digit and one that does not. For true temperatures between 15.0°C and 24.0°C in 1.6°C increments, what will each of the two displays show? What is your advice to the manufacturer?
3. What is your estimate of the standard reading error on the temperature indicator in Example 2.16? Justify your answer and indicate the values you determine for sensitivity, resolution, and fluctuations.
4. What is the typical reading error for a 500.0-ml volumetric flask? You may consult the literature, if necessary. Justify your answer and indicate the values you determine for sensitivity, resolution, and fluctuations.

5. What is the standard reading error for the Cannon–Fenske viscometers shown schematically in Figure 2.23? Justify your answer and indicate the values you determine for sensitivity, resolution, and fluctuations.

6. What is the standard reading error for the temperature indicator shown in Figure 3.2? Justify your answer and indicate the values you determine for sensitivity, resolution, and fluctuations.

7. What is the standard reading error for the Bourdon gauge shown in Figure 3.2? Justify your answer and indicate the values you determine for sensitivity, resolution, and fluctuations.

8. What is the standard reading error for the manometer shown in Figure 3.2? Large-scale markings on the ruler are in centimeters. Justify your answer and indicate the values you determine for sensitivity, resolution, and fluctuations.

9. What is the standard reading error for the weighing scale readout shown in Figure 1.2? Justify your answer and indicate the values you determine for sensitivity, resolution, and fluctuations.

10. What is the standard reading error for the rotameter shown in Figure 3.8? Justify your answer and indicate the values you determine for sensitivity, resolution, and fluctuations.

11. What is the standard reading error for the pneumatic valve shown in Figure 3.8? Justify your answer and indicate the values you determine for sensitivity, resolution, and fluctuations. Note that the scale shown is created in analogy to a ruler marked in inches, with markings at the 1/4 full scale shown as long dashes and markings at the in-between 1/8 full scale as short dashes. The top line represents the fraction 8/8 open, and the next line down is 7/8 open.

12. What is the standard reading error for the digital multimeter measuring electric current shown in Figure 3.9? Justify your answer and indicate the values you determine for sensitivity, resolution, and fluctuations.

13. What is the standard reading error for the automotive tachometer shown in Figure 3.16? Justify your answer and indicate the values you determine for sensitivity, resolution, and fluctuations.

14. What is the standard reading error for the automotive velocity display scale in Figure 3.16? Justify your answer and indicate the values you determine for sensitivity, resolution, and fluctuations.

15. What is the standard reading error for the stopwatch shown in Figure 3.16? Justify your answer and indicate the values you determine for sensitivity, resolution, and fluctuations.

16. When calibrating the frictional losses from a manual valve (see Figure 3.16), a student notes how many "180° turns" she makes and then

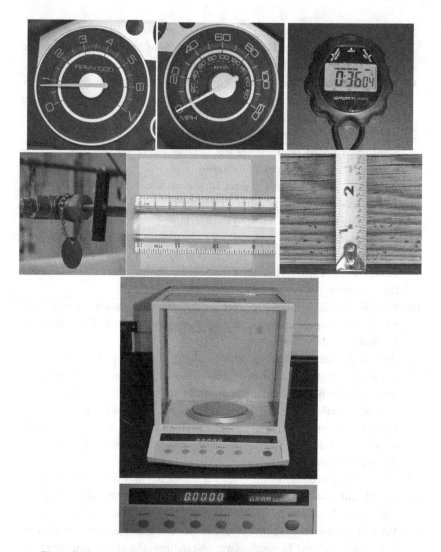

Figure 3.16 An automotive tachometer, an automotive speedometer, a stopwatch, a metering valve with a T handle, a ruler, a tape measure, and an analytical balance. Each provides a reading, which has an associated reading error. See Problems 3.13–3.18.

measures the flow rate and pressure drop across the valve. What is standard reading error on 180° valve turns?

17. What is the standard reading error for the tape measure shown in Figure 3.16? Justify your answer and indicate the values you determine for sensitivity, resolution, and fluctuations.

18. What is the standard reading error for the ruler shown in Figure 3.16? Give an answer for both measuring scales, *cm* and *inches*. Justify your answer and indicate the values you determine for sensitivity, resolution, and fluctuations.

19. What is the standard reading error for the analytical balance shown in Figure 3.16? Justify your answer and indicate the values you determine for sensitivity, resolution, and fluctuations.

20. A fitness app on a cell phone gives the following data on a morning walk:

$$
\begin{aligned}
\text{time:} \quad & 00 : 37 : 15 \text{ hh:mm:ss} \\
\text{air speed:} \quad & 2.4 \text{ mph} \\
\text{distance:} \quad & 1.4 \text{ miles}
\end{aligned}
$$

What are the standard reading errors on these data points? Justify your answer.

21. Eun Young uses her cellular phone to record her commuting time. The reading she obtained was 37:04.80 mm:ss. What is the standard reading error for this device? Comment on your result and its role in determining a good value for commuting time measured with the cell phone.

22. In Appendix E, we discuss the *Guide to the Expression of Uncertainty in Measurement* (GUM) [21] and explain the definitions of Type A and Type B standard errors. Is reading error Type A or Type B? What about replicate error? Discuss your answer.

23. For the rectangular probability distribution $f(w)$ (Equation 3.8) associated with reading error, what is the probability captured in the interval $\bar{w} \pm \sigma$? What is it for $\bar{w} \pm 2\sigma$? $\bar{w} \pm 3\sigma$? How do these compare with the same quantities for the normal distribution?

# 4

# Calibration Error

To begin this chapter, we present an example in which neither replicate nor reading error is able to explain the uncertainty that seems to be present. The uncertainty in this case is due to calibration error. For more on calibration, see the literature [31, 35, 36, 37].

**Example 4.1: Variability of laboratory temperature indicators.** *A laboratory has 11 digital temperature indicators equipped with Fe-Co (iron–constantan[1]), type-J thermocouples. As shown in Figure 4.1, the thermocouples are placed to measure the temperature of a water bath. The temperature of the bath is controlled to be 40.0°C (this value is assumed to be accurate). The readings shown on the indicators are given in Table 4.1. Do the different temperature indicators agree within acceptable limits?*

**Solution:** The thermocouples attached to the 11 digital temperature indicators are placed into a common bath. The readings on each of the indicators are recorded (see Table 4.1); these readings are steady and range from 36.7°C to 41.0°C. The temperature indicators do not seem to agree. Some of the indicators read only to the ones digit, while others read to the tenths digit.

First we consider if the discrepancy could be due to replicate error. In the setup in Figure 4.1, all 11 thermocouples are reading, simultaneously, the temperature of the same thermostated bath. The temperature of the bath is stable at the set temperature, so there is no issue of random temperature error in this case.

We could consider that the discrepancy is due to reading error. For each type of indicator, the reading error may be worked out with the help of the reading error worksheet, considering sensitivity, resolution, and fluctuations. For sensitivity we do not know much about this error, but we can use the

---

[1] Constantan is an alloy of copper (55%) and nickel (45%).

Table 4.1. *Temperatures reflected by 11 thermocouples submerged in a* 40°C *water bath.*

| Index | $T$ (°C) |
|-------|----------|
| 1 | 39.3 |
| 2 | 40.0 |
| 3 | 36.7 |
| 4 | 39.9 |
| 5 | 39.8 |
| 6 | 39 |
| 7 | 38 |
| 8 | 39.9 |
| 9 | 39 |
| 10 | 40.1 |
| 11 | 40.0 |

**Experiment:**

- 11 temperature indicators with iron–constantan thermocouples; all measuring bath temperature
- Water bath controlled to 40.0°C

Figure 4.1 Eleven digital temperature indicators equipped with Fe-Co thermocouples are set to measure the temperature of a foil-covered tank of water controlled to 40.0°C.

optimistic rule of thumb (ROT) of plus/minus the last digit (see Figure 3.5). For many commercial instruments sensitivity is not the dominant reading error, making this choice less critical. The optimistic ROT would give sensitivities of $1°C$ and $0.1°C$ for the two types of meters shown. For resolution, applying the definition in the worksheet we obtain $0.5°C$ and $0.05°C$ for the two types of meters. Finally, for fluctuations, the signals do not fluctuate, so the third source of reading error is not applicable for either indicator. From these reflections and following the worksheet, we obtain the standard reading errors for the two types of indicators.

Reading error:        $e_R = \max(\text{sensitivity}, \text{resolution}, \text{fluctuations})$

Less precise indicator: $e_R = \max(1, 0.5, \text{NA}) = 1°C$

Standard
reading error:        $e_{s,reading} = \dfrac{1°C}{\sqrt{3}}$

$e_{s,reading} = 0.57735°C$

Error limits:     $\boxed{2e_s = 1.2°C}$     Reading error only
                                              (optimistic ROT
                                              for sensitivity)
                                              95% level of confidence

More precise indicator: $e_R = \max(0.1, 0.05, \text{NA}) = 0.1°C$

Standard
reading error:        $e_{s,reading} = \dfrac{0.1°C}{\sqrt{3}}$

$e_{s,reading} = 0.057735°C$

Error limits:     $\boxed{2e_s = 0.12°C}$     Reading error only
                                              (optimistic ROT
                                              for sensitivity)
                                              95% level of confidence

The temperature readings along with error limits based on reading error[2] are shown in Figure 4.2. The error bars on 7 of the 11 indicators (64%) capture the true value of temperature; this is less agreement than we expect, as we are using 95% level of confidence error limits ($\pm 2e_s$). It also turns out that these error limits are much smaller than can be justified by the process used to manufacture and calibrate thermocouples (see Wikipedia for more on the physics of thermocouples).

---

[2] Both of these numbers are based on our guess of the sensitivity, which is optimistic for both calculations. One hypothesis for the lack of agreement could therefore be that we were too optimistic about sensitivity. This could be investigated; it turns out not to be the problem.

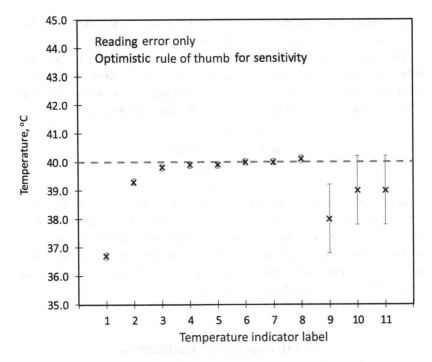

Figure 4.2 When only reading error is considered, the error bars on the more-precise indicators are quite small and many do not capture the true value of temperature. For the less-precise indicators, the error bars are broad. Even with broad error bars, however, one of the three does not capture the true temperature value. These are error limits at the 95% level of confidence, and we would expect that 95% of the error limits so constructed would capture the true value. This is not the case; perhaps some error has been missed in the determination of the error limits (Hint: the calibration error is missing).

In this chapter we consider limitations in the calibration of thermocouples as a source of additional uncertainty. Calibration is the process through which the accuracy of a device is established. Calibration error turns out to contribute strongly to the uncertainty in thermocouples and in many other measurements; the error bars on these measurements are revisited in Example 4.5.

In Figure 4.1, we showed 11 devices displaying their versions of water temperature in a controlled, 40.0°C bath. The temperatures shown on these devices ranged from 36.7°C to 40.1°C. These differences are not random errors – the bath temperature is constant, and all devices are registering simultaneously. In Example 4.1, we examined reading error as a possible explanation of these discrepancies and found that the more precise devices should exhibit no more than $\pm 0.12°C$ variation due to reading error; this variability is not enough to explain all the observations.

The observed differences are due to the functioning of the temperature indicators. Apparently, the devices are not sufficiently accurate for different instruments to give the same values when measuring the same thing. The issue of the *accuracy* of measurement devices is an issue of calibration.

> *Accuracy* is associated
> with calibration.

In this chapter we discuss calibration and calibration error, which is uncertainty left over after a device has been calibrated. Calibration error is the final of the three independent error sources that are part of our error-analysis system. Once calibration error is determined, we combine this error with replicate and reading error (in quadrature) to determine the combined standard uncertainty associated with a measurement.

## 4.1 Introduction to Calibration

How does a manufacturer of a device ensure that the device accurately displays what it is purported to measure? As we discussed in Chapter 3, measurement devices function through exploitation of material behavior and physics. For example, at the beginning of Section 3.1.1.1 we described the workings of a Bourdon gauge. In that device, a flexible hollow metal tube unfurls in response to pressure; this motion is linked to a pointer; and a reading is produced by noting the pointer position relative to a scale. For the scale on a Bourdon gauge to read pressure in some convenient units, however, the device must be *calibrated*. The calibration step is the essential last step in sensor design, as it makes the connection between the device's operation and the user's desire to infer an actual measured quantity from the device.

> Calibration makes the connection between the device's operation
> and the ability to infer a physical quantity from the device.

Calibration is a step made to establish the correctness and utility of a device. In calibration, the performance of a *unit under test* (the device) is put up against a *standard*, which is a device or material whose correctness or properties are known. In Figure 4.3 a calibration installation is shown schematically, in which

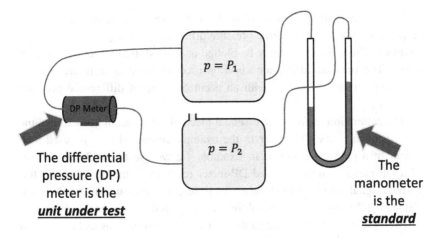

Figure 4.3  Calibration is a process of validating a device (*unit under test*, the DP meter shown here) against an accurate *standard* (in this figure, the manometer). The DP meter and the manometer both measure $(\Delta p)_{std} = p_1 - p_2$.

a differential-pressure (DP) meter is configured to respond to the difference in pressure between two reservoirs of pressure difference, $\Delta p = p_1 - p_2$. To know what that pressure difference between the two reservoirs actually *is*, the same reservoirs are simultaneously connected to a manometer; the reading on the manometer serves as a differential-pressure standard. A manometer works by reflecting a pressure difference in terms of a height difference in the working fluid of the manometer. If the density of the manometer fluid is known accurately, and if the height difference in the manometer can be recorded accurately, then the pressure difference causing the response of the manometer is known and can be associated with the reading on the DP meter.

The DP meter calibration setup in Figure 4.3 functions as follows. (1) Set the pressures in the reservoirs such that there is a stable pressure difference between them (This pressure difference is the independent variable during calibration). (2) Record the response of the DP meter to the pressure difference – the form of this signal and how it comes to correlate with the pressure difference is a consequence of the design and limitations of the DP meter. (3) Measure the response of the manometer to the *exact same* pressure difference (that is, measure the manometer-fluid heights difference; this value leads to a differential pressure standard value, $(\Delta p)_{std}$). (4) Repeat steps 1–3 with different pressure differences, exploring the entire range of operation of the unit under test (15–20 data points recommended).

The data from the calibration of the DP meter are a set of data pairs of meter response (signal)$_{cal}$ versus true pressure difference $(\Delta p)_{std}$ measured by the standard. These data pairs may be plotted and transformed into a calibration curve. The finished calibration curve is used to associate a future reading from the device (signal)$_{raw}$ with an accurate value of differential pressure $(\Delta p)_{meter}$.

The calibration curve is the essential piece of information needed to allow an instrument – the DP meter in the example discussed – to be used as a sensor of a physical quantity. In Example 4.2 we use the calibration curve of a DP meter to turn observed DP-meter readings in milliamps of electric current signal into measured values of differential pressure in psi (lb$_f$/in$^2$). Example 4.2 is the first of several linked examples discussed in Chapters 4–6. In Example 4.2 the DP meter calibration curve is given to us as known. In a series of discussions later in this text, we show all the steps to determining that calibration curve. For more context on how the linked examples fit together, see the overview in Section 1.5.2 and Figure 1.3.

**Example 4.2: \*LINKED\* Using a calibration curve: Differential-pressure meter.** *Water driven by a pump is made to flow through copper tubing at a variety of flow rates depending on the position of a manual valve (Figure 4.4). Along a 6.0-ft section of copper tube, the pressure drop is measured with a differential-pressure meter as a function of water volumetric flow rate, $Q(gal/min)$. The raw differential-pressure signals from the DP meter are in the form of electric current $(I, mA)_{raw}$ ranging from 4.0 to 20.0 mA (Table 4.2). These signals may be translated to differential pressures in psi by using the calibration curve for the device:*[3]

| DP-meter calibration curve: | $$(\Delta p, \text{psi})_{meter} = 0.23115(I, \text{mA})_{raw} - 0.92360$$ | (4.1) |
|---|---|---|

*For the raw electric-current data provided in Table 4.2, what are the associated differential-pressure measurements in psi? What are the appropriate error limits (95% level of confidence) to associate with each differential pressure data point? Plot the final results of the flow experiment for pressure drop versus flow rate; include the appropriate error bars on differential pressure.*

**Solution:** Designing a successful sensor is a creative and practical activity that puts together knowledge of material properties and physics, and, through a clever design, produces a reading that can be matched, through calibration, to

---

[3] We develop this calibration equation in Examples 4.8, 5.6 5.7, 6.1, 6.2, and 6.6; see the linked examples overview in Section 1.5.2 and Figure 1.3. At this stage, we simply use the result.

Table 4.2.  *Raw data signals from a differential-pressure meter in terms of current in milliamps as a function of fluid volumetric flow rate in gallons per minute (gal/min). Extra digits are shown for flow rate to avoid round-off error in downstream calculations.*

| Water flow rate ($Q$, gal/min) | DP meter raw reading ($I$, mA)$_{raw}$ |
|---|---|
| 0.5664 | 5.2 |
| 0.5664 | 4.9 |
| 0.7552 | 5.8 |
| 0.8496 | 6.3 |
| 1.0148 | 7.1 |
| 1.1564 | 8.3 |
| 1.2508 | 9.2 |
| 1.3216 | 9.7 |
| 1.5104 | 12.0 |
| 1.6048 | 12.25 |
| 1.7936 | 14.0 |
| 1.9588 | 15.65 |
| 2.0532 | 16.7 |
| 2.2892 | 19.4 |

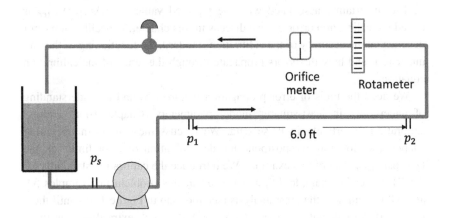

Figure 4.4  The pumping loop considered in Example 4.2. The DP meter is set up to measure the pressure difference across a 6-ft section of copper tubing.

known physical quantities. As is true for the DP meter in this example, devices are often designed so that their calibration curves are linear (Equation 4.1). A device with a linear calibration curve has a proportional response: if you double the signal, you double the response.[4] A proportional response is intuitive to the user of the sensor; also, the error limits on the calibration curve are more straightforward to determine with a linear calibration curve (see Chapter 6).

The DP meter used in this example has been calibrated, and the calibration curve forms a straight line when fit with an ordinary least-squares algorithm. To convert raw DP-meter signals $(I, \text{mA})_{raw}$ to calibrated differential-pressure values $(\Delta p, \text{psi})_{meter}$, we substitute the experimental electric current signal values from Table 4.2 into the calibration equation (Equation 4.1) and calculate the differential pressures. Note that the calibration equation, like all calibration curves, only has meaning for a particular set of units; reporting the units on a calibration equation is essential. The data from Table 4.2 converted to $\Delta p$ in psi are shown in Table 4.3.

We are also asked to obtain the appropriate error limits on the values of $(\Delta p, \text{psi})_{meter}$. Thus far in this text we have seen how to evaluate data replicates as well as how to assess uncertainty that results from reading error. The values in Table 4.3 do not represent data replicates, nor do they represent direct readings from an instrument; rather, these values were obtained from the operation of a device and the subsequent use of a calibration curve. The uncertainties in these values of $(\Delta p, \text{psi})_{meter}$ are associated with the details of how they were obtained.

The uncertainty associated with the reported values of $(\Delta p, \text{psi})_{meter}$ is called the calibration error. As we discuss in this chapter, the calibration error depends on the accuracy of the calibration standard, the functioning of the unit under test, and how the errors propagate through the terms of the calibration equation.

We need the tools of error propagation (Chapter 5) and an understanding of uncertainty in least-squares model parameters (Chapter 6) to address calibration uncertainty in these data. We discuss these tools in upcoming chapters, so for now we postpone the determination of error limits on the $(\Delta p, \text{psi})_{meter}$ data of this example. We introduce the calibration standards for the DP meter in Example 4.8, and we revisit this problem in Examples 5.9 and 6.8 to complete the error analysis (see the map in Figure 1.3). Until then, we report the calculated values of $(\Delta p, \text{psi})_{meter}$ with extra digits, as these are intermediate values (Table 4.3). Once the uncertainty in $(\Delta p, \text{psi})_{meter}$ is

---

[4] This is only strictly true when the calibration line is transformed to coordinates that include the point (0, 0); it is always possible to produce this shift (see Appendix F).

Table 4.3. *Pressure drops measured with a calibrated differential-pressure meter along a 6.0-ft section of 0.315-in inner-diameter copper tubing, as a function of volumetric flow rate. Extra digits on $\Delta p$ and $Q$ values are shown to avoid round-off error in downstream calculations.*

| Water flow rate ($Q$, gal/min) | DP meter results ($\Delta p$, psi)$_{meter}$ |
|---|---|
| 0.5664 | 0.2784 |
| 0.5664 | 0.2091 |
| 0.7552 | 0.4171 |
| 0.8496 | 0.5327 |
| 1.0148 | 0.7176 |
| 1.1564 | 0.9950 |
| 1.2508 | 1.2030 |
| 1.3216 | 1.3186 |
| 1.5104 | 1.8502 |
| 1.6048 | 1.9080 |
| 1.7936 | 2.3125 |
| 1.9588 | 2.6939 |
| 2.0532 | 2.9367 |
| 2.2892 | 3.5608 |

known, we can determine the appropriate number of digits to present in the error limits and in the results.

Calibration is an essential step in sensor design, installation, and use. In this chapter we discuss the issues behind calibration, and we present a methodology to identify a standard error related to calibration for a device of interest. The independent standard calibration error thus determined, or estimated, is added in quadrature to replicate and reading error to determine a final, overall, combined standard uncertainty, $e_{s,cmbd}$, for values obtained from a device.

## 4.2 Determination of Calibration Error

Accuracy of an instrument is limited, even for a brand-new device. As we learn to determine calibration error, we must keep mind several issues that may affect device accuracy.

1. *Limitations due to the quality of the calibration standard.* The calibration accuracy of a device can be no better than the accuracy of the standard. The rule of thumb in calibration is that the accuracy of the standard should be at least four times the desired accuracy in the device being calibrated. For example, to calibrate the DP meter of Example 4.2, a manometer is used as the standard (Example 4.8). The quality of the differential pressure determined by the manometer depends on how accurately the density of the fluid is known, how accurately the height of the fluid is read during calibration, and, more subtly, how well temperature is controlled during the calibration, as density is a function of temperature. An essential choice during calibration is the choice and quality of the standard.
2. *Installation issues.* Sensors must be installed as specified by the manufacturer if they are to perform to the levels specified by the manufacturer. Installation issues include the quality of electrical connections (including electrical grounding) and physical placement (in terms of exposure to the elements, radiation sources, whether the sensor is level, how close the device is to vibrating equipment, etc.).
3. *Proper use.* Some devices have multi-step operating instructions that must be followed to obtain the expected performance. All aspects of operating the device may affect the reliability of the calibration.
4. *Device age.* In addition to the accuracy limitations just discussed, if the unit is not brand-new or has not been well maintained, this status also affects the accuracy of the device. Physical devices can degrade over time, due to either chemical effects (metal surface oxidation is a common culprit), material fatigue (moving parts cause physical changes in metals and plastics), or electronic signal drift, among other reasons. An older device may not perform to the calibration limits of a brand-new device.

These issues and perhaps others may affect the accuracy of a device. The users of the measurements from a device are the ones ultimately responsible for any decisions made that rely on those results. Our understanding of the net accuracy of a device is encoded in the calibration error limits.

Calibration error of a device is a determined result that we must either estimate, find from the manufacturer, or measure ourselves. We organize our options by once again using a worksheet (Appendix A); a portion of the calibration error worksheet is shown in Figure 4.5. As with reading error, there are three suggested paths for determining the value for the calibration error, and, as before, we describe them from easiest (and least reliable) to hardest (and most reliable). The three paths to determining calibration error for a device of interest are (1) estimate by a rule of thumb (ROT); (2) obtain

| Device name: | | | |
|---|---|---|---|
| **Measured quantity:** | **Symbol:** | | **Representative value:** (include units) |
| | | | *Estimate of $e_s$:* (or **Not Applicable**) |
| Rule of Thumb Method: Least significant digit on provided value | Least significant digit varies by at least $\pm 1 = \pm 2e_s$ | | |
| Rigorous Method: Manufacturer maximum error allowable | $2\,e_s \approx$ | | |
| Method 3: User calibration | $2e_s \approx$ | | |

| | | | 95% CI, Calibration error only: quantity$\pm 2e_s$ |
|---|---|---|---|
| | Maximum of Methods 1 – 3 | $e_s =$ | |
| | | $2e_s =$ | (units) |

Figure 4.5  An excerpt of the calibration error worksheet in Appendix A.

the calibration-error information from the manufacturer; and (3) calibrate the device in-house. Whichever is largest is identified as the calibration error.

$$\text{Calibration error:} \quad e_{s,cal} = \max(\text{ROT, lookup, user})$$

We discuss each path in turn.

### 4.2.1  Calibration Error from Rule of Thumb

The way to obtain a value for calibration error with the least amount of work is to **estimate it.** When making an estimate, we can be optimistic or pessimistic; we provide rules of thumb for both cases.

*Optimistic Rule of Thumb*: We may reason that if the manufacturer provides the digit on a display, it must mean that the displayed value is accurate to within $\pm 1$ of that digit. This is the most optimistic estimate of the calibration error.

$$\begin{array}{l}\text{Optimistic} \\ \text{rule of thumb} \\ \text{for estimated } e_{s,cal}\end{array} \qquad \boxed{e_{s,est} = 1 \text{ times last digit}} \qquad (4.2)$$

*Pessimistic Rule of Thumb*: To determine a conservative (pessimistic) estimate of calibration error, we obtain a worst-case high value and a worst-case low value for the measurement quantity. Since we write error limits as $\pm 2e_s$ (see Appendix E), this means that the estimate for $e_s$ would be one fourth of the interval between the worst-case high and worst-case low values:

Rough estimate
of standard calibration error
(easy, but not necessarily accurate)

$$e_{s,est} \approx \frac{(\text{max} - \text{min})}{4}$$     (4.3)

If we feel we have grounds for picking less pessimistic estimates, we can use the estimation method with less than worst-case estimates (see Example 4.3). The reliability of this estimate depends on the user and the user's familiarity with the performance and condition of the device. For instance, when a highly accurately device such as an analytical balance is not well cared for (for example, placed on a shaky surface, not leveled, often used by inexperienced personnel, rarely calibrated against standards), it would make more sense to estimate the calibration error as worse than that certified for a brand-new instrument, since the conditions to which it has been subjected are likely to degrade performance.

We provide these rules of thumb for convenience, but they do not come with any guarantees. Assuming that all displayed digits are accurate is a particularly risky choice – a more pessimistic rule of thumb is a safer choice.

### 4.2.2 Calibration Error from Manufacturer

A good way to get the calibration error is to **look it up.** If the instrument is functioning the way the manufacturer warranties it to perform, then the number the manufacturer gives for calibration error should be a good number (see Examples 4.4 and 4.5). Calibration information may be found on the Internet by searching for the model number of the device. The calibration error may be posted as a $\pm$error limit, in which case the standard calibration error is calculated as

Standard calibration error
from manufacturer's error limits

$$e_{s,est} = \frac{(\text{error limit})}{2}$$     (4.4)

The number found from the manufacturer will be the best (smallest error) that the manufacturer could justify. Because more accurate devices are more valuable than less accurate ones, manufacturers may be expected to claim the best number they can justify, earning them the best price for their instruments. Some devices, including low-cost thermocouples, have surprisingly high calibration errors, even when measured values are displayed with many digits (we shall see that this is true for type-J thermocouples in Example 4.5). The manufacturer's specification sheets (specs, spec sheets, or datasheets) make this clear.

### 4.2.3 Calibration Error Determined by User

If the calibration information is not obtainable (as occurs sometimes when using an older instrument, for example), or if there is reason to believe that the manufacturer's calibration numbers may not apply (perhaps the instrument has been modified or has not been well maintained, or it was built in-house), then the solution is to **calibrate it in-house.** To carry out a calibration on a unit under test, one must have accurate standards and sufficient experimental expertise to carry out the calibration. To reveal systematic effects during calibration, we recommend to randomize trials when calibrating (see Section 2.3.4). Calibration error $e_{s,cal}$ must capture all the variability in accuracy associated with the use of the device. User calibration and the determination of calibration error is an involved process, which requires the tools of Chapter 5 and 6 (error propagation, model fitting). At the end of this chapter (in Example 4.8), we begin a user calibration in which we tackle the calibration of the DP meter featured in Example 4.2 (see the linked-examples overview in Figure 1.3).

As just noted, user calibration of an instrument is an involved process; however, we also use this term, sometimes in quotes ("user calibration"), for a less formal practice. When we know that results from a device are likely to be less accurate than they should be, we can add an estimated amount of error to the $e_{s,cmbd}$ to account for the lost accuracy. For example, if an inexperienced technician runs the instrument as a substitute, the results are expected to be less accurate. In this case we might "user calibrate" by estimating a large $e_{s,cal}$ in the worksheet. As always, we keep track of our estimates and revisit them when we consider the error limits on the final reported quantities. Example 4.3 includes a "user calibration."

All three potential sources for calibration error are included in the calibration error worksheet in Appendix A, part of which is shown in Figure 4.5. The $e_{s,cal}$ determined is the largest of these three values.

Calibration error $\quad \boxed{e_{s,cal} = \max(\text{ROT, lookup, user})} \qquad (4.5)$

To complete this introduction to calibration error, we turn now to one final topic – the special case of error effects that appear near the outer limits of sensor validity. We present several examples of working with calibration error in Section 4.4.

## 4.3 Special Considerations at Lower and Upper Performance Limits

All instruments have a limited range over which they provide accurate readings. Here we discuss special error considerations that affect the low and high ends of an instrument's range of operation.

### 4.3.1 Lower Limit

At low signal strength, there is a *limit of determination* (LOD), defined as the lowest value that can accurately be measured by the device [19]. To determine the LOD for a device, we begin with its calibration curve. Consider the low-signal range of the calibration curve shown schematically in Figure 4.6. The

**What's the lowest accurate signal the instrument is capable of measuring?**

$$value = (slope)signal + (intercept)$$

$$answer = value \pm 2e_{s,cal}$$

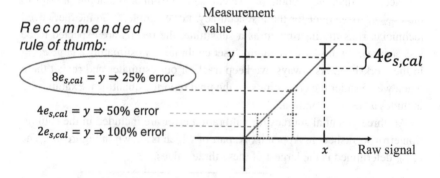

Figure 4.6 When signals are low, the size of the signal can be the same as the size of the error limits $y = 2e_{s,cal}$ (100% error). We make a choice as to how much error we tolerate, and this determines the limit of determination (LOD).

calibration error limits on the device's measurements may be determined from the methods of this chapter. As shown in Figure 4.6, we can display these error limits by creating error bars that show the values $\pm 2e_{s,cal}$ on each possible $y$ value on the calibration curve.

As we consider raw signal values $x$ that are smaller and smaller, the corresponding measurement values $y$ also approach zero. At some low value of $x$, the lower-limit error bar on $y$ will arrive at zero – it will bottom out. When the lower limit error bar hits the $x$-axis ($y = 0$), the measurement value $y$ is just equal to its error-bar amount.

$$\text{Device measurement value: } y \pm 2e_{s,cal}$$

$$\text{Lower error limit bottoms out: } y - 2e_{s,cal} = 0$$

$$
\begin{array}{l}
\text{100\% error on } y \\
\text{(95\% level of confidence):}
\end{array}
\qquad
\boxed{y_{100\%error} = 2e_{s,cal}}
\qquad (4.6)
$$

Since the value is equal to the uncertainty at this point, this represents a point with 100% error.

It would be unusual if we could tolerate 100% error in measurements we use. To avoid using data with intolerably high uncertainty, we must reject data points with this much, or larger, percentage error. While 100% error is often an intolerably high amount of error, perhaps this is too loose of a standard. What is the maximum amount of error that we can tolerate? There is no one answer to this question. We choose for ourselves the lower limit we can tolerate. We call this limit the *limit of determination* for the instrument. Values below this limit are essentially undetectable, since they are too uncertain to retain.

As we established in the preceding discussion, if we choose $2e_{s,cal}$ for the LOD, we are choosing to reject measurements with 100% error or higher. If we raise the LOD to $4e_{s,cal}$, we reject points with 50% error or higher (Figure 4.6); at LOD $= 8e_{s,cal}$, the rejected data have 25% error or higher:

$$\text{Measurement value} = y \pm 2e_{s,cal}$$

$$\%\text{Error} = \frac{(\text{error})}{(\text{value})} = \frac{2e_{s,cal}}{(\text{LOD})} \qquad (4.7)$$

$$\text{100\% error on } y: \quad \text{LOD} = 2e_{s,cal}$$

$$\text{50\% error on } y: \quad \text{LOD} = 4e_{s,cal}$$

$$\text{25\% error on } y: \quad \text{LOD} = 8e_{s,cal}$$

For many uses, data with 25% uncertainty are worth retaining. If 25% error is too much to tolerate for an application, we can continue to explore lower and lower acceptable error percentages, associated with higher multiples of $e_{s,cal}$ (higher LOD; see Example 6.12). As a rule of thumb, we recommend that the LOD be taken as no less than $8e_{s,cal}$.

$$
\begin{array}{c}
\text{Recommended rule of thumb} \\
\text{for limit of determination (LOD)} \\
\text{(maximum of 25\% error on } y\text{)}
\end{array}
\qquad \boxed{LOD = 8e_{s,cal}} \qquad (4.8)
$$

The choice is, as in many circumstances associated with error limits, up to the user and should be made carefully and with justification. The amount of time to allocate to thinking about a number's error estimate is driven by the end-use of the number and how accurate that value needs to be (if a high-quality number is essential, we should spend a great deal of time thinking about precision and accuracy).

To calculate the LOD for a device we need $e_{s,cal}$, which, as discussed, we can either determine ourselves or rely on the manufacturer's value. For thermocouples, we rely on the manufacturer, as discussed in Example 4.5; a related LOD calculation appears in Example 5.3. In Chapter 6 we learn how to obtain $e_{s,cal}$ when using an ordinary least-squares method to fit a model to calibration data; see Example 6.7 for a discussion of determining the LOD for a DP meter (part of the linked DP-meter examples; see Figure 1.3.). When relying on the manufacturer's calibration, the LOD is supplied with the instrument's documentation or is made available online. To find the LOD value, check the specification sheets, where LOD appears with the specification of the instrument's *range*. A lower limit to the range of 5% would be typical, meaning that the LOD recommended is 5% of full scale.

### 4.3.2 Upper Limit

At the upper end of a meter's range we encounter a second special considera-tion, that of *saturation* (Figure 4.7). The physics that makes a meter work has a limited range. Consider the Bourdon gauge. At high pressure, the curved metal tube that forms the sensing element straightens out completely, and increasing the pressure still further causes no change in output signal. Such a device is said to be saturated. To ensure that the top point of the scale remains a valid measurement, the gauge display would typically be designed to top out below the saturation of the physical response.

Measurements above the upper limit of the calibration curve – above $(x, y)_{upper\ limit}$ – are not interpretable: the meter response may continue to follow the linear relationship above $(x, y)_{upper\ limit}$, but the meter's response

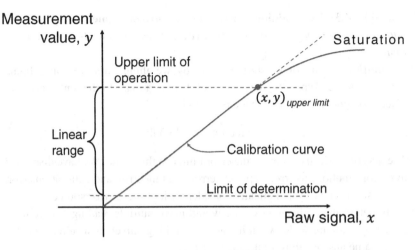

Figure 4.7 Most sensors and meters are designed to exhibit linear behavior. The range of operation spans from the limit of determination at low signal to an upper bound at high signal, which is usually a limit at which the signal response is no longer linear. If the signal "tops off" and the device fails to respond linearly to additional stimulation (as shown here), the response is called saturation. Some devices will give signals above their upper operation limit; it is up to the user to recognize that these are not valid signals.

may also be considerably above or below the linear relationship. It is not advisable to use any number at or above the top of the scale. For some devices it may even be advisable to refrain from using data that come near the top of the scale. Unless we specifically know how the instrument's manufacturer handles numbers near the top of the display range, it is prudent to avoid using these numbers in our results.

## 4.4 Working with Calibration Error

The various sources of calibration error discussed in Section 4.2 – estimate with a rule of thumb, obtain from manufacturer's spec sheets, and perform a user calibration – are incorporated into the calibration worksheet in Appendix A. Whichever is largest is identified as the calibration error.

Calibration error:     $e_{s,cal} = \max(\text{ROT, lookup, user})$

To practice with calibration error, we present several examples.

**Example 4.3: Determining mass on a bathroom scale.** *Using a home bathroom scale, what is your weight? Give an answer with well-reasoned error limits.*

**Solution:** Many of us start our day by weighing ourselves on a home bathroom scale. Typically, the scale will display a mass to the tenths decimal place. Imagine that we read that answer to be

$$\text{scale reading} = 145.3 \text{ lb}_m$$

The question asks us to determine error limits on this value. We have discussed three contributions to error: replicate error, reading error, and calibration error. In this solution we systematically consider each of these error sources.

To determine replicate error, we would need multiple readings. If we have a well-functioning scale, weighing and reweighing ourselves would give close to the same answer; thus, replicate error is zero.

To determine reading error, we begin with the reading error worksheet from Appendix A. For sensitivity (what is the smallest amount of mass that would cause the reading to change?), we have no information, so we skip this for now; for resolution, half the smallest decimal place is $0.05$ lb$_m$; for fluctuations, we see none (that would be typical). The reading error is thus $e_R = \max(\text{sensitivity, resolution, fluctuations}) = \max(\text{NA}, 0.05, \text{NA}) = 0.05$ lb$_m$ or whatever number we put for sensitivity, if that number is larger. If we assume for sensitivity that the last digit will toggle when just enough weight is added (this is the most optimistic assumption we can justify), that would imply that the sensitivity is $0.1$ lb$_m$. Following the worksheet, the value we use for $e_R$ is the largest of the three contributions to reading error; thus $e_R = \max(0.1, 0.05, \text{NA}) = 0.1$ lb$_m$ (the sensitivity). The standard reading error is then

$$e_{s,reading} = \frac{0.1}{\sqrt{3}} = 0.0577 \text{ lb}_m \quad \text{(sensitivity, tentative)}$$

Since the number we used for reading error is an optimistic estimate, it would be a good idea to explore a less optimistic estimate to see how it would affect the error limits we establish. If we assume that the scale can only sense a half-pound change in weight (a lower sensitivity), the reading error would be $e_R = \max(0.5, 0.05, \text{NA}) = 0.5$ lb$_m$, and the standard reading error is

$$e_{s,reading} = \frac{0.5}{\sqrt{3}} = 0.2887 \text{ lb}_m \quad \text{(lower sensitivity)}$$

The choice is up to us. Since this is not a high-precision device, it is possible that it had a low sensitivity, so we proceed with this less optimistic choice.

For calibration error, the calibration error worksheet guides us to consider three pathways. First, using a suggested rule of thumb based on the last displayed digit, toggling by one:

$$\text{calibration ROT} \qquad 2e_{s,est} = 0.1 \text{ lb}_m$$
$$e_{s,est} = 0.05 \text{ lb}_m$$

Second, we may find the manufacturer's specifications, which we will not take the time to seek out since the need for accuracy in this case is not very high. The third suggestion from the worksheet is to obtain the calibration error from a user calibration; the question in this example is not so urgent or important to warrant that effort; thus user calibration is not applicable in this case. The calibration error worksheet guides us to choose the largest of our estimates. Hence,

$$e_{s,cal} = \max(\text{ROT, lookup, user})$$
$$= \max(0.05, \text{NA}, \text{NA})$$
$$= 0.05 \text{ lb}_m$$

which is the estimate from the rule of thumb; this is not a particularly reliable choice. Then again, since the question in this example is not a measurement of great import, we press on with our calculations and revisit this issue later.

Having determined estimates of the three error sources, the combined standard error is the sum of the three, added in quadrature.

$$e_{s,cmbd}^2 = e_{s,random}^2 + e_{s,reading}^2 + e_{s,cal}^2$$
$$= 0^2 + (0.2887)^2 + (0.05^2) \qquad (4.9)$$
$$e_{s,cmbd} = 0.293 \text{ lb}_m$$
$$2e_{s,cmbd} = 0.6 \text{ lb}_m$$

We obtain that the weight is $145.3 \pm 0.6$ lb$_m$, dominated by reading error (pessimistic ROT on sensitivity).

In addition to the answer requested, we obtain insight into the factors that we have taken to govern the precision and accuracy of the answer. We found (mostly we assumed) the following:

1. The device is reproducible (there is no replicate error).
2. We estimated that sensitivity dominated the reading error at $e_R = 0.5$ lb$_m$. This was a guess, and for our combined standard uncertainty calculation it is the dominant error contribution. If we desire a more accurate error estimate, we need to measure sensitivity. This would be done by obtaining a set of standard weights and testing to see how the device responds. In

particular, we are interested in the sensitivity of the device in the range of the measurement, when it is measuring 140 lb$_m$ or so.

3. For calibration error, we chose to make an optimistic assumption rather than either looking up or determining the calibration. As a result, calibration error was irrelevant to the final answer (Equation 4.9). If we desire a more accurate error estimate, we would need to calibrate the device ourselves (or look up the manufacturer's calibration). If we have the standard weights to measure sensitivity as discussed earlier, we could also check the calibration, although larger standard weights would be required.

The final answer we obtained is an optimistic reporting of the accuracy of an inexpensive weighing device. If we know that the device is old and has been poorly maintained, we may find that we would be more comfortable with a less optimistic error estimate, reflecting that history. We could think of this as a type of "user calibration." We could follow the steps discussed around Equation 4.3 and in Example 1.2; that is, we estimate high and low values and calculate the error $e_{s,est}$ as

$$e_{s,est} = \frac{(\text{high value} - \text{low value})}{4}$$

We might estimate in the current case that although the device read 145.3 lb$_m$, we only trust the device to register to within a couple of pounds:

$$\text{user calibration} = \frac{(\text{high value} - \text{low value})}{4}$$
$$= \frac{(147 - 143)}{4}$$
$$= 1 \text{ lb}_m$$

Calibration error is the largest of the three values obtained when following the calibration error worksheet.

$$\text{Calibration error:} \quad e_{s,cal} = \max(\text{ROT, lookup, user})$$
$$e_{s,cal} = \max(0.05, \text{NA}, 1)$$
$$= 1.0 \text{ lb}_m$$

The combined standard uncertainty for the data from the bathroom scale is the random, reading, and calibration error combined in quadrature.

$$e_{s,cmbd}^2 = e_{s,random}^2 + e_{s,reading}^2 + e_{s,cal}^2$$
$$e_{s,total}^2 = 0^2 + (0.2887)^2 + (1.0)^2 = 1.083 \qquad (4.10)$$
$$= 1.0 \text{ lb}_m \quad (\text{calibration error dominated})$$

(Compare Equations 4.10 and 4.9.) With a 95% level of confidence, we report the weight as $145 \pm 2$ lb$_m$. This is an answer dominated by calibration error (estimated by the user).

The ability to estimate error well comes from having experience with an instrument's performance in a wide variety of situations. This was a simple (and trivial) example of using the tools of this book to estimate error limits. Note that we were constantly balancing the desire for precise error limits against the effort of chasing down all the contributions to error. This is a common feature of error analysis. In the current problem, the trivial nature of the question tipped the balance toward putting in less effort. For more impactful calculations, the balance will tip the other way. In addition to providing error limits, the discussion of this example tells us which error source is the one we believe dominates (Equation 4.10). In the case of the bathroom scale, we judge that calibration accuracy (from a user estimation) limits the results obtained with the scale.

**Example 4.4: Calibration error of a pycnometer.** *A pycnometer is a piece of specialized glassware used to measure fluid density; it is manufactured to have a precise, known volume (see Figure 4.8 and Example 3.8). The pycnometer has a glass cap with a capillary through its center. The vessel is full when fluid*

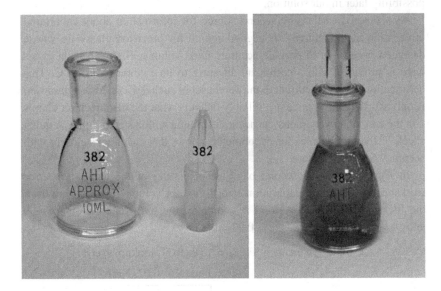

Figure 4.8 Pycnometers are carefully manufactured to hold precise volumes of liquid; they are used to measure the densities of fluids.

*completely fills that capillary. What is the standard error for the volume of fluid contained by the pycnometer?*

**Solution:** We were introduced to pycnometers in Example 3.8, and in that example we considered reading error for fluid volume. As discussed there, pycnometers do not have reading error since there is no "reading" connected to using fixed-volume glassware.

Another source of error would be from calibration error. Consulting the calibration error worksheet, we are guided by three issues: (1) a rule of thumb based on a reading scale; (2) error ranges provided by the manufacturer (this will turn out to be essential for pycnometers); and (3) an error range determined by the user. Whichever is largest is identified as the calibration error.

$$\text{Calibration error:} \quad e_{s,cal} = \max(\text{ROT, lookup, user}) \quad (4.11)$$

The first calibration-error issue is based on a reading, but this is irrelevant since this piece of glassware does not give its value by way of a reading. The third calibration-error issue is user calibration, which is a possible pathway to calibration error for the pycnometer: it would require that we calibrate the pycnometer by weighing it before and after carefully filling it with fluids of known densities. This is always possible, but recalibrating the device is not our first choice. User calibration could also involve an estimate of a "user calibration error" based on the process of using the device; we discuss this possibility later in our solution.

As we noted earlier, the second source for information about calibration error – the manufacturer – is a good source for precision glassware whose designed purpose is to provide accurate fixed volume. High-precision glassware is provided with a certificate attesting to its calibration at 20°C. This certification is very important to the purchaser of such devices, as the glassware is only worth its high price if it reliably delivers what the manufacturer claims – in the case of a pycnometer, an accurate volume within stated limits. A quick check of Internet sources for pycnometers shows that cost tracks closely with accuracy for pycnometers.

In Appendix A, the second page (reverse side) of the calibration error worksheet gives typical calibration error-limits for volumetric glassware from the literature [12]. For a 10-ml pycnometer, the usual calibration limit is $\pm 2e_{s,cal} = \pm 0.04$ ml or $e_{s,cal} = 0.02$ ml.

$$\text{Calibration error:} \quad e_{s,cal} = \max(\text{ROT, lookup, user})$$
$$= \max(\text{NA}, 0.02, \text{NA})$$

$$\begin{array}{l} \text{Pycnometer} \\ \text{calibration error} \\ \text{(from manufacturer):} \end{array} \quad \boxed{e_{s,cal} = 0.02 \text{ ml}} \quad (4.12)$$

Note that achieving this accuracy is only possible if the pycnometer is used as intended. If the operator does not follow good laboratory technique, the manufacturer's accuracy may not be achieved. If this is the case, it may be wise to "user calibrate" by making an adjustment to lower the claimed accuracy, taking into account your estimate of the expertise used when making the measurement. In any case, the manufacturer error limit is a lower bound – that is, the least expected error associated with the device.

We have not considered replicate error in this example. One way to assess the expertise of the individual using the pycnometer would be to ask for replicates. If the replicate error is large, it should be added to the calibration error in quadrature. A replicate error that exceeds the calibration error would indicate that the pycnometer is not being used correctly; if this is observed, more training would be called for.

Assuming that the person using the pycnometer is sufficiently well trained, the final error limits are

$$e_{s,cmbd} = e_{s,random}^2 + e_{s,reading}^2 + e_{s,cal}^2$$
$$= 0^2 + 0^2 + (0.02)^2$$
$$= 0.02 \text{ ml}$$

Volume of fluid
in pycnometer:
$$\boxed{V = 10.00 \pm 0.04 \text{ ml}}$$
(calibration error)

Individual pycnometers would be expected to have a volume between 9.96 and 10.04 ml. The error in volume for any given pycnometer would be a systematic error in any densities measured with that individual device. To eliminate the effect of the random deviation in pycnometer volume for density measurements, density replicates involving multiple pycnometers could be performed and averaged. This is a form of systematic-error randomization, as discussed in Section 2.3.4. Each pycnometer will have its individual error in volume, with the true volume being sometimes larger than 10.00 ml and sometimes smaller, usually (that is, with a 95% level of confidence) within the $\pm 0.04$ ml error range. Replicates with different pycnometers would be true replicates, and the systematic volume effect would average out.

**Example 4.5: Variability of laboratory temperature indicators, revisited.**
*A laboratory has 11 digital temperature indicators equipped with Fe-Co thermocouples (these data were considered previously in Example 4.1). As shown in Figure 4.1, the thermocouples are placed to measure the temperature of a water bath. The temperature of the bath is controlled to be $40.0^oC$*

*(this value is assumed to be accurate). The readings shown on the indicators are given in Table 4.1. Do the different temperature indicators agree within acceptable limits?*

**Solution:** Thermocouples attached to 11 digital temperature indicators are placed into a common bath. The readings on each of the indicators are recorded (see Table 4.1); these readings are steady and range from 36.7°C to 41.0°C. The temperature indicators do not seem to agree.

We examined this situation in Example 4.1, beginning with the effect of replicate error. The signals are stable, however, and all the indicators are measuring the temperature of the same thermostated bath. Thus, the effect is not a result of stochastic variability (random error).

As discussed in Example 4.1, the variability among the 11 thermocouples shown in Figure 4.1 is not within error limits based solely on reading error, which we estimated in Example 4.1 to be (extra digits provided)

$$\text{Less-precise indicator:} \quad e_{s,reading} = 0.577°C$$
$$\text{More-precise indicator:} \quad e_{s,reading} = 0.0577°C$$

Both estimates use an optimistic rule of thumb for sensitivity. Reading error alone does not account for the amount of variability observed in the data (Figure 4.2).

The third error source is calibration error (Figure 4.9). From the calibration error worksheet, we have three potential sources of information on the magnitude of calibration error; whichever is largest is identified as the calibration error.

$$\text{Calibration error:} \quad e_{s,cal} = \max(\text{ROT, lookup, user})$$

Beginning with measuring the calibration error, we are not planning to recalibrate and there is no issue of the user not operating the instrument well, so we discount user calibration as something to consider. The calibration rule of thumb works a bit like the reading-error calculation (like reading error, the calibration rule of thumb is based on the precision of the display). This rule of thumb is highly optimistic, however, so we discount it as well, since we plan to pursue the first information source, the manufacturer. As we hinted at in earlier discussions involving thermocouples, there is a great deal of uncertainty in temperature values obtained from thermocouples. Thus our best path is to consult the manufacturer's posted error limits to determine the calibration error.

We begin by describing how thermocouples function. These devices exploit the small voltage difference that is produced in a conductive material when

| Device name: | Digital Temperature Indicator w/t-couple | | |
|---|---|---|---|
| **Measured quantity:** | **Symbol:** | | **Representative value:** <br> **(include units)** |
| Temperature | $T$ | | 39.2°C |
| | | | *Estimate* of $e_s$: <br> (or **Not** Applicable) |
| Rule of Thumb Method: <br> Least significant digit on provided value | Least significant digit varies by at least $\pm 1 = \pm 2e_s$ | | 0.1°C |
| Rigorous Method: <br> Manufacturer maximum error allowable | $2e_s \approx 1.1°C$ | | 0.55°C |
| Method 3: <br> User calibration | $2e_s \approx$ | | *NA* |

| | | | **95% CI, Calibration error only:** <br> quantity$\pm 2e_s$ |
|---|---|---|---|
| | Maximum of Methods 1–3 | $e_s = 0.55°C$ | |
| | | $2e_s = 1.1°C$ | $39.2 \pm 1.1°C$ <br> (units) |

Figure 4.9 Filling out the calibration error worksheet guides us to a good estimate of calibration error and helps us keep track of our assumptions.

there is a temperature gradient in the conductor (Figure 4.10). Wires made of different materials respond differently to temperature gradients, and we can make a temperature sensor by creating an electrical circuit from such wires. The two wires are joined at one end, and the joint or junction of the two wires is exposed to the temperature we would like to measure, $T_{measure}$. The other ends of the two wires are exposed to a second, common temperature called the reference temperature $T_{ref}$. The voltage difference between the wire ends at temperature $T_{ref}$ is measured, and this voltage, along with material information about the wires, can be used to back-calculate the temperature of the junction, $T_{measure}$ (see Wikipedia for more on thermocouples).

The performance of a thermocouple is strongly influenced by the material properties of the wires from which it is manufactured. The purity and morphology of the conductors must be carefully controlled during manufacture. Batch-to-batch composition variation among wires must also be minimized. In addition, the accuracy of the sensor device is strongly affected by any changes taking place in the wires' properties over the lifetime of the device.

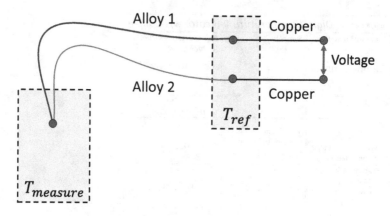

Figure 4.10 A thermocouple consists of wires made of two different metal alloys connected at one end and exposed to $T_{measure}$, and not connected at the other ends but exposed to a common reference temperature $T_{ref}$. The temperature $T_{measure}$ can be back-calculated from $T_{ref}$, the voltage difference between the two wires at $T_{ref}$, and information on previously determined material functions associated with the two alloys in use.

To make high-quality thermocouple wires, special care must be taken. When extra care has been taken to achieve high accuracy, the wire is called *special limits of error* (SLE) wire. Thermocouples made with such wire typically have calibration error limits reported by the manufacturer of $\pm 2e_s = \pm 1.1°C$ [48]. When standard care is taken in manufacturing the wire, the wire is categorized as having *standard limits of error*, which are usually $\pm 2e_s = \pm 2.2°C$ [48]. Error limits are equal to $\pm 2e_s$, and thus $e_{s,cal,T}$ for a thermocouple is half the manufacturer's stated error limits:

$$\text{Manufacturer specs:} \quad 2e_{s,est} = (\text{error limits})$$

The calibration error worksheet guides and organizes the determination of calibration error for both types of thermocouples.

$$\text{SLE wire:} \quad e_{s,cal} = \max(\text{ROT, lookup, user})$$
$$= \max(\text{NA}, 0.55, \text{NA})$$

$$\boxed{\text{SLE wire:} \quad e_{s,cal,T} = 0.55°C} \qquad (4.13)$$

Standard wire:    $e_{s,cal} = \max(\text{ROT, lookup, user})$

$$= \max(\text{NA, 1.1, NA})$$

$$\boxed{\text{Standard wire:} \quad e_{s,cal,T} = 1.1°C} \tag{4.14}$$

We now have estimates of all contributions to error for the various temperature measurements in Figure 4.1. The next step is to combine them to obtain combined standard uncertainty. In this example we have two types of temperature indicators, low precision and high precision. We also do not know which thermocouple wire is in use. To see the implications of error for the four possible situations, we calculate the error for all four cases (Figure 4.11). For each case, the overall standard error is the combination of the estimates of replicate error (zero), reading error (depends on the precision of the display), and calibration error (depends on which wire is used). The errors are added in quadrature. As an example, we show the calculation for special limits wire

| Special limits, high–precision display | $e_s$ | $e_s^2$ |
|---|---|---|
| replicate | 0 | 0 |
| reading | 0.0577 | 0.0033 |
| calibration | 0.5500 | 0.3025 |
| sum= | | 0.3058 |
| $e_{s,\,total}$= | | 0.55 °C |

| Standard limits, high–precision display | $e_s$ | $e_s^2$ |
|---|---|---|
| replicate | 0 | 0 |
| reading | 0.0577 | 0.0033 |
| calibration | 1.1000 | 1.2100 |
| sum= | | 1.2133 |
| $e_{s,\,total}$= | | 1.10 °C |

| Special limits, low–precision display | $e_s$ | $e_s^2$ |
|---|---|---|
| replicate | 0 | 0 |
| reading | 0.5770 | 0.3329 |
| calibration | 0.5500 | 0.3025 |
| sum= | | 0.6354 |
| $e_{s,\,total}$= | | 0.80 °C |

| Standard limits, low–precision display | $e_s$ | $e_s^2$ |
|---|---|---|
| replicate | 0 | 0 |
| reading | 0.5770 | 0.3329 |
| calibration | 1.1000 | 1.2100 |
| sum= | | 1.5429 |
| $e_{s,\,total}$= | | 1.24 °C |

Figure 4.11 We can explore the implications of using SLE wire versus standard wire and temperature indicators with two different levels of precision (ones digit only versus tenths digit displayed). When using the high-precision display (top row), reading error has no impact on the combined standard uncertainty. However, when the low-resolution display (bottom row) is used, the standard error for the SLE wire thermocouple degrades from $e_{s,cmbd} = 0.55°C$ to $0.8°C$ and the standard limits wire accuracy degrades from $e_{s,cmbd} = 1.1°C$ to $1.24°C$.

thermocouples attached to the more precise display (extra digits included to avoid round-off error):

$$e^2_{s,cmbd} = e^2_{s,random} + e^2_{s,reading} + e^2_{s,cal}$$
$$= 0^2 + (0.05774)^2 + (0.55^2)$$
$$= 0.30584$$
$$e_{s,cmbd} = 0.55303 = 0.55°C$$

When we use the high-precision display, reading error is very small and the combined error only reflects calibration error. Figure 4.11 shows that this is true for both SLE and standard wires (top row). For the low-precision display (bottom row), the reading error impacts the quality of the data. With the low-precision display, there is a small degradation of assigned precision for the standard wire thermocouple data from $e_{s,cmbd} = 1.1°C$ to $1.24°C$. For the SLE wire thermocouple, using the low-precision display decreases data accuracy significantly, from $e_{s,cmbd} = 0.55°C$ to $0.80°C$. The final error limits on the indicators in this problem are obtained from twice the standard error.

Thermocouple temperature indicator error limits
(manufacturer's calibration dominant,
95% level of confidence)

| | |
|---|---|
| SLE wire, high-precision display: | value ± 1.1°C |
| Standard wire, high-precision display: | value ± 1.6°C |
| SLE wire, low-precision display: | value ± 2.2°C |
| Standard wire, low-precision display: | value ± 2.5°C |

One cause of the high calibration error for these devices is variations in the composition and uniformity of their wires. When a manufacturer reports a general calibration error, it must represent any of the thousands of units made by the company. Random differences among units (due to material variations or production-process variations, for example) will cause unit performance differences and greater calibration error on average for the product.

We return now to the question asked in this example. Eleven digital temperature indicators with thermocouples attached were placed into a common bath. The readings recorded (see Figure 4.1) varied from 38°C to 41.0°C. Is this variation to be expected? The error limits just determined indicate that the high-precision indicators, assuming the thermocouples are special error limits wire, have an uncertainty of ±1.1°C, while the low-resolution indicators have an uncertainty of ±2.2°C (Figure 4.12). As seen in Figure 4.12 with SLE wire error limits shown, 10 of the 11 readings, when compared with the true value $T = 40.0°C$, are consistent with the true value. The results from one of the temperature indicators does not capture the true value between its error

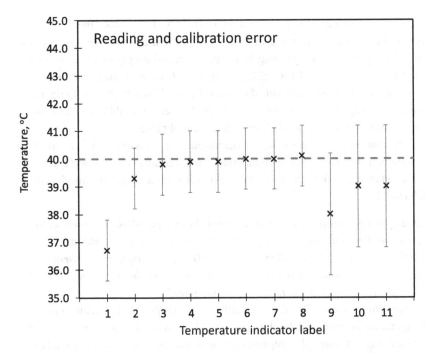

Figure 4.12 Eleven temperature indicators equipped with thermocouples (SLE wire assumed) all display a measured temperature from a common bath thermostated to 40.0°C (assumed accurate). Within the error limits determined by considering reading and calibration error, 10 of the 11 are consistent with the accurate temperature. The indicators with the larger error bars have a lower-precision display and hence larger reading error (Indicators 9 through 11).

limits; this may be because it falls into the fraction of responses that cause us to have only a 95% level of confidence rather than 100%; alternatively, the indicator or the thermocouple may be malfunctioning. Overall, the error limits we determined appear to explain the observed variability of the readings. It also suggests that the thermocouple in Indicator 1 may need to be replaced.

Thermocouples, as we see in this example, have limited accuracy. If a higher accuracy with temperature measurement is desired, a different measurement method must be employed. Resistance temperature detectors (RTDs), also called resistance thermometers (see Wikipedia or [9] for more information), are both more accurate than thermocouples and more stable over time; they are a common choice when the need for accurate temperature can justify the added expense of these devices.

In the previous example, even though the high-precision device displayed to the tenths digit, the uncertainty was at least an order of magnitude larger than plus or minus 1 in the last digit ($\pm 0.1°C$). This shows the danger in presuming accuracy from the number of digits that a manufacturer supplies in a display. The extra digits may have been included simply to allow two-digit uncertainties such as $\pm 1.6°C$, $\pm 2.2°C$, or $\pm 2.5°C$ to be applied to a measured value with a minimum amount of round-off error. When high-quality data are sought, a complete error analysis must be performed, and calibration error with information from the manufacturer is an essential piece.

The thermocouple error limits determined in Example 4.5 can provide insight into the reliability of temperature measurements in a wide range of circumstances. In the next example we revisit such a situation, introduced in Chapter 1.

**Example 4.6: Temperature displays and sig figs, revisited.** *A thermocouple is used to determine the melting point of a compound. If the compound is pure, the literature indicates it will melt at $46.2 \pm 0.1°C$. A sample of the chemical is melted on a hot plate and a thermocouple placed on the hot plate reads $45.2°C$ when the compound melts. Is the compound pure?*

**Solution:** We considered this situation in Example 1.1. The tenths place of temperature is displayed by the apparatus. Following the rules of significant figures, this optimistically implies that the device is "good" to $\pm 0.1°C$. Thus, the measured temperature is between $45.1°C$ and $45.3°C$. The lowest melting temperature consistent with the literature is $46.1°C$. Using sig figs for the measurement uncertainty, we would conclude that the compound is not pure, as the measured value with error limits is inconsistent with the literature value.

We now have another option for assessing uncertainty other than relying on significant figures alone – we can address this problem with error analysis. This work was done in Example 4.5. We have a high-precision display (it displays to the tenths digit), and we can initially assume that special limits wire was used (determining melting points requires as much accuracy and precision as possible, and it would be sensible for the manufacturer to have used the highest-accuracy thermocouples in the device). Thus, the error limits on the temperature measurement are no better than (see Example 4.1)

$$\text{SLE wire, high precision:} \quad \text{reading} \pm 1.1°C$$

The reading and its error limits become

$$T = 45.2 \pm 1.1°C$$
$$44.1°C \leq T \leq 46.3°C$$

There is overlap between the measured value and the literature value of the compound, so the measurements do not allow us to reject the hypothesis that the compound is pure (the pure compound's melting point is within the expanded uncertainty of the observed melting point, at a 95% level of confidence). If the wire is not SLE, there would be even wider error limits and even more possibility that the compound is pure. Therefore, with current instruments, we cannot rule out that the compound is pure.

To be more definitive in our conclusions about the compound's purity, we need a more precise measurement, and thus thermocouples should not be used. RTDs provide greater precision. RTDs are recommended for very high temperatures (higher than 500°C), and for applications requiring better than ±1.1°C accuracy. RTDs have a slow response time, however, and may be physically larger than thermocouples.

The final two examples in this chapter address electric-current measurement with a digital multimeter and user calibration of a differential-pressure meter (one of the linked examples identified in Figure 1.3).

**Example 4.7: Error on fluctuating and non-fluctuating measurements of electric current, resolved.** *In Examples 3.6 and 3.9, we considered the reading error of a digital multimeter (DMM) display presenting fluctuating and nonfluctuating electric current signals. We did not consider calibration error in those examples. What is the combined standard uncertainty for the current signal in both cases, taking all error contributions into account?*

**Solution:** In Example 3.6, we explored the reading error for a fluctuating current signal registered with a Fluke Model 179 DMM. Following the reading error worksheet, we neglected sensitivity, calculated resolution, and evaluated the fluctuations. The standard reading error was found to be (extra digits shown)

Reading error
(nonstandard)
$$e_R = \max(\text{sensitivity, resolution, fluctuations})$$
$$= \max(\text{NA}, 0.05, 0.15) = 0.15 \text{ mA}$$

Single, fluctuating
current measurement
$$e_{s,reading} = \frac{0.15}{\sqrt{3}} = 0.0866 \text{ mA} \qquad (4.15)$$

To evaluate the combined standard uncertainty, we must consider calibration error.

Calibration error: $\quad e_{s,cal} = \max(\text{ROT, lookup, user})$

Rather than relying on the rule of thumb (or performing a measurement), now we will look it up.

In Example 3.9, we searched out the datasheet for the Fluke Model 179 DMM. The calibration error is indicated on the datasheet under "detailed specifications," which are listed in a table for each of the functions displayed by the DMM. We are interested in the mA readings of current. The manufacturer specifies the calibration error this way:

$$\text{accuracy} = \pm[\% \text{ Reading}] + [\text{counts}]$$

The "accuracy" is equivalent to our error limits due to calibration, $\pm 2e_{s,cal}$. For the Fluke Model 179 DMM reading mA electric current:

Fluke
model 179    $2e_{s,cal} = (1\%)[\text{Reading}] + 3[\text{counts}]$    (4.16)
DMM

To understand the calibration errors given here, we must first understand what a "count" is, and we turn to the manufacturer's literature for the answer. The Fluke Model 179 is a "6000 count" DMM. The largest number that a 6000 count meter is capable of displaying is the number of counts minus 1: 5999. The leftmost display only shows digits 0, 1, 2, 3, 4, and 5; the other three display slots show digits from 0 to 9. The decimal point floats to wherever it is needed on the display.

To see how display design affects measurements of 4–20 mA, we reason as follows. The largest signal we will view is 20 mA. To display this number we use the first two slots, leaving two more for decimal places. The meter will show 20.00 mA or down to the hundredths place in mA. When set up to read this signal, then, one count is the toggling of "1" of the least significant digit; one count is 0.01 mA for this case. The language of counts is needed with digital multimeters to accommodate the wide variety of functions they are capable of displaying.

We are now ready to interpret the calibration error. The spec sheet says the accuracy of the Fluke 179 is 1% of the reading plus 3 counts. For a reading of 14.60 mA, we calculate (assuming the limits are at 95% level of confidence)

Manufacturer's specs:    $2e_{s,est} = (1\%)(14.60) + 3(0.01)$

$$= 0.1760 \text{ mA}$$

Calibration error:    $e_{s,cal} = \max(\text{ROT, lookup, user})$

$$= \max(\text{NA}, 0.0880, \text{NA})$$

Calibration error
for Fluke 179,
electric current       $e_{s,cal} = 0.0880 \text{ mA (extra digits shown)}$    (4.17)
of 14.6 mA:

To obtain the combined standard uncertainty for a Fluke 179 DMM reading the fluctuating signal considered here, we must combine the reading and calibration error in quadrature.

$$\text{(with fluctuations):} \quad e_{s,cmbd}^2 = e_{s,reading}^2 + e_{s,cal}$$
$$= (0.0866)^2 + (0.0880)^2$$

single, fluctuating
electric current measurement      $e_{s,I,true} = 0.12347 \text{ mA}$
of 14.6 mA:

$$\text{electric current} = 14.60 \pm 0.25 \text{ mA}$$

The uncertainty is dominated by the calibration error, but due to large fluctuations, there is an augmentation of the error. Note that in Example 3.9 we considered reading error on the non-fluctuating DMM and obtained a small number ($e_{s,reading} = \frac{0.05}{\sqrt{3}} = 0.02887$ mA). Adding the calibration error in quadrature to this non-fluctuation reading error gives the $e_{s,I}$ for the non-fluctuating case.

$$\text{(no fluctuations):} \quad e_{s,cmbd}^2 = e_{s,reading}^2 + e_{s,cal}$$
$$e_{s,I}^2 = (0.02887 \text{ mA})^2 + (0.0880 \text{ mA})^2$$

single
electric current
measurement        $\boxed{e_{s,I} = 0.09261 \text{ mA}}$              (4.18)
(no fluctuations)
of 14.6 mA:

Error limits:    $2e_{s,I} = 0.18523 \text{ mA}$
$$= 0.19 \text{ mA}$$

At this value for the current reading, the error limits turn out to be plus or minus two times the smallest digit displayed. The calibration error is the most significant error for this measurement, indicating that for some devices it is important to consult the manufacture's datasheets to know the extent to which one can rely on the results. Over the entire range of 4.0–20.0 mA, the calibration error (Equation 4.16) ranges from 0.035 mA $\leq e_{s,cal} \leq 0.115$ mA, and the total combined error (reading and calibration error) ranges from 0.045 mA $\leq e_{s,cmbd} \leq 0.1186$ mA.

Some devices are notoriously subject to calibration issues; we have met one of them – the thermocouple. Note also that calibration error was significant for milliamp electrical current signals measured by a DMM. The versatility of a

DMM means that the display is not optimized for individual settings, making it important to check the specifications from the manufacturer to determine the actual accuracy for the measurement of interest. Some weighing scales have limited calibration accuracy, while others are quite precise, if handled correctly. The manufacturer's documentation is the best source for sorting out these issues.

The final example in this chapter shows how a formal user calibration is performed. This is one of the linked examples identified in Figure 1.3.

**Example 4.8: \*LINKED\* User calibration of a differential pressure (DP) meter.** *A differential-pressure meter produces a 4–20 mA electrical current signal that correlates with sensed pressure difference (inner workings are described in Figure 4.13). For the electrical current output of the DP meter to be interpreted in terms of differential pressure $\Delta p$ in psi, the instrument must be calibrated. The calibration data using an air-over-Blue-Fluid manometer as the standard (Figure 4.13) are given in Table 4.4. What is the calibration error associated with the DP meter?*

**Solution:** We have established three pathways to calibration error, using a rule of thumb, looking up the calibration, and measuring it. In this example, we will measure the calibration function using a manometer as the differential-pressure standard.

$$\text{Calibration error:} \quad e_{s,cal} = \max(\text{ROT, lookup, user})$$
$$= \max(\text{NA, NA, user})$$

We begin with a description of the DP meter's functioning.

The DP meter is designed to register the pressure difference between two locations, perhaps between two points in a flow (as in Example 4.2). When installed to measure a flow pressure drop, appropriate tubing connects the locations of the pressures of interest, with one line connecting to the high-pressure location and the other line connecting to the low-pressure location. The lines are attached to the DP meter, which has an internal structure that exposes the pressures of interest to opposite sides of a stiff diaphragm (Figure 4.14). Inside the housing of the device, imprinted on the diaphragm, is a strain sensor, a pattern made from piezoelectric material – a material that produces electric current when subject to strain (the piezoelectric effect [9]). When the pressure is higher on one side of the diaphragm than the other, the diaphragm and the piezoelectric pattern deform. The deformation of the pattern that occurs when the diaphragm is pushed by the high-pressure side creates a piezoelectric current that is amplified and conditioned by the DP meter's electronics to become an output signal ranging from 4.0 to 20.0 mA (current loop; see [9]).

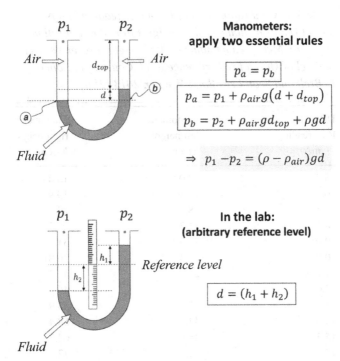

**Manometers:
apply two essential rules**

$$p_a = p_b$$

$$p_a = p_1 + \rho_{air}g(d + d_{top})$$

$$p_b = p_2 + \rho_{air}gd_{top} + \rho gd$$

$$\Rightarrow\ p_1 - p_2 = (\rho - \rho_{air})gd$$

**In the lab:
(arbitrary reference level)**

*Reference level*

$$d = (h_1 + h_2)$$

Figure 4.13 When the pressures on the two sides of a manometer are different, we can relate the height differences to the pressure difference using two simple rules. First, within the same fluid, fluid at the same height is at the same pressure; this tells us that $p_a = p_b$. Second, the pressure at the bottom of a column of fluid is equal to the pressure at the top plus $\rho g$ multiplied by the height, where $\rho$ is the density of the fluid and $g$ is the acceleration due to gravity [45].

To associate the output currents of the DP meter to pressure differences, the unit must be calibrated (the setup appears schematically in Figure 4.3). We choose the $\Delta p$ standards to be values determined by an air-over-Blue-Fluid manometer (Figure 4.13). To calibrate the DP meter, we produce a variety of differential pressures with two tanks: we use a tank vented to atmospheric pressure as the low-pressure source, and we install a tank that we can pump up to various pressures to serve as the high-pressure source. Calibration consists of pumping up the high-pressure tank to an arbitrary pressure, connecting the tank to the high-pressure side of the DP meter and atmospheric pressure to the low-pressure side, and recording the electrical-current signal output by the DP meter, $(I, \mathrm{mA})_{cal}$. The accurate differential-pressure value associated with the DP meter signal is determined simultaneously by connecting the high- and

Table 4.4. *Calibration data of DP meter reading in* mA
*output versus manometer fluid heights from the calibration
standard. Measured manometer fluid heights are above ($h_1$)
and below ($h_2$) a fixed reference level. Courtesy of Robert
LeBrell and Sean LeRolland-Wagner, 2013 [25].*

| DP meter Raw reading $(I, \mathrm{mA})_{cal}$ | Manometer left height $h_1$, cm | Manometer right height $h_2$, cm |
|---|---|---|
| 5.2 | –3.0 | 13.4 |
| 6.6 | 3.5 | 19.9 |
| 7.5 | 8.3 | 24.7 |
| 8.2 | 11.3 | 27.7 |
| 9.0 | 15.3 | 31.7 |
| 10.0 | 20.4 | 36.8 |
| 11.2 | 25.8 | 42.2 |
| 12.0 | 29.8 | 46.2 |
| 13.7 | 37.5 | 53.9 |
| 14.6 | 42.0 | 58.4 |
| 15.5 | 45.9 | 62.3 |
| 17.8 | 56.3 | 72.7 |
| 18.7 | 60.3 | 76.7 |
| 19.5 | 64.3 | 80.7 |
| 19.9 | 66.0 | 82.4 |

low-pressure tanks to the two sides of an air-over-Blue-Fluid manometer [the $\Delta p$ standard values, determined from ($h_1 + h_2$)]. We repeat this process for 15–25 data points corresponding to different differential pressures.

The manometer height differences for each calibration data point are converted to an accurate pressure difference ($\Delta p$, psi)$_{std}$ using an analysis of the manometer reading (Figure 4.13).

$$(\Delta p, \mathrm{psi})_{std} = \alpha_0 \rho g (h_1 + h_2) \qquad (4.19)$$

where $\rho$ is the density of the manometer fluid at the measurement temperature, $g$ is the acceleration due to gravity, and $h_1$ and $h_2$ are the heights of the fluid on the two sides of the manometer above (and below) an arbitrary reference level. The constant $\alpha_0$ in Equation 4.19 is a unit conversion used to obtain the desired differential-pressure units (psi).

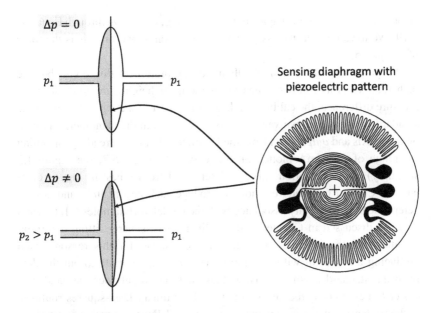

Figure 4.14  A differential-pressure (DP) meter is typically built around a chamber that allows two sides of a metal sensing-diaphragm to be exposed to different pressures. The sensing diaphragm has a piezoelectric pattern deposited on it that generates current when deformed. The design of the pattern is chosen to maximize the deformational sensitivity of the instrument. For more on DP meter design, see the literature [9].

The manometer differential-pressures $(\Delta p, \text{psi})_{std}$ are being used as the standard to allow us to calibrate the DP meter. Thus, this $(\Delta p, \text{psi})_{std}$ must be accurate, and a rule of thumb is that the accuracy of the standard must be at least four times the accuracy of the unit under test. In other words, the error in the $\Delta p$ associated with the manometer is to be no more than one-quarter the anticipated calibration error $e_{s,cal}$ for the DP meter. We cannot attribute a high accuracy to a device that has been calibrated using low quality standards.

To determine the accuracy of the manometer-determined $\Delta p$, we must obtain the uncertainties in all the quantities on the right side of Equation 4.19 and propagate the errors to an error associated with differential pressure $\Delta p$. We have these pieces: we can determine the reading errors of $h_1$ and $h_2$ as outlined in Chapter 3; a good value of Blue Fluid density and its error have been determined through replication (Example 2.8); and the gravitational acceleration constant $g$ and the unit-conversion constant $\alpha_0$ are universal constants with zero error (for our purposes). The next step in determining the errors on $(\Delta p)_{std}$ is to propagate the errors in $\rho$, $g$, $h_i$, and $\alpha_0$ through Equation 4.19. We have not yet discussed error propagation, however; this is the topic of Chapter 5.

We pause the discussion of DP meter calibration at this point, but the path forward is as follows. Once the manometer heights are converted to a pressure difference, the calibration data will consist of the *input data* from the standard [accurate, known values of $(\Delta p, \text{psi})_{std}$ from the manometer] plotted on the $x$-axis and *output data* from the DP meter $(I, \text{mA})_{cal}$, readings spanning the range of 4–20 mA plotted on the $y$-axis (Example 5.7; see Figure 1.3 summary). The calibration curve is determined from a best-fit line to these data. In Chapter 6 we address how to produce such best-fit lines and how to determine the uncertainty associated with the model fit (Examples 6.1, 6.2, and 6.6). As discussed in Chapter 6, the ordinary least-squares fitting algorithm assumes that the $x$ values of a fit have zero error; for this reason, when producing a model fit with calibration data, it is appropriate to put the data from the standard on the $x$-axis, as these data are highly accurate and thus closer to being error-free, as assumed by the ordinary least-squares routines. Ordinary least-squares is implemented with the LINEST function in Excel or with MATLAB.

Producing a user calibration is an involved process, as we see from the description just given. The care with which we calibrate devices determines the accuracy of our measurements with these devices.

## 4.5 Summary

In this chapter we discuss a third source of uncertainty in experimental data, the uncertainty that is a consequence of the limitations in the calibration of a measuring device. Calibration is what determines the accuracy of measurements with a device. We must be mindful of the need to calibrate and recalibrate devices at appropriate intervals. Three potential sources of information on device calibration error are included in the calibration error worksheet in Appendix A: rule of thumb (ROT), looking up the manufacturer's calibration, and performing our own calibration. The $e_{s,cal}$ determined is the largest of these three values.

$$\text{Calibration error} \quad \boxed{e_{s,cal} = \max(\text{ROT, lookup, user})}$$

For most determinations of calibration error, our best source is the literature provided by the device's manufacturer. In addition, calibration error, reading error, and replicate error may all be present in a measurement; if so, they are determined independently and combined in quadrature.

We explored special considerations at lower and upper performance limits. At low signal strength, we disregard data with more than 25% error; this establishes the limit of determination.

$$
\begin{array}{l}
\text{Recommended rule of thumb} \\
\text{for limit of determination (LOD)} \\
\text{(maximum of 25\% error on } y\text{)}
\end{array}
\qquad \boxed{\text{LOD} = 8e_{s,cal}} \qquad (4.20)
$$

At high signal strength, to avoid saturation effects we must stay away from the upper end of the device's display range.

We discussed devices in which calibration error dominates a measurement (thermocouple, electric current with a multimeter); it is also common for calibration error to be negligible. An ordinary stopwatch, for instance, may be used to time the fall of a meniscus in a Cannon–Fenske viscometer (Examples 2.19 and 5.8). In that application, the calibration error in the stopwatch will be irrelevant compared to the large random error introduced by the variable reaction time of the operator; we can use our judgment to choose to neglect the calibration error of the stopwatch in that case. We can see more formally how various quantities influence the final error in a measurement when we propagate error in calculations (Chapter 5). Error propagation affords us the opportunity to assess the impact of estimates of error as well: estimates of little consequence can be tolerated, while high-consequence estimates should be reexamined before affected results are used in important decisions.

## 4.6 Problems

1. Explain calibration in your own words. What is the purpose of calibration?
2. Is calibration error a Type A or a Type B error? Explain your answer.
3. Name a measuring device/situation in which the calibration error dominates. Name a measuring device/situation in which calibration error is negligible. Explain your reasoning.
4. Name a device or measurement situation in which replication error is dominant and reading and calibration error are negligible. Explain your reasoning.

5. The scale at the gym is used to settle a dispute about who weighs more. Chris weighs 172 lb$_f$ and Lee weighs 170 lb$_f$. The dispute then turns to the issue of whether the scale is accurate enough to settle the dispute. What is the possible role of calibration error in the dispute?

6. A chemist uses a 500-ml volumetric flask to prepare an aqueous sugar solution. What are the reading and calibration errors on the volume of such a flask? Explain your reasoning.

7. For 14 days in a row, Milton fills a 1-liter volumetric flask with water, weighs it, and then empties the flask and stores it. The mass of the full flask, nominally of volume one liter, is a little different each day. What is the expected calibration error for the volume metered by this device? How would you explain Milton's observations?

8. For the 250-ml graduated cylinder in Figure 3.12, what is the calibration error? What is the reading error? Give error limits (95% level of confidence) based on calibration error and based on combined standard error. Which error dominates?

9. A 25-ml volumetric pipet is used to transfer 25 ml of fluid to a reaction vessel. What are the error limits on the amount of volume transferred?

10. Eun Young uses her cellular phone to record her commuting time. The reading she obtained was 37:04.80 mm:ss. What is the calibration error for this device? Comment on your result and its role in determining a good value for commuting time measured with the cell phone.

11. What is your estimate of calibration error for the automotive tachometer in Figure 3.16? Include worksheets for your estimates and explain your reasoning.

12. What is your estimate of calibration error for the automotive speedometer in Figure 3.16? Include worksheets for your estimates and explain your reasoning.

13. What is your estimate of calibration error for the stopwatch in Figure 3.16? Include worksheets for your estimates and explain your reasoning.

14. What is your estimate of calibration error associated with the position of the manual metering valve in Figure 3.16? Include worksheets for your estimates and explain your reasoning.

15. What is your estimate of calibration error for the tape measure in Figure 3.16? Include worksheets for your estimates and explain your reasoning.

16. What is your estimate of calibration error for the ruler in Figure 3.16? Include worksheets for your estimates and explain your reasoning.

17. What is your estimate of calibration error for the analytical balance in Figure 3.16? Include worksheets for your estimates and explain your reasoning.

18. What is your estimate of calibration error for your kitchen oven? Include worksheets for your estimates and explain your reasoning.

19. For a DMM found in your lab or on the Internet (give a citation for the one you choose), fill in the reading and calibration worksheets for electrical resistance readings in the tens of ohms range. Which error is dominant?

20. For a DMM you find on the Internet (give a citation for the one you choose), fill in the reading and calibration worksheets for voltage readings in the millivolt range. Which error is dominant?

21. What is the combined standard uncertainty when determining temperatures between 0°C and 100°C with RTDs? Be guided by an actual instrument located on the Internet (give a citation for the one you choose). Include worksheets for your estimates and explain your reasoning.

22. We measure a reactor temperature to be 50.0°C using a digital temperature indicator using thermocouples to sense temperature. What is the likely standard calibration error for the thermocouple? The readings of room temperature are unvarying, and thus the replicate error is zero. What is the reading error? What are the error limits (95% level of confidence)?

23. A student measures replicates of reactor surface temperature in Example 2.16. What are the likely reading and calibration errors in Example 2.16? What is the combined standard uncertainty? What is the expanded uncertainty at a level of confidence of 95%?

24. For an analytical balance that reads in grams out to four decimal places, what is the standard reading error? What is the calibration error? If the replicate error is negligible, what is the combined standard uncertainty for this device? What limitations do you attach to your answers? Include worksheets for your estimates and explain your reasoning.

25. Inexperienced workers use an analytical balance to weigh sugar to use in making a sugar-water solution. The balance is in good operating order. The intended weight is 32.200 g of sugar. What are reasonable error limits with a 95% level of confidence on the workers' reported mass? Include worksheets for your estimates and explain your reasoning. Which error dominates?

26. In Chapter 6 we obtain the calibration error on the DP meter that is discussed throughout the text, $e_{s,cal,DP} = 0.021$ psi (Equation 6.80). What is the reading error for the DP meter? What is the combined standard uncertainty in a measurement with this device? Determine

plausible error limits at a 95% level of confidence; include filled-in
worksheets.

27. A digital multimeter measuring electric current has a calibration error of
$e_{s,cal,I} = 0.088$ mA $\approx 0.09$ mA. What is your estimate of the LOD –
that is, what is the lowest signal that you can retain from this instrument?
What is wrong with data signals that are below this threshold?

28. What is your estimate of the LOD of your home bathroom scale? In other
words, what is the lowest signal that you can reliably retain from this
device? What is wrong with data signals that are below this threshold?

29. What is the shortest interval that you can time with a stopwatch? Devise
an experiment to measure the shortest accurate time interval that is
advisable. Explain your reasoning, including your definition of
"advisable."

# 5

# Error Propagation

In the preceding chapters, we established a foundation for understanding uncertainty in measurements. We categorize the various effects that contribute to uncertainty in measurements as random (replicate), reading, and calibration error. For quantities obtained by direct measurement, the worksheets in Appendix A and the methods of Chapters 2–4 guide us to determine values for error contributions. We have shown how independent errors may be expressed in standard form and summed in quadrature to produce a combined standard uncertainty, $e_{s,cmbd}$. We construct error limits at the 95% level of confidence on measured values using $\pm 2e_{s,cmbd}$. We now arrive at the need to assess how uncertainty in measurement impacts calculations that interest us.

Most scientific measurements are used in follow-on calculations, with the uncertainty for the final result reflecting contributions from several measured quantities. In this chapter, we show how to obtain error limits on a final calculated result by propagating error through the equations used. Error propagation is based on variances: standard error is defined as the standard deviation of the quantity's sampling distribution, standard deviation squared is the variance, and the method of error propagation is based on how variances of stochastic variables combine.

Although error propagation may seem involved, it is easily implemented in software, and the error propagation process itself yields valuable insight into identifying both the dominant errors in a process and the impact of certain error assumptions. The insight gained during error propagation may, in some circumstances, even be more valuable to the investigator than the numerical value of the propagated error itself. Reliable scientific results are established from well-designed experiments, with results presented and interpreted within their known limitations. Error propagation is an important part of this process.

# 5.1 Introduction

The need to combine various measured quantities arises quite often in science and engineering. We introduce a simple example to illustrate the context of propagating error.

**Example 5.1: Estimate density error without replicates, revisited.** *We seek to determine the density of a sample of Blue Fluid 175, but there is only a small amount of fluid available, and we can only make the measurement once. The technique used to measure density is mass-by-difference with a $10.00\pm0.04$ ml pycnometer (Figure 4.8) and an analytic balance (the balance is accurate to $\pm10^{-4}$ g). What are the appropriate 95% level of confidence error limits on this single measurement?*

**Solution:** We first considered this example in Chapter 2 (Example 2.11); there, we were unable to determine error limits because, at that time, our only error tool was replicate error and there are no replicates to evaluate.

Although we are limited by not having replicates, we can estimate error limits using another approach. We know the methods and calculations we used to obtain our single result for density, and this is our starting place. Fluid density is measured with a pycnometer by first weighing the clean, dry, empty pycnometer and cap; this yields mass $M_E$. The pycnometer is carefully filled with solution and the cap is inserted, allowing the capillary in the pycnometer cap to fill with fluid. Any excess is wiped off, and the filled pycnometer is weighed, yielding mass $M_F$. The density $\rho$ is calculated from the measurements as

$$\begin{array}{cc} \text{Fluid density from} \\ \text{mass-by-difference} \end{array} \qquad \rho = \frac{(M_F - M_E)}{V} \qquad (5.1)$$

where $V$ is the volume of the pycnometer. This is the mass-by-difference technique.

We can determine the uncertainties for all the quantities on the right side of Equation 5.1. For the masses, we know from the problem statement that the balance is accurate to $\pm10^{-4}$ g; thus, for the analytic balance, standard errors for masses $M_E$ and $M_F$ may be obtained from the error limits $\pm2e_{s,M} = \pm10^{-4}$ g. For the pycnometer volume, the manufacturer certifies that the calibration error limits (there is no reading error) are $2e_{s,V} = 0.04$ ml (see Example 4.4). We thus determine

$$\text{Analytic balance: } e_{s,M} = 5 \times 10^{-5} \text{ g}$$
$$\text{Pycnometer: } e_{s,V} = 0.02 \text{ ml}$$

We have the formula for density, and we have the standard errors of all the numbers that go into the formula. What we lack is the knowledge of the correct technique for propagating the error through the formula to find a final error on density. This is the technique we are about to discuss.

We resolve this problem once we establish the error propagation process (see Example 5.4).

## 5.2 How to Propagate Error

The circumstance we now consider is this: we take a measurement, obtain a value for some quantity, and subsequently use that number in a calculation. How does the uncertainty in the measured value affect the uncertainty in the calculated quantity?

We may answer this question by considering measurements as stochastic variables. As discussed in Section 3.3, a stochastic variable is a variable that can take on different values every time it is observed or sampled. Specifically, measurement results are numbers composed of the sum of three contributions: the true value of the quantity that interests us; a stochastic contribution from random errors; and stochastic contributions from systematic errors such as reading and calibration error.

$$\begin{array}{c}\text{observed value}\\\text{(stochastic)}\end{array} = \text{true value} + \begin{array}{c}\text{random error}\\\text{(stochastic)}\end{array} + \begin{array}{c}\text{systematic errors}\\\text{(stochastic)}\end{array} \quad (5.2)$$

Consider the case where three stochastic variables are manipulated in a function to produce a quantity $y$. Let $x_1$, $x_2$, and $x_3$ be continuous stochastic variables with variances $\sigma_{x_1}^2$, $\sigma_{x_2}^2$, and $\sigma_{x_3}^2$, respectively. The function $y = \phi(x_1, x_2, x_3)$ is a function of these stochastic variables.

$$y = \phi(x_1, x_2, x_3) \quad (5.3)$$

Because $x_1$, $x_2$, and $x_3$ are stochastic, $y$ will also be stochastic and will have an associated variance $\sigma_y^2$. To put this discussion in terms of the quantities in the fluid density example (Example 5.1), $x_1$, $x_2$, and $x_3$ are $M_F$, $M_E$, and $V$, which are all quantities with uncertainties, and the standard errors of these quantities are the standard deviations $\sigma_{x_1}$, $\sigma_{x_2}$, and $\sigma_{x_3}$. The density $\rho$ is the calculated quantity $y$, and the formula in Equation 5.1 is the function $y = \phi(x_1, x_2, x_3)$. We are asking how to calculate $e_{s,\rho} = \sigma_y$ from the individual standard errors of the variables $M_F$, $M_E$, and $V$.

The variance of a continuous stochastic variable is defined with respect to an integral over the variable's probability density function (Equation 3.10, repeated here):

Variance of a
continuous
stochastic variable
$$\sigma_y^2 = \int_{-\infty}^{\infty} \left(y' - \bar{y}\right)^2 f(y')dy' \qquad (5.4)$$

where $\sigma_y^2$ and $\bar{y}$ are the variance and mean value of the stochastic variable $y$, and $f(y)$ is the probability density function (pdf) of $y$. We are looking to determine $\sigma_y$, but, unfortunately, we rarely know the pdf $f(y)$; thus, Equation 5.4 is not a helpful starting place. Instead, we follow an indirect approach based on knowing the function $y = \phi(x_1, x_2, x_3)$ and the variances for the three dependent variables, $x_1$, $x_2$, and $x_3$.

The simplest case to consider is error propagation in a linear function. If $\phi(x_1, x_2, x_3)$ is a linear function given by

$$y = \phi(x_1, x_2, x_3) = c_1 x_1 + c_2 x_2 + c_3 x_3 \qquad (5.5)$$

where the $c_i$ are constant coefficients, then the variance of $y$ may be shown (using a first-order Taylor series approximation) to be [38, 3]

$$\sigma_y^2 = c_1^2 \sigma_{x_1}^2 + c_2^2 \sigma_{x_2}^2 + c_3^2 \sigma_{x_3}^2$$
$$+ 2c_1 c_2 \text{Cov}(x_1, x_2) + 2c_1 c_3 \text{Cov}(x_1, x_3) + 2c_2 c_3 \text{Cov}(x_2, x_3) \qquad (5.6)$$

where $\text{Cov}(x_i, x_j)$ is the covariance of the variables $x_i$ and $x_j$. The first three terms are a linear combination of the three variances; the last three terms are the somewhat complex covariance terms, which account for interactions between the variables. The covariance statistic $\text{Cov}(x, z)$ measures the interrelationship between a pair of stochastic variables, and its formal definition is an integral similar to Equation 5.4, involving the joint probability density function for the two variables.[1]

The presence of the covariance terms in Equation 5.6 makes the calculation of $\sigma_y^2$ rather complex; it would be nice if the covariance terms would go away. Luckily, quite often this is the case: the covariance is zero when the variables $x_i$ are mutually independent.[2] An example of independent or

---

[1] The formal definition of covariance is

Covariance of a
pair of continuous
stochastic variables
$$\text{Cov}(x, z) \equiv \int_{-\infty}^{\infty} \left(z' - \bar{z}\right)\left(x' - \bar{x}\right) f(x', z')dx'dz' \qquad (5.7)$$

where $f(x, z)$ is the joint probability density function for the variables $x$ and $z$ [5, 38].

[2] Or if they are simply *not correlated*; see the literature for the distinction between independent and uncorrelated.

uncorrelated stochastic variables would be the temperature of the room in which the experiments are performed and the weight of an empty pycnometer – uncertainties in one are not linked to the values and uncertainties in the other. For a function of variables with zero covariances and with $y$ given by the linear function in Equation 5.5, we obtain the following rule for combining errors [38, 3]:

Variance of
linear function $y = \phi(x_1, x_2, x_3)$,
with independent variables $x_i$
($y$ given by Equation 5.5)
$$\sigma_y^2 = c_1^2 \sigma_{x_1}^2 + c_2^2 \sigma_{x_2}^2 + c_3^2 \sigma_{x_3}^2 \quad (5.8)$$

This is just the quadrature rule we have been using to combine random, reading, and calibration errors; we have been assuming random, reading, and calibration errors are independent and linearly additive with coefficients $c_i$ equal to 1.

As a first error propagation calculation, we turn to the question of how uncertainty is affected when a length is determined by adding two measurements rather than by measuring the same length with one, longer ruler. This example is followed by another from engineering that involves a linear combination of variables.

**Example 5.2: Length from one measurement versus two.** *A meter stick is used to measure the length of a long pipe. A mark is made on the pipe and two lengths are recorded, $l_1 = 85.0$ cm from one end to the mark and $l_2 = 55.0$ cm from the mark to the other end of the pipe. The standard error in a length measurement from the meter stick is $e_{s,l} = 0.0577$ cm (this error value is determined in Example 5.6, Equation 5.34). Extra error digits are employed here to minimize subsequent round-off error. What is the length of the pipe, including error limits? If a longer tape measure (with the same standard error $e_{s,l}$) could have been used, what would have been the error limits?*

**Solution:** The total length $l_T$ of the pipe was determined from the following equation:

$$l_T = l_1 + l_2 = 140.0 \text{ cm} \quad (5.9)$$

We seek error limits for this final, calculated value. Equation 5.9 is a linear equation with constant coefficients with independent variables $l_1$ and $l_2$; we therefore use a version of Equation 5.8 to calculate the uncertainty through error propagation, and the error limits at the 95% level of confidence will be given by $\pm 2e_{s,l_T}$.

$$\sigma_{l_T}^2 = c_1^2 \sigma_{l_1}^2 + c_2^2 \sigma_{l_2}^2$$
$$e_{s,l_T}^2 = (1)^2 e_{s,l_1}^2 + (1)^2 e_{s,l_2}^2$$
$$= (0.0577)^2 + (0.0577)^2$$
$$= 0.006659 \text{ cm}^2$$
$$e_{s,l_T} = 0.0816 \text{ cm} = 0.08 \text{ cm}$$

Adding lengths from
two measurements:=                $l_T = 140.0 \pm 0.16 \text{ cm}$
(95% level of confidence)

If we had measured the length directly without having to combine two separate measurements, the uncertainty in length would have been

$$\pm 2e_{s,l} = \pm(2)(0.0577)$$
$$= \pm 0.1154 \text{ cm}$$

Length from one measurement
(95% level of confidence)          $l = 140.0 \pm 0.12 \text{ cm}$

There is a 25% increase in uncertainty in the case when two measurements are added compared to making a single measurement.

**Example 5.3: Limit of determination (LOD) on temperature difference.** *At steady state, we can determine the sensible heat $\dot{q}$ given up by a water stream in a double-pipe heat exchanger using the following equations:*

$$\dot{q} = \dot{m}\hat{C}_p \Delta T \tag{5.10}$$
$$\Delta T = (T_{out} - T_{in}) \tag{5.11}$$

*where $\dot{m}$ is the mass flow rate of water, $\hat{C}_p$ is the water heat capacity, and $T_{out}$ and $T_{in}$ are the outlet and inlet temperatures of the water, respectively. These equations come from a steady-state macroscopic energy balance on the water stream. If the temperatures are measured with a thermocouple constructed of standard limits wire, what is the lowest temperature difference $\Delta T$ we can accurately measure? What is that lowest accurate temperature difference if special limits wire is used ?*

   **Solution:** The lowest accurate value of a measured quantity such as temperature difference is related to the LOD of the device, or technique, used to measure it. The LOD is recommended to be $8e_s$ (Equation 4.8). When the measured value is the result of a calculation, such as of $\Delta T$ in Equation 5.11, we determine the standard error through error propagation.

First, we need the standard calibration error for a thermocouple using standard wire. We determined this in Example 4.1, $e_{s,cal} = 1.1°C$. Second, we calculate the standard error for $\Delta T$ from error propagation (Equations 5.11 and 5.8):

$$\sigma_{\Delta T}^2 = c_1^2 \sigma_{T_{out}}^2 + c_2^2 \sigma_{T_{in}}^2$$
$$e_{s,\Delta T}^2 = (1^2)(e_{s,T_{out}}^2) + (-1)^2(e_{s,T_{in}}^2)$$
$$= 1.1^2 + 1.1^2$$

Standard-limits wire: $\quad e_{s,\Delta T} = 1.5556°C$ $\qquad$ (5.12)

Finally, the LOD is $8e_{s,\Delta T}$ (maximum 25% error):

Standard-limits wire
(maximum 25% error)

$$\boxed{LOD_{\Delta T} = 12.4°C}$$ $\qquad$ (5.13)

For special limits wire, the calibration error is smaller, $e_{s,cal} = 0.55°C$ (Equation 4.13). Following the same calculation procedure, we obtain the LOD for $\Delta T$ determined with special-limits-wire thermocouples.

Special limits wire: $\quad e_{s,\Delta T} = 0.7778°C$ $\qquad$ (5.14)

Special limits wire
(maximum 25% error)

$$\boxed{LOD_{\Delta T} = 6.2°C}$$ $\qquad$ (5.15)

With our LOD limit set at a maximum of 25% error, the smallest $\Delta T$ interval we can accurately determine is 6.2°C (special limits wire) or 12.4°C (standard limits wire).

The linear error propagation calculation is quite straightforward, and we readily determine the increased uncertainty that is a consequence of combining two stochastic variables in this way. Most functions we work with are not linear combinations, however, including the equation for determining density from pycnometer weights and volume (Equation 5.1). For the general case, including that of a nonlinear function, the individual variances combine with pre-factors calculated from the function [38]. For general nonlinear function $y = \phi(x_1, x_2, x_3)$, one can write the variance of $y$ as[3]

$$\sigma_y^2 = \left(\frac{\partial \phi}{\partial x_1}\right)^2 \sigma_{x_1}^2 + \left(\frac{\partial \phi}{\partial x_2}\right)^2 \sigma_{x_2}^2 + \left(\frac{\partial \phi}{\partial x_3}\right)^2 \sigma_{x_3}^2 + \begin{pmatrix} \text{covariance} \\ \text{terms if} \\ x_i \text{ are} \\ \text{correlated} \end{pmatrix} \quad (5.17)$$

Note that Equation 5.17 reduces to Equation 5.8 for $\phi$ given by a simple linear combination of the variables (Equation 5.5). Often we are working with uncorrelated variables, in which case we set the covariance terms to zero. Thus, the equation we use most often for error propagation is Equation 5.17 with the covariance terms set to zero and the variances written as the squared standard errors of the stochastic variables.

Error propagation
equation
(3 variables,
all uncorrelated)

$$e_{s,y}^2 = \left(\frac{\partial \phi}{\partial x_1}\right)^2 e_{s,x_1}^2 + \left(\frac{\partial \phi}{\partial x_2}\right)^2 e_{s,x_2}^2 + \left(\frac{\partial \phi}{\partial x_3}\right)^2 e_{s,x_3}^2$$

$$(5.18)$$

Equation 5.18 is extended to a function of more than three variables in a straightforward way.

$$y = \phi(x_1, x_2, \ldots x_n) \quad (5.19)$$

Error propagation
equation
($n$ variables,
all uncorrelated)

$$e_{s,y}^2 = \sum_{i=1}^{n} \left(\frac{\partial \phi}{\partial x_i}\right)^2 e_{s,x_i}^2 \quad (5.20)$$

It is worth repeating that for both Equations 5.18 and 5.20 we have neglected the covariance terms, and this is not always an appropriate assumption. A frequently encountered case of correlated variables occurs when evaluating the error on the value of a function predicted from a least-squares model fit; we discuss this case in Section 6.2.4, where we determine and incorporate the appropriate covariance terms.

---

[3] The complete equation for the case of three variables is [38]

$$\sigma_y^2 = \left(\frac{\partial \phi}{\partial x_1}\right)^2 \sigma_{x_1}^2 + \left(\frac{\partial \phi}{\partial x_2}\right)^2 \sigma_{x_2}^2 + \left(\frac{\partial \phi}{\partial x_3}\right)^2 \sigma_{x_3}^2 + 2\left(\frac{\partial \phi}{\partial x_1}\right)\left(\frac{\partial \phi}{\partial x_2}\right)\text{Cov}(x_1, x_2)$$

$$+ 2\left(\frac{\partial \phi}{\partial x_1}\right)\left(\frac{\partial \phi}{\partial x_3}\right)\text{Cov}(x_1, x_3) + 2\left(\frac{\partial \phi}{\partial x_2}\right)\left(\frac{\partial \phi}{\partial x_3}\right)\text{Cov}(x_2, x_3) \quad (5.16)$$

This is obtained from a first-order Taylor series approximation [21].

| $\phi(x_1, x_2, x_3, x_4, x_5)$: | Formula for $\phi$:* *Note: units must work as written. | | Representative value of $\phi$: (include units) | 95% C.I. of $\phi$: $(\phi \pm 2e_{s,\phi})$ (include units) |
|---|---|---|---|---|
| **Measured quantities, $x_i$** | | | $\dfrac{\partial \phi}{\partial x_i}$ | $e_{s,x_i}$ | $\left(\dfrac{\partial \phi}{\partial x_i}\right)^2 e^2_{s,x_i}$ |
| $x_i$ | Symbol | Representative value | | | |
| $x_1$ | | units | | | |
| $x_2$ | | units | | | |
| $x_3$ | | units | | | |
| $x_4$ | | units | | | |
| $x_5$ | | units | | | |

$$e^2_{s_\phi} = \left(\frac{\partial \phi}{\partial x_1}\right)^2 e^2_{x_1} + \left(\frac{\partial \phi}{\partial x_2}\right)^2 e^2_{x_2} + \left(\frac{\partial \phi}{\partial x_3}\right)^2 e^2_{x_3} + \left(\frac{\partial \phi}{\partial x_4}\right)^2 e^2_{x_4} + \left(\frac{\partial \phi}{\partial x_5}\right)^2 e^2_{x_5}$$

$e^2_{s,\phi} =$ 　　　 $e_{s,\phi} =$ 　　units

*All variables $x_i$ must be independent for this equation to hold.

Figure 5.1 An excerpt of the error propagation worksheet provided in Appendix A.

We provide a worksheet that applies Equation 5.18 to error propagation problems (an excerpt of the worksheet is given in Figure 5.1; the full worksheet is available in Appendix A). In the examples that follow, we show how to use the error propagation worksheet. We begin by completing the error calculation that we initiated in Example 5.1.

**Example 5.4: Estimate density error without replicates, resolved.** *We seek to determine the density of a sample of Blue Fluid 175, but there is only a small amount of fluid available, and we can make the measurement only once. The technique used to measure density is mass-by-difference with a $10.00 \pm 0.04$ ml pycnometer and an analytic balance (the balance is accurate to $\pm 10^{-4}$ g). What are the appropriate 95% level of confidence error limits on this single measurement?*

**Solution:** We first considered this example in Chapter 2 (Example 2.11) and then again at the start of this chapter (Example 5.1). Now, having determined the mathematics of error propagation (Equation 5.18), we have the tools we need to calculate the requested uncertainty in density.

As discussed previously, we determine fluid density with a pycnometer by measuring two masses, one of the empty pycnometer ($M_E$) and one of the full pycnometer ($M_F$). The density is calculated from these measurements and from the known volume of the pycnometer, $V$:

$$\rho = \frac{(M_F - M_E)}{V} \tag{5.21}$$

We previously established the uncertainties of all the quantities in Equation 5.21 (see Example 5.1).

$$\text{Analytic balance: } e_{s,M} = 5 \times 10^{-5} \text{ g}$$
$$\text{Pycnometer: } e_{s,V} = 0.02 \text{ ml}$$

To find the uncertainty in the density from the errors of the measurements, we use the error propagation equation for uncorrelated variables, Equation 5.18. To complete the calculation, we take the indicated partial derivatives of $\phi(x_1, x_2, x_3) = \rho(V, M_F, M_E)$, evaluate and carry out the squares of each $\left( \frac{\partial \phi}{\partial x_i} e_{s,x_i} \right)$ term, and then sum. The final answer for error is the square root of the sum.

$$e_{s,\phi}^2 = \left( \frac{\partial \phi}{\partial x_1} \right)^2 e_{s,x_1}^2 + \left( \frac{\partial \phi}{\partial x_2} \right)^2 e_{s,x_2}^2 + \left( \frac{\partial \phi}{\partial x_3} \right)^2 e_{s,x_3}^2$$

$$e_{s,\rho}^2 = \left( \frac{\partial \rho}{\partial M_F} \right)^2 e_{s,M_F}^2 + \left( \frac{\partial \rho}{\partial M_E} \right)^2 e_{s,M_E}^2 + \left( \frac{\partial \rho}{\partial V} \right)^2 e_{s,V}^2 \tag{5.22}$$

$$= \left( \frac{1}{V} \right)^2 e_{s,M_F}^2 + \left( -\frac{1}{V} \right)^2 e_{s,M_E}^2 + \left( -\frac{(M_F - M_E)}{V^2} \right)^2 e_{s,V}^2$$

$$e_{s,\rho} = \sqrt{\left[ \left( \frac{1}{V} \right)^2 e_{s,M_F}^2 + \left( -\frac{1}{V} \right)^2 e_{s,M_E}^2 + \left( -\frac{(M_F - M_E)}{V^2} \right)^2 e_{s,V}^2 \right]} \tag{5.23}$$

The squared error (variance) in density is the weighted sum of the squared errors of the variables involved in the calculation (Equation 5.22). The weights used in the sum are the squared partial derivatives. The weights determine the impact of the various errors.

We carry out this calculation in the error propagation worksheet (Figure 5.2). We have listed already most of the quantities called for in the worksheet. To complete the calculation of $e_{s,\rho}$, we require values for $V$, $M_F$, and $M_E$; for this example we use the following measured values:

$$V = 10 \text{ ml}$$
$$M_F = 30.800 \text{ g}$$
$$M_E = 13.410 \text{ g}$$

| $\phi(x_1, x_2, x_3, x_4, x_5)$: | Formula for $\phi$: $\rho_{BF} = \dfrac{M_F - M_E}{V}$ *Note: units must work as written.* | | Representative value of $\phi$: (include units) 1.739 g/cm³ | 95% C.I. of $\phi$: $(\phi \pm 2e_{s,\phi})$ (include units) $1.739 \pm 0.007$ g/cm³ |
|---|---|---|---|---|
| **Measured quantities, $x_i$** | | | $\dfrac{\partial \phi}{\partial x_i}$ | $e_{s,x_i}$ | $\left(\dfrac{\partial \phi}{\partial x_i}\right)^2 e_{s,x_i}^2$ |
| $x_i$ | Symbol | Representative value | | | |
| $x_1$ | $M_F$ | 30.800 g $\quad$ units | $1/V$ | $5.0 \times 10^{-5}$ g | $2.5 \times 10^{-11}$ g²/cm⁶ |
| $x_2$ | $M_E$ | 13.410 g $\quad$ units | $-1/V$ | $5.0 \times 10^{-5}$ g | $2.5 \times 10^{-11}$ g²/cm⁶ |
| $x_3$ | $V$ | 10.00 ml $\quad$ units | $-(M_F - M_E)/V^2$ | 0.02 ml | $1.21 \times 10^{-5}$ g²/cm⁶ |
| $x_4$ | | $\quad$ units | | | |
| $x_5$ | | $\quad$ units | | | |

*Table note columns:*

$$e_{s_\phi}^2 = \left(\frac{\partial \phi}{\partial x_1}\right)^2 e_{x_1}^2 + \left(\frac{\partial \phi}{\partial x_2}\right)^2 e_{x_2}^2 + \left(\frac{\partial \phi}{\partial x_3}\right)^2 e_{x_3}^2 + \left(\frac{\partial \phi}{\partial x_4}\right)^2 e_{x_4}^2 + \left(\frac{\partial \phi}{\partial x_5}\right)^2 e_{x_5}^2$$

* All variables $x_i$ must be independent for this equation to hold.

$e_{s,\phi}^2 = 1.21 \times 10^{-5} \frac{g^2}{cm^6}$

$e_{s,\phi} = 0.0035$ $\quad$ units g/cm³

Figure 5.2 The error propagation worksheet guides the error propagation process. The values in the rightmost column indicate the relative impact of the uncertainties of each stochastic variable. As shown, the uncertainty of the volume of the pycnometer dominates the uncertainty in the density. Values shown are from Example 5.4.

The standard error on density is calculated to be (Figure 5.2)

$$\text{Density standard error} \quad e_{s,\rho} = 3.5 \times 10^{-3} \text{ g/cm}^3 \qquad (5.24)$$

The results of the error propagation allow us to determine the error limits on density, which may be written as $\pm 2e_{s,\rho}$:

$$\begin{array}{c} \text{Blue Fluid density} \\ \text{with 95\% level} \\ \text{of confidence} \end{array} \quad \rho = 1.739 \pm 0.007 \text{ g/cm}^3 \qquad (5.25)$$

This value along with its error limits is our best estimate of the density (95% level of confidence) based on the single measurement and the known uncertainties of the measurement method.

The final result for Blue Fluid density error limits shows that we are able to determine density to between three and four significant figures. Three sig figs would imply error limits of $\pm 0.01$ g/cm³, which is too conservative, but four sig figs would imply $\pm 0.001$ g/cm³, which is overstating the precision.

Through error propagation we are able to determine a 95% level of confidence on the density uncertainty, given our understanding of the uncertainties in $M_F$, $M_E$, and $V$. Error propagation is simple and powerful.

We give several additional examples of error propagation in Section 5.4. In Section 5.3 we discuss how to use the error propagation worksheet to explore the relative impacts of various error contributions in a calculation, and we discuss automating error propagation with Excel and MATLAB. One final note: care must be taken with units in error propagation calculations. Equation 5.18 is dimensionally consistent, but the terms have complex units due to the partial derivatives. As always when substituting numbers, the appropriate unit conversions must be included to maintain unit consistency (see this effect in Example 5.6).

## 5.3  Error Propagation Worksheet

The error propagation worksheet (Figure 5.1) is a handy tool for carrying out and documenting an error propagation. The worksheet also helps with the valuable task of testing the impact of choices made when determining the individual errors, $e_{s,x_i}$. This is needed since, as discussed in earlier chapters, we often resort to estimates or rules of thumb to determine errors. If we determine through error propagation that a particular error has minimal impact on the calculated quantities we care about, then we can rest assured that our estimates are doing no harm to our final answers. Conversely, if an error estimate dominates the calculation, we would rightfully worry about the impact of error choices we made.

When trying out different estimated values for $e_{s,x_i}$, it is advantageous to automate error propagation calculations with Excel, MATLAB, or another program. We take a moment now to show how the paper worksheet guides, organizes, and records an error propagation. Once organized into the worksheet format, the calculation is readily transferred to computer software for further explorations.

Consider the worksheet in Figure 5.2. Equation 5.18 appears at the bottom, written in the form that includes up to five independent stochastic variables, $x_i$. The error propagation worksheet has places for the formula for $\phi(x_i)$ and for the names of the variables. There is also a column for the standard errors associated with each variable (second from right). The middle column of the table records the formulas for the necessary partial derivatives. The expressions for the partial derivatives contain variables – in this problem, $V$ and the two masses – and included in the table are representative values of these quantities. The final column on the right of the worksheet represents each of the squared terms that make up Equation 5.18. These entries are summed near the bottom of the column, and, finally, the square root is taken at the very bottom to give the answer for $e_{s,\phi}$. As indicated previously, the final answer for Example 5.4 is $e_{s,\phi} = e_{s,\rho} = 3.5 \times 10^{-3}$ g/cm$^3$.

With the calculation laid out in the worksheet, it is easy to test the impacts of the errors associated with each variable. Each variable has a row in the worksheet, and in the rightmost column we can view the associated individual terms that make up the final sum for squared error. Looking at the magnitudes in that column in Figure 5.2, we see that the term associated with the volume error is six orders of magnitude larger than the terms associated with the masses. If our estimates of the uncertainties in the mass measurements were increased by an order of magnitude (or even two orders of magnitude), these contributions would still be dwarfed by the volume uncertainty. From this comparison, we conclude that the small weighing errors have no impact on the overall error; the uncertainty in volume is the dominant error.

Exploring logic of this type is very powerful: by putting these numbers before us, the error propagation worksheet guides us to pay attention to which errors dominate the error in the calculated variable. Knowing these relationships allows us to assess choices we have made – of which pycnometer or balance to use, for example, or how many replicates to perform, or to whom we assign the experimental work.

The error propagation worksheet is readily automated in Excel. An Excel spreadsheet version of the pycnometer problem is shown in Figure 5.3. The column headings indicate the formulas that must be entered to reproduce the worksheet; for the partial derivatives, we code the manually derived partial derivative formulas. Each row is labeled with the name (symbol) of the appropriate variable. The squared weighted errors in the last column are summed to calculated the variance $e_{s,\phi}^2$, which leads to the final values of $e_{s,\phi}$.

Excel results
(dominated by $e_{s,V}$)

$e_{s,\rho} = 0.0035$ g/cm$^3$

Once coded, the Excel worksheet becomes an active version of the error propagation. If we wish to explore the impact of the error estimates $e_{s,x_i}$ ("what if" scenarios), we can change the value in the cell that contains the estimate and immediately see the effect on the propagated values. In the software we would also see, dynamically, the relative magnitude of the $\left(\frac{\partial \phi}{\partial x_i}\right)^2 e_{s,x_i}^2$ terms, which

**Error propagation worksheet**

| $\phi(x_1,x_2,x_3)$ | | $\phi=$ $\rho_{BF}=$ | | 1.739 g/cm³ | | $2e_{s\phi}=$ | | 0.007 g/cm³ |

| | $x_i$ | value | units | $d\phi/dx_i$ | $(d\phi/dx_i)^2$ | $e_{xi}$ | $e_{xi}^2$ | $(d\phi/dx_i)^2 e_{xi}^2$ | |
|---|---|---|---|---|---|---|---|---|---|
| $x_1$ | $M_F$ | 30.800 | g | 0.10 | 0.010 | 5.0E–05 | 2.5E–09 | 2.50E–11 | g²/cm⁶ |
| $x_2$ | $M_E$ | 13.410 | g | –0.10 | 0.010 | 5.0E–05 | 2.5E–09 | 2.50E–11 | g²/cm⁶ |
| $x_3$ | $V_{pyc}$ | 10.000 | cm³ | –0.174 | 0.0302 | 0.02 | 4.0E–04 | 1.210E–05 | g²/cm⁶ |
| | | | | | | | $e_{s,\phi}^2$ | 1.21E–05 | g²/cm⁶ |
| | | | | | | | $e_{s,\phi}$ | 0.0035 | g/cm³ |

Figure 5.3 Excel or other spreadsheet software is an effective platform for error propagation calculations. The data from Example 5.4 are shown in this excerpt from an Excel spreadsheet. Once set up in this way, it becomes easy to try out alternative values for the $e_{s,x_i}$ to see the impact on the overall error calculated.

are shown in the last column. Spending a few moments running scenarios would tell us all we need to know about the various potential improvements we might consider for our processes.

We can carry out the same automation of error propagation with MATLAB, although the solution looks a bit different from the Excel approach (Figure 5.4). MATLAB is array based, and quantities are input by the user and manipulated by MATLAB as vectors or matrices. The quantities created exist within a user's MATLAB Workspace and may be displayed, exported, or manipulated further. In MATLAB, there is a command-line interface, and commands are typed in sequentially. Alternatively, we can store a series of commands into a function, which is itself stored for repeated use in a file with the ".m" extension. The home-grown MATLAB functions that we use in this text are provided in full in Appendix D.

Our solution to automating the error propagation worksheet with MATLAB is to create a function *error prop* that carries out the error propagation (section 1). The function *error prop* requires the following arguments:

$$results = error\,prop(fcn, nominal, names, SE, dx)$$

- *fcn:* the name of the function $\phi(x_1, x_2, x_3)$ when coded in MATLAB
- *nominal*: a vector of nominal values (representative values) of the variables
- *names*: a vector of the names of the variables

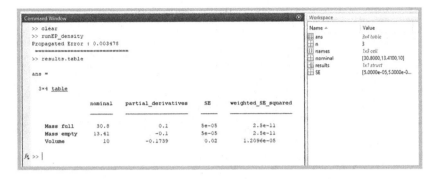

Figure 5.4 Using our user-written MATLAB function *runEP density* (code provided in Appendix D), along with the conditions presented in Example 5.4 (function @*phi density*; vectors *names, nominal, SE*), we obtain a MATLAB structure array *results* that contains three fields: *results.table, results.SEfcn,* and *results.fcn.* The propagated error $e_{s,\phi}$ is displayed in the Workspace when we run *runEP density* and it is also stored in *results.SEfcn.* The name of the user-programmed function for the equation $\rho(M_F, M_E, V)$ (@*phi density*) is stored in *results.fcn,* and the table shown in this figure is stored in *results.table.* The Workspace window (right) lists all the variables that have been created and some information about them. Typing the names of the variables causes their contents to be displayed. For more on using these functions, see Example 5.6 and the appendix.

- *SE*: a vector of the standard errors of the variables
- *dx*: an optional step size used to numerically calculate the needed partial derivatives

The entire error propagation calculation, therefore, consists of creating these four (or five) inputs and running *error prop*. Those steps – that is, creating the inputs and running *error prop* – can themselves be written into a function, which for the fluid density error propagation in Example 5.4 we called *runEP density* (Figure 5.5 and Appendix D).

For the fluid density example, the various arguments of *error prop* are as follows:

- function $\rho(M_F, M_E, V)$ when coded in MATLAB, @*phi density* (see Appendix D)
- vector *nominal* = [ 30.800, 13.410, 10.00]
- vector *names* = {'Mass full', 'Mass empty', 'Volume' }
- vector *SE* = [5.0e − 5, 5.0e − 5, 0.02 ]

In *runDP density*, once the input vectors are created, we call *error prop* with the arguments, yielding the information in the error propagation worksheet.

```matlab
function results = error_prop(fcn,nominal,names,SE,dx)
%
%   function results = error_prop(fcn,nominal,names,SE,dx)
%
%   A function to perform the error propagation worksheet calculations
%   from Morrison, "Uncertainty Analysis for Engineers and Scientists:
%   A Practical Guide," Cambridge University Press
%   ==================================================
%       where fcn :  user-supplied function
%           nominal : representative values of all variables
%           names: names of variables
%           SE: standard errors of variables
%           dx (optional):  user-supplied differentials for determining
%                            partial derivatives (default:
%                            1e-8 times nominal values)
%
% (c) 2019 by Faith Morrison and Tomas Co
% @ Michigan Technological University

% Make inputs to be columns.

    nominal = nominal(:);
    names = names(:);
    SE = SE(:);

% Set the default dx for calculating the partial derivatives;
% protect it from negative values of dx and values of zero.

    if nargin<5
        dx = max((1e-8)*abs(nominal),1e-18);
    end

% Calculate partial derivatives; function dphidx is below.

    z = dphidx(fcn,nominal,dx);
    partial_derivatives = z;

% Build the table.

    weighted_SE_squared = (z.*SE).^2;
    tab = table(nominal,partial_derivatives,SE,weighted_SE_squared);
    tab.Properties.RowNames = names; %remove semicolon to echo table

% Obtain the final result.

    SEfcn = sqrt(sum(weighted_SE_squared));
%    disp(['Propagated Error : ', num2str(SEfcn)]); %displays SE result
%    disp(' =============================== ')

% Build a structure array to save all results together

    results.table = tab;
    results.SEfcn = SEfcn;
    results.fcn = fcn;

end

function z = dphidx(fcn,x,dx)

    n = length(x);

    for i=1:n
        nx = x;
        nx(i) = nx(i)+dx(i);
        z(i) = (feval(fcn,nx)-feval(fcn,x))/dx(i);
    end
    z = z(:);

end
```

Figure 5.5 MATLAB code for the function *error prop*; also provided in Appendix D.

The output of function *runEP density* is $e_{s,\phi} = 0.03478$ (see Figure 5.4), and the error propagation table can be displayed by typing *results.table* in the MATLAB Workspace.

| MATLAB results (dominated by $e_{s,V}$) | $e_{s,\rho} = 0.0035$ g/cm$^3$ |
|---|---|

One feature of using the supplied MATLAB functions is that inputting a set of manually obtained partial derivatives is not necessary,[4] as we can readily program a digital version of the required partial derivatives (see MATLAB function *error prop*, which contains function *dphidx*).[5] This saves us the trouble of calculating and programming the formulas for the partial derivatives, which can be a significant savings in effort when the relationship in the function is complex.[6] Changing the value of a standard error to run a different scenario may be accomplished by swapping out the target standard error within the vector *SE* and rerunning *errorprop* (see Example 5.6 to see how this could be done systematically).

The error propagation worksheet and computer software are handy tools in making error calculations. In the next section we present several examples that demonstrate the value of these tools.

## 5.4 Working with Error Propagation

We have the fundamental tools we need for error propagation; now we apply them to a variety of situations to demonstrate the value and added understanding obtained by examining the uncertainty in our calculations and experiments.

---

[4] Nor is it necessary in Excel, if one is willing to do some VBA programming.
[5] The definition of the partial derivative of the function $\phi(x, y, z)$ with respect to $x$ is

$$\frac{\partial \phi}{\partial x} = \lim_{\Delta x \to 0} \frac{\phi(x + \Delta x, y, z) - \phi(x, y, z)}{\Delta x}$$

We can approximate this quantity at a point $(x_0, y_0, z_0)$ as

$$\left.\frac{\partial \phi}{\partial x}\right|_{x_0, y_0, z_0} \approx \frac{\phi(x_0 + \Delta x_0, y_0, z_0) - \phi(x_0, y_0, z_0)}{\Delta x_0}$$

This approximation is coded into our MATLAB function *dphidx* (see Appendix D).
[6] An example of a complex relationship would be the friction factor–Reynolds number data correlation that determines pressure drop/flow rate information in smooth pipes.

For our first example (Example 5.5), we compare two methods of estimating uncertainty in fluid density. In Example 5.4 we obtained a density for Blue Fluid 175 from a single pycnometer measurement; previously (Example 2.8), we discussed some replicates of the same density measurement, with the same fluid. The density values obtained via the two pathways are similar, but the 95% level of confidence error limits are quite different. We address the implications of these results now.

**Example 5.5: Replicate density error versus error from propagation.** *In Example 5.4 we obtained a density for Blue Fluid 175 from a single pycnometer measurement; previously, in Example 2.8, we discussed some replicates of the same measurement, with the same fluid. The density values obtained are similar, but the error limits are quite different in the two examples. Which answer and which error limits should we use in follow-on calculations?*

**Solution:** In Example 2.8, 10 replicate values of Blue Fluid 175 density were presented, and the appropriate statistical analysis was performed to obtain

$$\begin{matrix} \text{Density from replicates} \\ \text{(mean of 10 measurements)} \end{matrix} \quad \rho_{BF175} = 1.734 \pm 0.003 \text{ g/cm}^3 \quad (5.26)$$

In Example 5.4, we used error propagation and a single measurement of density of the same fluid to determine

$$\begin{matrix} \text{Density from} \\ \text{error propagation} \\ \text{(one measurement)} \end{matrix} \quad \rho_{BF175} = 1.739 \pm 0.007 \text{ g/cm}^3 \quad (5.27)$$

Comparing these two numbers and their error limits (Figure 5.6), we see that the two error ranges overlap, indicating that both methods may be correct in their estimate of the density. Looking at the error limits of the two answers, however, we see that the single measurement has a much larger uncertainty than the replicate result. We would like to claim the small error as the correct error – that number is much more precise in expressing the density. But if that number is correct, how do we explain the much wider error limits obtained in the single-measurement determination? Is the error propagation calculation wrong?

This confusion may be resolved by considering the effect of systematic error in the two experiments. The error propagation example calculated both density and the error limits based on a single use of a pycnometer. The error limits came from assuming our best estimates of the uncertainties in the mass and volume measurements and completing an error propagation. This calculation is correct. We also learned from that example that the error in the pycnometer experiment is dominated by the uncertainty in the volume of the pycnometer.

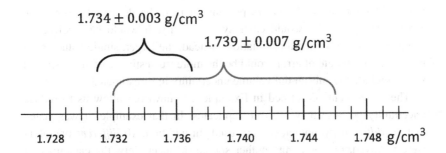

1.734 ± 0.003 g/cm$^3$

1.739 ± 0.007 g/cm$^3$

1.728    1.732    1.736    1.740    1.744    1.748 g/cm$^3$

Figure 5.6 We determined the density of Blue Fluid 175 two ways: (1) through averaging true replicates and (2) from a single measurement and error propagation. Both are correct; the answer from the error propagation calculation has wider error limits, which is appropriate, as discussed in the text.

The replicate-error calculation is also correct. In that example, 10 student groups used 10 different pycnometers to measure the density of Blue Fluid 175. When we obtain repeated measurements that are only subject to random errors, it is correct to calculate the best estimate of error from replicate error, $e_{s,random} = s/\sqrt{n}$, where $s$ is the standard deviation of the sample set and $n$ is the number of observations in the sample set. With replicate error, because of the $\sqrt{n}$ in the denominator, large values of $n$ will greatly diminish the amount of error associated with estimating the mean of the dataset. The mean of the dataset is the best value we can determine for the density of Blue Fluid 175. The more *true replicates* we take, the better and better will be the estimate of the true mean value of the density.

How can both calculations and error estimates be correct? The key to understanding this situation resides in the meaning of the phrase "true replicates." A *true replicate* is exposed to all the error sources that bring uncertainty into a measurement. It turns out that not all repeated experiments produce true replicates. Consider the calculation described in Example 5.4. There, the largest source of uncertainty in using a pycnometer for a density measurement was found to be the uncertainty in the volume of the pycnometer. If one obtains 10 replicates of density *with the same pycnometer*, the calibration error of the volume would be the same, systematically, for all the replicates. If the true volume of the particular pycnometer in use were 10.05 ml, for example, then for *all* of the replicate measurements, the pycnometer volume would be systematically larger than the assumed value of 10.00 ml. These replicates would therefore not be "true." In the case of repeating the experiments over and over with the same pycnometer, we would not produce a sample set that randomly sampled all the errors present; in fact, the dominant error – the error in the volume – would enter systematically. It would be incorrect to use $e_s = s/\sqrt{n}$ for such a calculation. Instead, the larger, single-value error-propagation estimate of error would be the more true estimate of error, since it would include the substantial volume uncertainty in its estimate.

The experiments described in Example 2.8, however, allow us to reduce uncertainty because they are true replicates. In that example, it was stated that *10 different* pycnometers were used. In addition, 10 *different* operators were involved, randomizing another source of uncertainty, the capability of the operator to carry out the density measurement. The replicates described in Example 2.8 were carried out with all sources of error sampled within the sample set, with each error randomly increasing or decreasing the value obtained. When this is the case, the replicates in the dataset truly represent samples subjected to random errors, and replicate error alone can characterize the set. Since the replicates are true, we may confidently use the more precise value of the density of Blue Fluid 175 obtained from the replicates in Example 2.8. By replicating the density measurements with various pycnometers, the systematic pycnometer volume error is *randomized* (see Section 2.3.4): each individual pycnometer has its actual volume, sometimes slightly higher than the 10 ml expected and sometimes lower, varying stochastically. These errors randomly influence the density values obtained, and the influence can be neutralized by calculating the mean value of density from replicates. Figure 2.19 demonstrates that the mean of true replicates can be very precisely known even when individual observations have a large amount of stochastic variation.

Thus, we conclude that in follow-on calculations, we should use the following value for the density of Blue Fluid 175 (from averaging true replicates):

Mean of true replicates
95% confidence interval
$$\rho_{BF175} = 1.734 \pm 0.003 \text{ g/cm}^3 \qquad (5.28)$$

To arrive at this conclusion, we needed to know that the dominant error in the calculation is the pycnometer volume and that this error was randomized in the trials, producing true replicates.

In our next example, we use MATLAB to calculate the standard error $e_{s,\phi} = e_{s,\Delta p,std}$ of a differential pressure calculated from data for manometer fluid heights. We subsequently run several scenarios to explore the impact on $e_{s,\phi}$ of each contributing error $e_{s,x_i}$.

**Example 5.6:** *LINKED* **Uncertainty in differential pressure from a manometer.** *An air-over-Blue-Fluid-175 manometer is used as a pressure standard when calibrating a differential-pressure (DP) meter (see Figure 4.3 for a schematic of the experiment). Two pressurized vessels are connected simultaneously to the manometer and the DP meter. This arrangement results in a fluid-height difference both between the two sides of the manometer (Figure 4.13) and in the generation of an electric-current reading on the DP meter. To produce the calibration curve, we need to convert the manometer-height readings to* $(\Delta p, psi)_{std}$. *The fluid-height differences on the manometer are measured with a meter stick as two separate heights, the fluid height above an arbitrary zero level plus the height below that same reference level; the raw data are given in Table 4.4. For the data point that produced a DP-meter current signal of* 12.0 mA, *calculate the value* $(\Delta p, psi)_{std}$ *and its 95% level of confidence error limits. Which contributing errors* $e_{s,x_i}$ *dominate or strongly influence the error propagation?*

**Solution:** The discussion of the calibration of the DP meter is one theme of the linked examples mapped in Figure 1.3. In Example 4.8, we introduced the raw calibration data, described the workings of the DP meter, and discussed the error propagation process needed to make progress on determining the calibration error from the raw data. We now return to this problem, having learned the error propagation steps.

To calculate a pressure difference that causes a fluid height difference in the manometer, we use the following relationship from physics, which comes from a momentum balance (Figure 4.13):[7]

$$\text{Manometer equation} \quad \Delta p = p_1 - p_2 = \rho g(h_1 + h_2) \quad (5.29)$$

where $h_1$ and $h_2$, respectively, are the heights of fluid above and below the arbitrary reference level on the high- and low-pressure sides of the manometer; $\rho = \rho_{BF175}$ is the density of the fluid in the manometer; $g = 980.66$ cm/s$^2$ is the gravitational force constant; and the density of air $\rho_{air}$ has been neglected relative to the fluid density. Measurement of the manometer fluid density and its uncertainty are discussed in Examples 2.8 and 5.5; the density is

$$\rho = \rho_{BF175} = 1.734 \pm 0.003 \text{ g/cm}^3 \quad \begin{array}{c} \text{(95\% confidence} \\ \text{interval)} \end{array} \quad (5.30)$$

This value was obtained from averaging true data replicates (replicate error, $e_{s,random} = e_{s,\rho} = 0.0015$ g/cm$^3$).

---

[7] The momentum balance on a stationary fluid may be summarized as follows: the pressure at the bottom of a column of fluid is equal to the pressure at the top plus $\rho g$(height) [41].

For the data point of interest, $(I, mA) = 12.0$ mA on the DP meter, $(\Delta p, psi)_{std}$ is ($\alpha_0$ is a unit conversion)

$$(\Delta p, psi)_{std} = p_1 - p_2 = \rho g(h_1 + h_2)\alpha_0 \tag{5.31}$$

$$= \left(\frac{1.734 \text{ g}}{\text{cm}^3}\right)\left(\frac{980.66 \text{ cm}}{\text{s}^2}\right)\left(\frac{76.0 \text{ cm}}{1}\right)\left(\frac{14.696 \text{ psi}}{1.01325 \times 10^6 \text{ g/cm s}^2}\right)$$

$$= 1.8744 \text{ psi} \qquad \text{(extra digits)} \tag{5.32}$$

To obtain the error limits, we carry out an error propagation on the function in Equation 5.31. The worksheet is shown in Figure 5.7.

Following the error propagation procedure, we need values of the standard errors $e_{s,x_i}$ for all the variables in the problem: density, gravity, and the fluid heights. We have the standard error for density from recent discussions (Example 5.5, $e_{s,\rho} = 0.0015$ g/cm$^2$). For the acceleration due to gravity, we write zero for the error, as this is a standard reference value whose uncertainty is presumed to be much less than the other uncertainties in the calculation. For the heights $h_1$ and $h_2$, we use a common standard error, $e_{s,h}$, which we must determine.

| $\phi(x_1, x_2, x_3, x_4, x_5)$: | Formula for $\phi$:[*] $\Delta p = \alpha_0 \rho g(h_1 + h_2)$ *Note: units must work as written.* | | Representative value of $\phi$: (include units) 1.874 psi | 95% C.I. of $\phi$: $(\phi \pm 2e_{s,\phi})$ (include units) $1.874 \pm 0.005$ psi |
|---|---|---|---|---|
| **Measured quantities, $x_i$** | | | $\dfrac{\partial \phi}{\partial x_i}$ | $e_{s,x_i}$ | $\left(\dfrac{\partial \phi}{\partial x_i}\right)^2 e_{s,x_i}^2$ |
| $x_i$ | Symbol | Representative value | | | |
| $x_1$ | $\rho$ | $1.734 \frac{\text{g}}{\text{cm}^3}$ units | $\alpha_0 g(h_1 + h_2)$ | $1.5 \times 10^{-3} \frac{\text{g}}{\text{cm}^3}$ | $2.63 \times 10^{-6}$ |
| $x_2$ | $g$ | 980.66 cm/s$^2$ units | $\alpha_0 \rho(h_1 + h_2)$ | 0 | 0 |
| $x_3$ | $h_1$ | 29.8 cm units | $\alpha_0 \rho g$ | 0.0577 cm | $2.03 \times 10^{-6}$ |
| $x_4$ | $h_2$ | 46.2 cm units | $\alpha_0 \rho g$ | 0.0577 cm | $2.03 \times 10^{-6}$ |
| $x_5$ | $\alpha_0$ | $1.45038 \times 10^{-5} \frac{\text{psi}}{\text{dynes/cm}^2}$ units | $\rho g(h_1 + h_2)$ | 0 | 0 |

$$e_{s,\phi}^2 = \left(\frac{\partial \phi}{\partial x_1}\right)^2 e_{x_1}^2 + \left(\frac{\partial \phi}{\partial x_2}\right)^2 e_{x_2}^2 + \left(\frac{\partial \phi}{\partial x_3}\right)^2 e_{x_3}^2 + \left(\frac{\partial \phi}{\partial x_4}\right)^2 e_{x_4}^2 + \left(\frac{\partial \phi}{\partial x_5}\right)^2 e_{x_5}^2$$

*All variables x_i must be independent for this equation to hold.*

$e_{s,\phi}^2 = 6.68 \times 10^{-6}$ psi$^2$

$e_{s,\phi} = 0.0026$ psi    units

Figure 5.7 The error propagation worksheet guides the error propagation process for Example 5.7. The unit-conversion constant $\alpha_0$ (Equation 5.35) is included to produce the appropriate units. Note that an optimistic value $e_{s,h} = 0.0577$ cm was used in these calculations. See the discussion in the text.

The standard error associated with measuring the heights with a wall-mounted meter stick is obtained by following the procedures we established earlier in the text. We consider three error sources: replicate error (not relevant, as we have a single value of each height), reading error (important), and calibration error (possibly important). For reading error, we must know something of how these heights were obtained. We investigate and find that a wooden meter stick was installed vertically and between the two arms of the manometer (Figure 3.2). The reading is the height of the fluid meniscus as determined with the meter stick. The scale on the meter stick is divided into 1-mm divisions. Following the reading-error worksheet, we estimate as follows:

1. Sensitivity:[8] (0.1 cm)
2. Resolution: $\frac{(0.1 \text{ cm})}{2}$
3. Fluctuations: not applicable, if the fluid level is steady.

The largest of the three is the sensitivity, and thus the standard reading error is

$$\text{Reading error} \quad e_R = \max(\text{sensitivity, resolution, fluctuations})$$
$$= \max(0.1, 0.05, NA) = 0.1 \text{ cm}$$
$$e_{s,reading} = \frac{0.1 \text{ cm}}{\sqrt{3}} = 0.0577 \text{ cm} \quad \text{(sensitivity)} \quad (5.33)$$

Turning to calibration error for the meter stick, the accuracy with which the markings have been etched into the stick is the responsibility of the manufacturer. When assembling the manometer apparatus, the designer probably chose an inexpensive ruler. Thus the accuracy in $h$ is unlikely to be exceptionally good, but it is probable that it is acceptable. There is also the issue of how well the meter stick was mounted vertically. In the end, we choose to neglect the stick's calibration error in the expectation that the accuracy of the meter stick is not the controlling error for our differential-pressure question. When we complete the error propagation, we can test this assumption by toggling the error on the measured fluid heights to larger values to see the impact of setting $e_{s,h}$ at various levels.

Having neglected calibration error on the meter stick, we now obtain the overall error for that device by combining the errors:

---

[8] The reasoning here is that sensitivity is the amount that the instrument can *sense*. If we are careful in reading the meter stick, perhaps we can sense a change when the height of the manometer fluid changes by about a millimeter. If the meter stick is not perfectly perpendicular, the sensitivity would be less, maybe more like 1.5–2.0 mm. These estimates are up to the individual investigator.

$$e_{s,h}^2 = e_{s,random}^2 + e_{s,reading}^2 + e_{s,cal}^2$$
$$= 0^2 + (0.0577)^2 + 0^2$$
$$= (0.0577)^2 \text{ cm}^2$$

Length, from a
single measurement
with ruler

$$e_{s,h} = 0.0577 \text{ cm} = 0.06 \text{ cm} \qquad (5.34)$$

(We used this three-sig-figs value for $e_{s,l}$ in Example 5.2.) The error is purely reading error, obtained from an estimate of sensitivity. We now have all the inputs to the differential-pressure error propagation, and we carry out the calculation using the paper worksheet (Figure 5.7) and MATLAB function *runEP DP* in Appendix D, whose output is shown in Figure 5.8.

We pause to warn once again about a common calculation mistake made in error propagation – one that is related to unit conversions. In the current problem, there is a unit conversion. Beginning with Equation 5.29, the units are

$$\Delta p = \rho g (h_1 + h_2)$$
$$\Delta p \text{[units]} [=] \left(\frac{\text{g}}{\text{cm}^3}\right) \left(\frac{\text{cm}}{\text{s}^2}\right) (\text{cm})$$
$$[=] \frac{\text{g}}{\text{cm s}^2} = \frac{\text{dynes}}{\text{cm}^2}$$

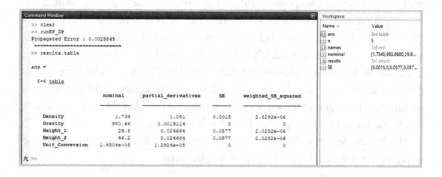

Figure 5.8 The error propagation worksheet guides the error propagation process for Example 5.7, which here is implemented in MATLAB. We ran the function *runEP DP*, which prints the output $e_{s,\phi}$ = Propagated Error = 0.0025845 (extra digits provided). To display the table, we type "results.table" in the command line. The Workspace window (right) lists all the variables that have been created and some information about them. Typing the names of the variables causes their contents to be displayed.

These are not the pressure units requested, so we use a unit conversion to obtain the requested units of psi (Equation 5.31). Under normal circumstances we leave unit conversions out of formulas and rely on the user to make sure that units are sorted out before terms are added or cancelled. When using the error propagation worksheet, however, the user is a bit removed from the calculations made in the worksheet rows, and this may allow unit-conversion errors to creep into the calculation. To avoid this, we recommend that unit conversions be included explicitly in the formula being propagated, as has been done with the conversion factor $\alpha_0$ in the worksheet in Figure 5.7.

$$\text{Unit-conversion factor} \quad x_5 = \alpha_0 \equiv \left( \frac{14.696 \text{ psi}}{1.01325 \times 10^6 \text{ dynes/cm}^2} \right)$$

$$= 1.4503824328 \times 10^{-5} \frac{\text{psi}}{\text{dynes/cm}^2} \qquad (5.35)$$

The unit conversion enters into all the partial derivatives. Since unit conversion $\alpha_0$ is a universal constant, it has zero uncertainty.

We carry out the error propagation in MATLAB, using the function *runEP DP* to store the steps (see Appendix D for the code). In that function, we call *error prop* with the following arguments:

- Function @*phi DP* (Equation 5.31; see Appendix D)
- Vector *nominal* = [1.734, 980.66, 29.8, 46.2, 1.4504e-5]
- Vector *names* = {'Density', 'Gravity', 'Height_1', 'Height_2', 'Unit_Conversion'}
- Vector $SE = [1.5e - 3, 0, 0.0577, 0.0577, 0]$

In Figure 5.8, we see the effect in the MATLAB Workspace of running the commands in *runEP DP*, including the call to function *error prop*. The error propagation calculation for manometer differential pressure $(\Delta p, \text{psi})_{std}$ at $I = 12.0$ mA is now complete, and the standard error on $(\Delta p, \text{psi})_{std}$ at DP-meter signal 12 mA is calculated to be

$$\text{Differential pressure from manometer} \quad \boxed{e_{s, \Delta p, std} = 0.0025845 \text{ psi}} \quad \text{(extra digits)} \quad (5.36)$$
$$(I, \text{mA}) = 12.0 \text{ mA}$$
$$(\text{tentative})$$

Figure 5.8 shows the error propagation table created by *runEP DP* as displayed in the MATLAB Workspace by typing the variable name *results.table*.

Before using the value of $e_{s, \Delta p, std}$ in Equation 5.36, we should examine its reliability, which we can determine by reviewing how the answer was obtained. Equation 5.36 gives the estimate of the standard error on differential pressure as measured by the manometer, based on the estimated standard errors that we entered into the error propagation (vector *SE*):

| $e_{s,x_i}$ | Value | Source |
|---|---|---|
| $e_{s,\rho}$ | 0.015 g/cm$^3$ | replicates |
| $e_{s,g}$ | 0 | assumed accurate |
| $e_{s,h_1}$ | 0.0577 cm | estimate, sensitivity |
| $e_{s,h_2}$ | 0.0577 cm | estimate, sensitivity |
| $e_{s,\alpha_0}$ | 0 | accurate |

A quick calculation with the error propagation worksheet results indicate (rightmost column, Figure 5.8) that 39% of the total squared error is associated with the error in density, while 60% is associated with the errors in obtaining the heights. The other two variables in the equation for $(\Delta p, \text{psi})_{std}$ are reference constants, which we assume to be accurate. The question of accuracy of $e_{s,\Delta p,std}$ therefore comes down to the accuracies of $e_{s,\rho}$ and $e_{s,h}$.

In the previous example (Example 5.5), we discussed $e_{s,\rho}$, which was obtained through the averaging of true replicates. This is a reliable way to assess uncertainty, so we are highly confident in this estimate.

The uncertainty in measured height $e_{s,h}$ was discussed earlier and emerged from an estimate of the sensitivity, with calibration error neglected. Since $e_{s,h}$ is the dominant error in the determination of $e_{s,\Delta p,std}$, it would be prudent to explore this estimate a bit more before accepting the result in Equation 5.36.

The manometer-fluid heights used in the calculation were taken with a conventional meter stick mounted vertically with the level of the fluid meniscus compared to the neighboring scale (see photograph in Figure 3.2). The value of height recorded will depend on how carefully the meter stick was mounted (is it truly vertical?). The heights are obtained by estimating the level of a fluid meniscus; the technique for doing this involves avoiding parallax error[9] and extending horizontally the reading of the fluid level to the nearby scale. We reasoned that $e_{s,h} \approx 0.06$ cm or a bit over half of a millimeter. On reflection, and given the possibility of the meter stick deviating from vertical, this is a bit optimistic. Perhaps we should try a larger number – a millimeter or two – and see what the effect is on the error of $(\Delta p, \text{psi})_{std}$.

Suggestions of this sort are easily addressed when the error propagation is set up in MATLAB, Excel, or equivalent software. In our MATLAB

---

[9] When reading a fluid meniscus against a mark, it is essential for the viewer's eye to be even (at the same horizontal level) with the meniscus. Looking at the meniscus from a bit above or a bit below horizontal introduces an error; see Wikipedia and [30] for more on parallax error.

calculation, we incorporated the number we originally estimated for $e_{s,h}$ (0.0577 cm) into a vector of standard errors (*SE*). The overall error on $(\Delta p, \text{psi})_{std}$ with this value used turned out to be $e_{s,\Delta p,std} = 0.00258$ psi. If we try other values for $e_{s,h}$ (by repeatedly running the function *runEP DP* or by using MATLAB function *runEP multipleDP*, both found in Appendix D), we obtain the following revised values for $e_{s,\Delta p,std}$:

| Trial | $e_{s,h}$ (cm) | $e_{s,\Delta p,std}$ (psi) | Comment |
|---|---|---|---|
| 1 | 0.0577 | 0.00258 | initial value |
| 2 | 0.0800 | 0.00323 | |
| 3 | 0.0900 | 0.00353 | |
| 4 | 0.1000 | 0.00385 | |
| 5 | **0.1500** | **0.00548** | chosen value |
| 6 | 0.2000 | 0.00716 | |
| 7 | 0.5000 | 0.01752 | |

Changing $e_{s,h}$ from approximately 0.06 cm to up to half a centimeter significantly changes the propagated standard error on differential pressure (increases it more than six-fold). The choice for $e_{s,h}$ is impactful, and in retrospect, our initial estimate was quite optimistic. We therefore resolve to use a less optimistic value than we had initially used. The step of estimating an error in this way is still a bit arbitrary, and it depends on the experimenter's judgment. The choice also reflects back on the experimenter, as the results of our work are shared with, and impact, our coworkers and collaborators. These decisions should be made with the highest ethical standards in mind; the error worksheets of this text allow us to record our decisions and the factors that went into them if we need need to justify or revisit our decisions.

Considering parallax, less-than-vertical mounting of the meter stick, as well as the need for horizontal extension from the fluid level to the scale, we choose to use a value of $e_{s,h} = 0.15$ cm (1.5 mm) for the uncertainty in manometer-fluid height measurements. Using this value increases the propagated $e_{s,\Delta p,std}$ from 0.00258 psi to 0.00548 psi.

Differential pressure
from manometer
$(I, \text{mA}) = 12.0$ mA          $\boxed{e_{s,\Delta p,std} = 0.00548 \text{ psi} = 0.005 \text{ psi}}$          (5.37)
95% level of confidence
(final, extra digits)

In the next example, we consider a larger dataset that needs error limits for each data pair. These calculations are again carried out in MATLAB, which is well designed for handling larger data arrays. The calculations may be carried out in Excel in a straightforward way.

**Example 5.7: *LINKED* Uncertainty in DP meter calibration dataset.**
*An air-over-Blue-Fluid-175 manometer is used as a pressure standard when calibrating a DP meter as discussed in Example 5.6. To produce the calibration curve, we need to convert the entire dataset of manometer-height readings (Table 4.4) to $(\Delta p, psi)_{std}$. What are the $(\Delta p, psi)_{std}$ values from the manometer for the complete dataset? What are the 95% level of confidence error limits on $(\Delta p, psi)_{std}$ for each data point? How can the calibration data be used to convert future DP-meter signals to $(\Delta p, psi)_{meter}$?*

**Solution:** In the previous problem we showed how an individual manometer height measurement could be converted to differential pressure, and we showed how to assess the uncertainty in $(\Delta p, psi)_{std}$ with the error propagation worksheet and MATLAB. In this problem, we use this technique on an entire calibration dataset. All the MATLAB commands discussed in this solution are collected into the function *runEP DPall* (Appendix D).

In Figure 5.9 we show the effect in the MATLAB Workspace of running *runEP DPall*. First, the data from Table 4.2 are read into MATLAB and $15 \times 1$ vectors are created containing the 15 data points of the DP meter signal $(I, mA)_{cal} = calDP\ I$, as well as two vectors containing the fluid heights, $h1 = h_1$ and $h2 = h_2$ (Table 5.1):

> *data = importdata('dp.csv');*
> *calDP I = data(:, 1);*
> *h1 = data(:, 2);*
> *h2 = data(:, 3);*

Subsequently, variables are created for the constants $\alpha_0$, $g$, and $\rho$ as well as for the $1 \times 5$ vector *SE* of standard errors $e_{s,x_i}$:

$$SE = [1.5e{-}3, 0, 0.15, 0.15, 0]$$

*SE* contains the values of $e_{s,i}$ for $\rho$, $g$, $h_1$, $h_2$, and $\alpha_0$ in their respective units (see worksheet in Figure 5.7 for units). We use the new, agreed-upon value of $e_{s,h} = 0.15$ cm. The function $\phi(x_i)$ used to calculate $(\Delta p, psi)_{std}$ (Equation 5.29) is programmed as MATLAB function *phi DP*, which we used in Example 5.6 when we performed a single-point error propagation. Finally,

Table 5.1. *Variables for Example 5.7 as referenced in the text and as they appear in MATLAB, created in the function runEP DPall (Appendix D).*

| Variable name | MATLAB name | Array size |
|---|---|---|
| $(I, \mathrm{mA})_{cal}$ | *calDP I* | $15 \times 1$ |
| $h_1$ | *h1* | $15 \times 1$ |
| $h_2$ | *h2* | $15 \times 1$ |
| $\alpha_0$ | *alpha0* | Scalar |
| $g$ | *g* | Scalar |
| $\rho$ | *rho* | Scalar |
| $e_{s,x_i}$ | *SE* | $1 \times 5$ |
| $(\Delta p, \mathrm{psi})_{std}$ | deltaP | $15 \times 1$ |
| $e_{s,\Delta p, std}$ | SEdeltaP | $15 \times 1$ |

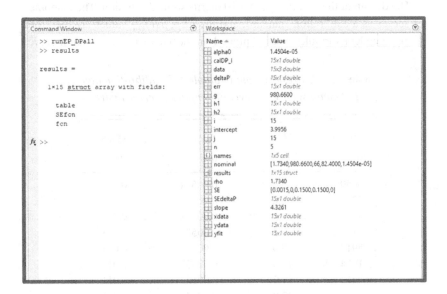

Figure 5.9 The MATLAB commands for Example 5.7 are collected in the function *runEP DPall*, which has no output to the Workspace, but which creates the solution described in the text. The MATLAB struct *results* stores the key output.

to calculate $(\Delta p, \mathrm{psi})_{std}$ and $e_{s,\Delta p, std}$ for all 15 data points, we use a "for loop" that runs through the 15 points and calls *error prop* to perform the error propagation for each data point.

```
for i=1:j
nominal=[rho, g, h1(i), h2(i), alpha0];
results(i)=error prop(@phi DP,nominal,names,SE);
SEdeltaP(i) = results(i).SEfcn;
deltaP(i) = feval(@phi DP,nominal);
end
```

The values of $(\Delta p, \text{psi})_{std}$ for each point are also calculated within this loop, employing the MATLAB function *feval*, which is used to evaluate function *phi DP*. The error-propagation results, including $e_{s, \Delta p_i, std}$ and error propagation tables for each point, are stored in the MATLAB struct *results(i)*. The values of the key calculated quantities are given in Table 5.2.

The function *runEP DPall* also contains code that serves to plot the data $(I, \text{mA})_{cal}$ versus $(\Delta p, \text{psi})_{std}$ along with horizontal error limits as on $(\Delta p, \text{psi})_{std}$ (Figure 5.10). In addition, Figure 5.10 includes a best-fit line with slope and intercept determined in Chapter 6.

The design of the plot in Figure 5.10 merits some discussion. The manometer serves as a calibration standard, and the $(\Delta p, psi)_{std}$ values are therefore the independent variables and are plotted on the $x$-axis of a calibration curve.

Table 5.2. *Key results from Example 5.7: calibration data for DP meter with 95% level of confidence error limits.*

| $(\Delta p)_{std}$ (psi) | $(I)_{cal}$ (mA) | $e_{s, \Delta p, std}$ (psi) | $2e_{s, \Delta p, std}$ (psi) |
|---|---|---|---|
| 0.25650 | 5.2 | 0.00524 | 0.010 |
| 0.57712 | 6.6 | 0.00526 | 0.011 |
| 0.81389 | 7.5 | 0.00528 | 0.011 |
| 0.96187 | 8.2 | 0.00530 | 0.011 |
| 1.15917 | 9.0 | 0.00533 | 0.011 |
| 1.41074 | 10.0 | 0.00537 | 0.011 |
| 1.67710 | 11.2 | 0.00543 | 0.011 |
| 1.87441 | 12.0 | 0.00548 | 0.011 |
| 2.25422 | 13.7 | 0.00558 | 0.011 |
| 2.47619 | 14.6 | 0.00565 | 0.011 |
| 2.66856 | 15.5 | 0.00572 | 0.011 |
| 3.18156 | 17.8 | 0.00591 | 0.012 |
| 3.37886 | 18.7 | 0.00599 | 0.012 |
| 3.57617 | 19.5 | 0.00608 | 0.012 |
| 3.66002 | 19.9 | 0.00612 | 0.012 |

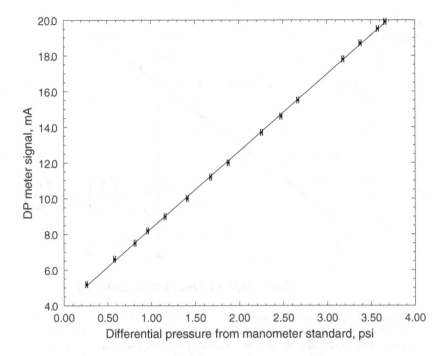

Figure 5.10 The raw manometer data presented in Example 5.7 have been converted to DP meter reading versus $(\Delta p, \text{psi})_{std}$ and plotted along with 95% level of confidence error limits, determined through manometer error propagation for $x$ and from the least-squares curve fit for $y$ (calculation discussed in Chapter 6 and Example 6.2). For the manometer error propagation calculations, $e_{s,h} = 0.15$ cm was used as a conservative (pessimistic) estimate of the error on manometer-fluid height measurements as recommended in Example 5.6.

This choice is supported, as discussed in Chapter 6, by the choice to use the ordinary least-squares algorithm to produce a model. The least-squares algorithm assumes that there is no uncertainty in the $x$-values of the data to be fit. The signal from the DP meter, which is the response of the meter to the imposed differential pressure standard, is the dependent variable and appears on the $y$-axis. The $x$-direction error bars $\pm 2e_{s,\Delta p,std}$ are from the MATLAB error propagation calculations of this example (Table 5.2); these error limits are quite small, as befits calibration standards. Appropriate $y$-direction error bars are discussed in Chapter 6 (Example 6.2) and are associated with the deviations of the data from the fit of a straight line. We discuss vertical error limits further in Chapter 6.

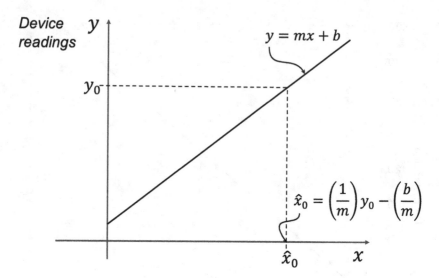

**Device readings**

*Calibrated values, from standard*

Figure 5.11 Once a calibration curve is established, we can associate a future reading with a calibrated value by locating the reading on the $y$-axis of the calibration curve and identifying the associated $x$ value. This process is most conveniently handled by manipulating a model that has been fit to the calibration data.

The final task we are asked to address in this example is determining how the calibration data could be used in future experiments to associate differential-pressure values with DP-meter readings. The relationship between DP-meter signal and $(\Delta p, psi)_{std}$ in Figure 5.10 appears linear. To use this plot for future DP-meter measurements, we would obtain a raw signal $y_0 = (I, mA)_{raw}$ from the DP meter, locate the signal value on the $y$-axis of Figure 5.10, and determine the corresponding $x_0 = (\Delta p, psi)_{meter}$ value from the $x$ value of the graph (Figure 5.11).

Alternatively, and more conveniently, we could create a calibration curve. The process would be as follows: (1) fit a line to the data in Figure 5.10; (2) rearrange the equation for the line to be explicit in $x$; and (3) use that equation $(\Delta p, psi)_{meter} = \phi[(I, mA)_{raw}]$ as our calibration equation. With such a calibration curve, we could simply substitute values of raw meter signals into the formula to obtain the associated differential pressure.

For the DP meter, we have most of what we need to follow these steps, as we now discuss. Using the model fit shown here (determined in Chapter 6), we are able to produce an accurate calibration curve.

$$\text{Model fit to calibration data} \quad (I, \text{mA})_{cal} = m(\Delta p, \text{psi})_{std} + b \qquad (5.38)$$

where for our DP-meter data $m = 4.326143$ mA/psi and $b = 3.995616$ mA (line shown in Figure 5.10; high-precision values of slope and intercept obtained in Example 6.1, extra digits included here). We invert this equation, solving for $(\Delta p, \text{psi})$, which we now associate with future DP-meter results [19].

$$(\Delta p, \text{psi})_{meter} = \frac{(I, \text{mA})_{raw} - b}{m}$$

$$\text{Inverted calibration curve fit} \quad \boxed{(\Delta p, \text{psi})_{meter} = \left(\frac{1}{m}\right)(I, \text{mA})_{raw} - \left(\frac{b}{m}\right)} \qquad (5.39)$$

In terms of the Chapter 6 values of $m$ and $b$, this becomes

$$\text{DP-meter calibration curve} \quad \boxed{(\Delta p, \text{psi})_{meter} = 0.23115(I, \text{mA})_{raw} - 0.92360} \qquad (5.40)$$

We used this equation in Example 4.2 (Equation 4.1).

An open question is how to determine the $(\Delta p, \text{psi})_{meter}$ error limits on future uses of the calibration curve; to answer that last question, we need to perform an error propagation of Equation 5.39. To carry out the error propagation, we first need to address uncertainty in the model fit (errors on parameters $m$ and $b$). We start on that error propagation in Example 5.9, but the question will only be resolved in Chapter 6 (Example 6.8). The questions surrounding the calibration curve are part of our linked DP meter examples. To recall how the linked DP meter examples fit together, see Figure 1.3.

We have completed all the tasks requested in this example: we calculated the $(\Delta p, \text{psi})_{std}$ values for the manometer data and their 95% level of confidence error limits, and we showed how to obtain a calibration curve from the resulting plot (Figure 5.11 and Equation 5.40). The least-squares model fit for $m$ and $b$ in Equation 5.38 is calculated in Chapter 6 both with Excel's LINEST and with MATLAB.

In the next example we use Excel to address error propagation in another measurement-calculation system, determining viscosities from efflux times in Cannon–Fenske viscometers. Each new system considered builds our experience and confidence.

Table 5.3. *Data for Example 5.8. Measured efflux times of a 30% aqueous sugar solution at 26.7°C in three different Cannon–Fenske viscometers. Extra digits provided for $\Delta t_{eff}$ to avoid calculation round-off error. The manufacturer's viscometer calibration constants in $(mm^2/s^2)$ are M267, $\tilde{\alpha}(40°C) = 0.01480$, $\tilde{\alpha}(100°C) = 0.01474$; Y614, $\tilde{\alpha}(40°C) = 0.01459$, $\tilde{\alpha}(100°C) = 0.01453$; and Y733, $\tilde{\alpha}(40°C) = 0.01657$, $\tilde{\alpha}(100°C) = 0.01650$.*

| Viscometer ID→ | M267 | Y614 | Y733 |
|---|---|---|---|
| trials, $i$ | $\Delta t_{eff,i}$ (s) | $\Delta t_{eff,i}$ (s) | $\Delta t_{eff,i}$ (s) |
| 1 | 158.035 | 160.156 | 140.599 |
| 2 | 157.878 | 160.387 | 140.984 |
| 3 | 158.134 | 160.515 | 141.323 |
| 4 | 158.371 | 160.586 | 141.516 |
| 5 | 158.614 | 160.826 | 141.763 |
| 6 | 158.708 | 161.141 | 141.938 |
| 7 | 158.970 | 161.381 | 142.092 |
| mean | 158.3871 | 160.7131 | 141.4593 |
| standard deviation | 0.39617 | 0.43090 | 0.53378 |

**Example 5.8: Uncertainty in viscosity determined with Cannon–Fenske viscometers.** *In Example 2.19, we introduced the use of Cannon–Fenske viscometers to measure fluid viscosity; their operation is described below. Calculate viscosity from the Cannon–Fenske efflux-time data given in Table 5.3 (30 wt% sugar aqueous solution; operated at 26.7°C, solution density = $1.124 \pm 0.018°C$ determined at 21°C). What are the 95% level of confidence error limits on the viscosity determined? Does the value agree with published values for the viscosity of this solution? Discuss your observations.*

**Solution:** A Cannon–Fenske (CF) ordinary viscometer is a specialized piece of glassware used to measure fluid viscosity. The viscometer is charged with a fixed volume of the fluid of interest and allowed to equilibrate in a thermostated bath (Figure 5.12) at the desired temperature, $T_{bath}$. For a CF viscometer measurement to give an accurate viscosity, the correct amount of fluid must be charged (a loading protocol is specified by the manufacturer). In addition, during operation, the upper, straight tubes of the viscometer must be vertical (aligned with gravity). Once thermal equilibrium has been reached, the fluid is drawn upward to an upper reservoir (schematic in Figure 5.13) and allowed to flow down to a lower reservoir under the pull of gravity. A stopwatch

Figure 5.12 To give accurate viscosities, Cannon–Fenske ordinary viscometers must be placed vertically in a bath at the desired temperature. The viscometer remains in the bath the entire time, with fluid drawn up to the upper reservoir with a pipette bulb and the passage of the meniscus through marks timed through the glass walls of the bath.

is used to time the progression of the fluid meniscus as it passes the start timing mark and the stop timing mark. The timing interval is called the efflux time.

$$\text{Viscometer efflux time:} \quad \Delta t_{eff}$$

Below the lower timing mark, the CF viscometer has a capillary tube of uniform bore; the gravity-driven flow through the capillary is the key piece of physics that allows this instrument to measure viscosity (see reference [41], her Example 7.4, p. 508). Appropriate analysis of gravity-driven flow through the angled capillary tube yields the following equation relating viscosity and efflux time:[10]

---

[10] Note that the usual symbol for viscosity is the Greek letter $\mu$; we have already used the symbol $\mu$, however, for the true value of the mean of a distribution, in accordance with the convention in statistics. To avoid any confusion between these two uses of this letter, we have added the tilde symbol ($\tilde{\mu}$) to the mu when referring to viscosity. We have done the same with the Cannon–Fenske calibration constant $\tilde{\alpha}$, to avoid confusion with significance level $\alpha$ (Equation 2.46).

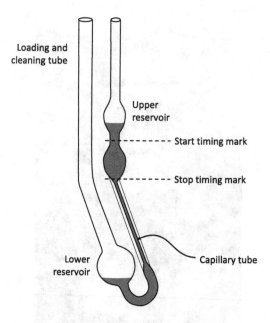

Figure 5.13 The Cannon–Fenske ordinary viscometer uses gravity to drive a fluid through a capillary tube; this device was introduced in Example 2.19.

$$\tilde{\mu} = \rho \tilde{\alpha} \Delta t_{eff} \qquad (5.41)$$

where $\tilde{\mu}$ is the fluid's viscosity at $T_{bath}$, $\rho$ is the fluid density at $T_{bath}$, $\tilde{\alpha}$ is the temperature-dependent calibration constant, evaluated at $T_{bath}$, and $\Delta t_{eff}$ is the fluid's efflux time. The manufacturer-provided calibration constant $\tilde{\alpha}$, associated with each individual viscometer, is given in units of $mm^2/s^2$, which, when used with CGS (centimeters–grams–seconds) units for density and time, results in a viscosity value in *centipoise*, cp (poise = g/(cm s) = 100 cp):

$$\tilde{\mu} = \rho \tilde{\alpha} \Delta t_{eff}$$
$$[=] \left[ \left( \frac{g}{cm^3} \right) \left( \frac{mm^2}{s^2} \right) (s) \right] \left( \frac{10^2 \, cp}{poise} \right) \left( \frac{poise}{g/(cm \, s)} \right) \left( \frac{cm}{10 \, mm} \right)^2$$
$$[=] \, cp$$

The viscosity of water at room temperature is about one centipoise.

Efflux time data are provided for a 30% aqueous sugar solution at 26.7°C in Table 5.3. The efflux times are from measurements with three different CF viscometers, each with its own temperature-dependent values of $\tilde{\alpha}$, supplied by the manufacturer. For each viscometer, the efflux time was measured seven times for a single loading of the glassware; we use the average efflux time to

Table 5.4. *Viscosities of 30% aqueous sugar solution at 26.7°C, determined for the Cannon–Fenske viscometers indicated. Extra digits are retained; the appropriate number of sig figs on viscosity is determined by the error analysis.*

| Viscometer ID | $\tilde{\alpha}(26.7°\text{C})$ $(\text{mm}^2/\text{s}^2)$ | $\tilde{\mu}(26.7°\text{C})$ (cp) |
|---|---|---|
| M267 | 0.01481 | 2.6373 |
| Y614 | 0.01460 | 2.6381 |
| Y733 | 0.01659 | 2.6372 |

calculate viscosity. The sample mean and standard deviation of $\Delta t_{eff}$ for each viscometer are given in Table 5.3.

To calculate viscosity from average efflux time, we need $\tilde{\alpha}$ at $T_{bath} = 26.7°\text{C}$. The function $\tilde{\alpha}(T)$ for each viscometer is a straight-line calibration function provided in the form of two values, one at 40°C and a second one at 100°C, listed for our viscometers in the caption of Table 5.3. To obtain $\tilde{\alpha}$ at the desired temperature, we calculate the line between these two points and evaluate $\tilde{\alpha}$ at $T_{bath} = 26.7°\text{C}$ (see Table 5.4).

The value of the density of the fluid at 21°C is provided in the problem statement: $\rho(21°\text{C}) = 1.124 \pm 0.018 \text{ g/cm}^3$. This temperature is different from the viscosity measurement temperature, but it is close. We use this density value in Equation 5.41, ignoring for the moment the slight temperature offset. We return to this discrepancy when we determine error limits on the final reported viscosity value.

We are now ready to calculate viscosity for the data provided, using Equation 5.41. For viscometer M267 we obtain

$$\tilde{\mu} = \rho \tilde{\alpha} \Delta t_{eff}$$
$$= \left(1.124 \frac{\text{g}}{\text{cm}^3}\right) \left(0.0148 \frac{\text{mm}^2}{\text{s}^2}\right) (158.3871 \text{ s}) \left(\frac{10^2 \text{ cp}}{\text{g}/(\text{cm s})}\right) \left(\frac{\text{cm}^2}{10^2 \text{ mm}^2}\right)$$
$$= 2.6373 \text{ cp} \tag{5.42}$$

This result and the values obtained for the other two viscometers appear in Table 5.4. The three values, taken with separate viscometers (and hence with separate sample loadings), with $\Delta t_{eff}$ determined from a mean of seven timing replicates, may reasonably be described as true viscosity replicates ($n = 3$). It is a testimony to the precision of the instruments and the robustness of the techniques used that we obtained three very close values for viscosity from three different viscometers (Table 5.4). Averaging the viscosity replicates, we obtain $\tilde{\mu} = 2.63753$ cp (extra digits provided) for the viscosity of the 30% aqueous sugar solution at 26.7°C.

The literature value for the viscosity of 30% aqueous sugar solution at 26.7°C is 2.73 cp ([26], interpolated). This is close to what we obtained (the values are 3% different), although the literature and experimental results agree to fewer than two significant figures. Is this the best we can do, given the intrinsic uncertainty of the measurement process? To answer this question, we must perform error analysis.

| Viscosity of 30wt% aqueous sugar solution, data from this example | $\tilde{\mu}$, cp, 26.7°C $= 2.63753 \pm$ ? | (extra digits provided) |
|---|---|---|

| Literature value [26] | $\tilde{\mu}$, cp, 26.7°C $= 2.73$ |
|---|---|

The uncertainty in CF-determined viscosity is assessed using the error propagation worksheet. To perform the error propagation, we need $e_s$ for each of the quantities in Equation 5.41: $\tilde{\alpha}(T_{bath})$, $\rho(T_{bath})$, and $\Delta t_{eff}$. For our preliminary error propagation, we estimate $e_{s,\tilde{\alpha}} = 0.00005$ mm$^2$/s$^2$ (half the last digit provided; optimistic). If this turns out to be the dominant error, we can revisit this assumption.

The error on the density of the fluid at $T_{bath}$ must be known to carry out the error propagation. We were told that the density error limits are $2e_{s,\rho} = 0.018$ g/cm$^3$, although the temperature at which density was reported (21°C) is 5.7°C lower than $T_{bath}$. We return to this uncertainty later in the discussion. For now we accept this density value and the provided error.[11]

The efflux time was measured by repeated timing of a fluid meniscus traveling between two timing marks. The reaction time of the operator affects the accuracy of the timing, and thus to obtain a reliable $\Delta t_{eff}$ we performed replicates on the timing. The error on efflux time is calculated from replicate error (stopwatch calibration is assumed to be accurate, and reading error is negligible).

Replicate error on efflux time

$$2e_{s,\Delta t_{eff}} = \frac{t_{0.025,n-1}s}{\sqrt{n}}$$

$$= \frac{t_{0.025,6}s}{\sqrt{7}} = \frac{(2.4469)(0.53378)}{\sqrt{7}}$$

$$e_{s,\Delta t_{eff}} = 0.2468 \text{ seconds}$$

---

[11] In our calculations we have used an untruncated value for measured density standard error, $e_{s,\rho} = 9.17 \times 10^{-3}$ g/cm$^3$.

| Error Propagation Worksheet | | | | | 30wt% aqueous sugar solution | | | |
|---|---|---|---|---|---|---|---|---|
| $\phi\,(x_1, x_2, x_3)$ | | | $\phi=$ | $= \mu =$ | 2.64 cp | | $2e_{s\phi}=$ | 0.05 cp |
| | $x_i$ | value | units | $d\phi/dx_i$ | $(d\phi/dx_i)^2$ | $e_{xi}$ | $e_{xi}^2$ | $(d\phi/dx_i)^2 e_{xi}^2$ |
| $x_1$ | $\rho$ | 1.124 | g/cm$^3$ | 2.35E+00 | 5.50E+00 | 9.17E–03 | 8.41E–05 | 4.63E–04 | cp$^2$ |
| $x_2$ | $\alpha$ | 0.0166 | mm$^2$/s$^2$ | 1.59E+02 | 2.53E+04 | 5.00E–05 | 2.50E–09 | 6.32E–05 | cp$^2$ |
| $x_3$ | $\Delta t_{\text{eff}}$ | 141.5 | s | 1.86E–02 | 3.48E–04 | 2.47E–01 | 6.09E–02 | 2.12E–05 | cp$^2$ |
| | | | | | | | $e_{s,\phi}^2$ | 5.47E–04 | cp$^2$ |
| | | | | | | | $e_{s,\phi}$ | 0.023 | cp |

Figure 5.14 Excel calculations for error propagation of $\tilde{\mu}$ determined with a Cannon–Fenske viscometer. The density error dominates (largest values in the rightmost column); 85% of the total squared error is attributable to the uncertainty in density.

Note that the error limits in replicate error involve the Student's $t$ statistic. When using these limits in the error propagation, we equate the error limits to $2e_{s,\Delta_{eff}}$ to obtain a standard error to combine with other errors. To be conservative, we have used the largest replicate error among the three viscometers (viscometer ID = Y733).

For the third viscometer (ID = Y733), the error estimates are combined in an Excel error propagation calculation as shown in Figure 5.14. The error estimate on viscosity is $e_{s,\tilde{\mu}} = 0.023$ cp and the dominant error is the error in the density (85% of the weighted squared errors sum), distantly followed by the error in $\tilde{\alpha}$ (12%), which was an optimistic estimate. The error on efflux time is quite small. The range for the viscosity is thus determined to be ($\pm 2e_{s,\phi} =$ 95% level of confidence)

$$2.60 \leq \tilde{\mu} \leq 2.70 \text{ cp} \quad (95\% \text{ level of confidence}) \tag{5.43}$$

$$\tilde{\mu} = 2.64 \pm 0.05 \text{ cp} \quad (\text{tentative}) \tag{5.44}$$

As stated previously, the literature value for the viscosity of 30% aqueous sugar solution at 26.7°C is 2.73 cp [26], which is not, unfortunately, captured by the 95% level of confidence error limits determined here.

It is a bit disappointing to discover that the measured value of viscosity does not capture the true value of viscosity for the solution in question. A significant contribution of the error calculation, however, is that it gives us a framework for investigating discrepancies such as this. We begin by enumerating the possible causes of the discrepancy:

1. *The density used may be incorrect.* We know that the density used was
   measured at 21°C rather than at $T_{bath} = 26.7°C$. This is a known
   systematic error. Looking at the value of viscosity determined, the value
   obtained is lower than the literature value. The density used was
   determined at a temperature 5.6°C too low, which would imply that the $\rho$
   value we used is likely higher than the value we should have used. Using a
   higher value of density than we should would lead to a higher value of
   viscosity (Equation 5.41); yet, we determined a $\tilde{\mu}$ that is too low. Thus,
   although we know there is an issue with the density in our solution (likely
   a bit too high), the systematic effect is in the wrong direction and thus
   cannot explain the mismatch with the literature value.
2. *The error on $\tilde{\alpha}$ may be underestimated.* We did not know what value to
   assign to the uncertainty on $\tilde{\alpha}$, and thus perhaps we need to adjust the
   estimated error on $\tilde{\alpha}$ upward to assess correctly the uncertainty on our
   value of $\tilde{\mu}$. The accuracy of $\tilde{\alpha}$ is affected by the accuracy with which $T_{bath}$
   is known and the accuracy with which the appropriate volume of fluid is
   charged to the viscometer. The manufacturer supplies instructions on how
   to charge a CF viscometer with fluid so that the amount of fluid charged to
   the viscometer in the customer's laboratory matches the fluid volume
   previously used during the manufacturer's determination of $\tilde{\alpha}$. With the
   error propagation calculation set up in Excel, we can try new values for
   this error and see the effect. The results of our simulation are as follows:

| $\tilde{\alpha}$ $(mm^2/s^2)$ | $e_{s,\tilde{\mu}}$ (cp) | $\tilde{\mu}_{meas} + 2e_{s,\tilde{\mu}}$ (cp) |
|---|---|---|
| $0.5 \times 10^{-4}$ | 0.023 | 2.68 |
| $1 \times 10^{-4}$ | 0.027 | 2.69 |
| $2 \times 10^{-4}$ | 0.039 | 2.71 |
| $2.5 \times 10^{-4}$ | 0.045 | **2.73** |
| $3 \times 10^{-4}$ | 0.053 | 2.74 |

The simulation shows that we can just capture the true value of the
viscosity (at the 95% level of confidence) by assuming that the uncertainty
in $\alpha$ is $e_{s,\alpha} = 2.5 \times 10^{-4}$ mm$^2$/s$^2$, which corresponds to $\pm 2.5$ times one
in the last digit of the supplied $\alpha$. Is this a plausible revised estimate for
uncertainty? The manufacturer may have provided more digits than are
strictly significant to protect our results from round-off error. This is a
plausible circumstance, which we accept, although even with this
assumption our measurement is at the outer range of the error limits on
viscosity.

3. *The value of $\tilde{\mu}$ from the literature may not correspond to our solution.* This could occur for at least two good reasons. First, for the literature value to correspond to the solution in the lab, the concentration of the solution in the lab must be 30%, which depends on the accuracy of the process for creating the solution (unknown to us). Second, the temperature measurement in the lab must be accurate at 26.7°C. The result we obtained agreed with the literature to almost two significant figures; a slight difference in concentration, or a temperature offset (from thermocouples, which we acknowledge have at least $2e_s = 1.1°C$), could possibly create the difference in viscosity observed.

The discrepancy in this example between the measured data and the literature data may have a variety of sources. We have identified two reasonable possibilities that could be investigated, the accuracy of the solution concentration and the accuracy of the bath temperature. The viscosity of the laboratory solution has been precisely determined. To verify the accuracy and meaning of the answer, we would need to investigate the remaining issues further. For now, with revised error limits based on our new assessment of $e_{s,\alpha} = 2.5 \times 10^{-4}$ mm$^2$/s$^2$, we report the viscosity as follows:

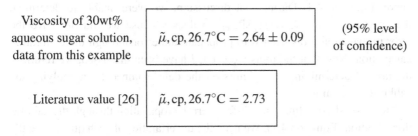

| Viscosity of 30wt% aqueous sugar solution, data from this example | $\tilde{\mu}, \mathrm{cp}, 26.7°C = 2.64 \pm 0.09$ | (95% level of confidence) |
| --- | --- | --- |
| Literature value [26] | $\tilde{\mu}, \mathrm{cp}, 26.7°C = 2.73$ | |

Example 5.8 shows how error analysis can help us to understand our results. To draw reliable conclusions about the meaning of the viscosity data obtained in the example, we needed to be able to investigate the sources of error. The sources of random and systematic errors in a dataset may be well known or they may be hidden. The essential practice is to carefully assess each component of the experimental process, noting assumptions and revisiting them as new information is determined. The estimates adopted and logic to support them should also be reported with compete transparency. As shown by the literature electron-charge measurements discussed in Figures 3.6 and 3.7, following a careful procedure does not guarantee an immediate correct answer every time, but the method is a step along the path to the right answer.

In our final example of this chapter, we return to the problem introduced in Example 4.2, one of the linked examples (Figure 1.3), in which we use

a DP-meter calibration curve to convert flow pressure-drop measurements
to final results, including error limits. Now that we know how to do error
propagation, we can make some progress in this problem.

**Example 5.9: \*LINKED\* Using a calibration curve: Differential-pressure
meter, revisited.** *Water driven by a pump is made to flow through copper
tubing [8] at a variety of flow rates depending on the position of a manual
valve (Figure 4.4). Along a 6.0-ft section of copper tube, the pressure drop is
measured with a calibrated differential-pressure meter as a function of water
volumetric flow rate, Q(gal/min). The differential-pressure signals from the
DP meter are in the form of electric current ranging from 4.0 to 20.0 mA
(Table 4.2). These signals may be translated to differential pressures in psi
by using the calibration curve for the device, which is given in Equation 4.1.
For the raw data provided in Table 4.2, what are the associated differential-
pressure measurements in psi? What are the appropriate error limits (95%
level of confidence) to associate with each differential pressure? Plot the final
results for pressure drop versus flow rate, including the appropriate error bars.*

**Solution:** We began this problem in Example 4.2, and we obtained the
desired values of $(\Delta p, \text{psi})_{meter}$ (Table 4.3) using a supplied calibration
curve, Equation 4.1. During that discussion, we were unable to determine
the error limits on the $(\Delta p, \text{psi})_{meter}$ values because we did not have the
error propagation tools needed to propagate the error through the calibration
calculation. Now that the tools we needed have been discussed, we return to
the task of determining error limits on the data obtained from applying the
calibration equation.

The desired error limits result from error propagation through the calibration equation (Equation 4.1). We need the uncertainties of each quantity in the
calibration equation.

DP-meter
calibration
curve

$$(\Delta p, \text{psi})_{meter} = 0.23115(I, \text{mA})_{raw} - 0.92360$$

(Equation 4.1)

The uncertainties in the coefficients in this equation are tied to how the
calibration equation was determined.

The calibration curve in Equation 4.1 is developed in Example 5.7 (Equation 5.40). The numerical values in Equation 5.40 were obtained from a
calculation that begins with a least-squares model fit to a straight line of
manometer calibration data (Equation 5.38, repeated here).

Model fit to
calibration data
$$(I, \text{mA})_{cal} = m(\Delta p, \text{psi})_{std} + b \qquad (5.45)$$

where $m = 4.326143$ mA/psi and $b = 3.995616$ mA are the slope and intercept of this line, respectively (values provided to us at this point; the full calculation is presented in Chapter 6). The differential-pressure values $(\Delta p, \text{psi})_{std}$ used during calibration are the calibration standards determined with a manometer; thus, they are the independent variables and appear on the $x$-axis during model fitting. To obtain the calibration curve in our preferred format, explicit in differential pressure, we rearranged Equation 5.45.

DP-meter
calibration curve
$$(\Delta p, \text{psi})_{meter} = \left(\frac{1}{m}\right)(I, \text{mA})_{raw} - \left(\frac{b}{m}\right) \qquad (5.46)$$

$$(\Delta p, \text{psi})_{meter} = 0.23115(I, \text{mA})_{raw} - 0.92360 \qquad (5.47)$$

where $(\Delta p, \text{psi})_{meter} = p_1 - p_2$ is differential pressure in $psi$ and $(I, \text{mA})_{raw}$ is the raw current signal from using the DP meter [25]. Equation 5.47 matches Equations 4.1 and 5.40.

To calculate the uncertainty in the DP-meter differential pressures $(\Delta p, \text{psi})_{meter}$ from the raw DP-meter readings $(I, \text{mA})_{raw}$ obtained, we need to carry out an error propagation with Equation 5.46 as shown in the worksheet in Figure 5.15. We have filled in the worksheet as much as we can at this time, including the particulars of this problem: the calibration equation, the variables, and their representative values. During this initial setup of the error propagation, we arbitrarily use 10 mA for the value of $(I, \text{mA})_{raw}$. To calculate error bars, we will need to do this calculation for each data pair in Table 4.2. The error propagation calls for values of the uncertainties of the three variables $m$, $b$, and $(I, \text{mA})_{raw}$.

Note that the uncertainty in $(I, \text{mA})_{raw}$ is the uncertainty in the DP-meter signal with units of mA; it must reflect all the uncertainties of the operation of the DP meter itself. The uncertainty we seek for $(I, \text{mA})_{raw}$ is *not* the general uncertainty in measuring arbitrary electric currents with a digital multimeter. During use of the meter, the values of $(I, \text{mA})$ read when $(\Delta p, \text{psi})$ is applied will exhibit a stochastic variation due to the operations of the meter, as installed. This variability is the intrinsic uncertainty of the DP-meter signal. The intrinsic meter-signal uncertainty may be determined from the least-squares curve fit to the calibration data and is given the symbol $s_{y,x}$, as shown in Figure 5.15. We discuss this issue in Chapter 6.

The other two variables in Equation 5.46 are $m$ and $b$, the slope and intercept determined during the fitting of a model to the calibration dataset (Equation 5.45). These values are not perfectly certain, as they were calculated from

| $\phi(x_1, x_2, x_3, x_4, x_5)$: | Formula for $\phi$:* $$\Delta p = \left(\frac{1}{m}\right)I - \left(\frac{b}{m}\right)$$ *Note: units must work as written. | | Representative value of $\phi$: (include units) | 95% C.I. of $\phi$: $(\phi \pm 2e_{s,\phi})$ (include units) |
|---|---|---|---|---|
| | | | psi | psi |

| Measured quantities, $x_i$ | | | $\dfrac{\partial \phi}{\partial x_i}$ | $e_{s,x_i}$ | $\left(\dfrac{\partial \phi}{\partial x_i}\right)^2 e_{s,x_i}^2$ |
|---|---|---|---|---|---|
| $x_i$ | Symbol | Representative value | | | |
| $x_1$ | $m$ | 4.326143 mA/psi <br> units | $\dfrac{\partial \Delta p}{\partial m} = (I - b)\left(-\dfrac{1}{m^2}\right)$ | $s_m =?$ | |
| $x_2$ | $(I, \text{mA})_{raw}$ | 10 mA <br> units | $\dfrac{\partial \Delta p}{\partial I} = \left(\dfrac{1}{m}\right)$ | $s_{y,x} =?$ | |
| $x_3$ | $b$ | 3.995616 mA <br> units | $\dfrac{\partial \Delta p}{\partial b} = \left(-\dfrac{1}{m}\right)$ | $s_b =?$ | |
| $x_4$ | | units | | | |
| $x_5$ | | units | | | |

| $e_{s_\phi}^2 = \left(\dfrac{\partial \phi}{\partial x_1}\right)^2 e_{x_1}^2 + \left(\dfrac{\partial \phi}{\partial x_2}\right)^2 e_{x_2}^2 + \left(\dfrac{\partial \phi}{\partial x_3}\right)^2 e_{x_3}^2 + \left(\dfrac{\partial \phi}{\partial x_4}\right)^2 e_{x_4}^2 + \left(\dfrac{\partial \phi}{\partial x_5}\right)^2 e_{x_5}^2$ | $e_{s,\phi}^2 =$ |
|---|---|
| | $e_{s,\phi} =$  units  psi |

Figure 5.15 Worksheet for error propagation to determine differential pressure from new DP meter observations $(I, \text{mA})_{raw}$ and a calibration curve $\Delta p = \left(\frac{1}{m}\right)I - \left(\frac{b}{m}\right)$; the worksheet shows the calculation for a single value of electric current, $10\ mA$, chosen at random. To complete the error propagation, we need to know the uncertainty in $m$ and $b$ as determined during calibration. See Example 6.6 for the continuation of this calculation. The error on the electric current reading from the DP meter is also discussed there.

a least-squares fitting of a linear model to the manometer calibration data. We must determine the uncertainties of the quantities $m(s_m)$ and $b(s_b)$ from the fitting calculation. In Chapter 6 we discuss ordinary least-squares curve fitting, and we show how to determine both the model fit to the calibration data and the related errors. We use Excel's program LINEST for ordinary least-squares curve-fitting calculations. We also discuss how to do the calculations with MATLAB.

Once again we suspend our work on this problem until we study error analysis a bit more. In Chapter 6 we see how to determine the three missing standard errors $s_m$, $s_b$, and $s_{y,x}$; once these errors are obtained, the error propagation in Figure 5.15 allows us to calculate the errors on each flow-DP-meter measurement in Table 4.2. In addition, as discussed in Section 6.2.4, the interdependency of the least-squares slope and intercept – they are calculated from the same calibration dataset $(x_i, y_i)$ – means that the error propagation worksheet in Figure 5.15 must be modified to include a nonzero covariance term (see Equation 5.16). We explain this issue in Example 6.6 and resolve the flow pressure-drop problem in Example 6.8 (see the linked examples map in Figure 1.3).

## 5.5 Summary

In this chapter, we have introduced and explained error propagation, a process that allows the uncertainties in individual quantities to be propagated through a subsequent calculation to give an estimate of the error associated with the result of the calculation. When the variables in an equation are uncorrelated, error propagation proceeds through Equation 5.18, an expression in which the squared errors are summed with weights that depend on how the variables are treated in the calculation (squared partial derivatives).

Error propagation
equation
(3 independent
variables)
$$e_{s,y}^2 = \left(\frac{\partial \phi}{\partial x_1}\right)^2 e_{s,x_1}^2 + \left(\frac{\partial \phi}{\partial x_2}\right)^2 e_{s,x_2}^2 + \left(\frac{\partial \phi}{\partial x_3}\right)^2 e_{s,x_3}^2$$

(Equation 5.18)

When the variables in a calculation are correlated, as we encounter in Chapter 6, we use Equation 5.16 for error propagation. This equation includes terms that account for the interactions between variables in the calculation.

Error propagation
equation
(3 variables, general)
$$\sigma_y^2 = \left(\frac{\partial \phi}{\partial x_1}\right)^2 \sigma_{x_1}^2 + \left(\frac{\partial \phi}{\partial x_2}\right)^2 \sigma_{x_2}^2 + \left(\frac{\partial \phi}{\partial x_3}\right)^2 \sigma_{x_3}^2$$

$$+ 2\left(\frac{\partial \phi}{\partial x_1}\right)\left(\frac{\partial \phi}{\partial x_2}\right) \text{Cov}(x_1, x_2)$$

$$+ 2\left(\frac{\partial \phi}{\partial x_1}\right)\left(\frac{\partial \phi}{\partial x_3}\right) \text{Cov}(x_1, x_3)$$

$$+ 2\left(\frac{\partial \phi}{\partial x_2}\right)\left(\frac{\partial \phi}{\partial x_3}\right) \text{Cov}(x_2, x_3) \quad \text{(Equation 5.16)}$$

In this chapter we provide an error propagation worksheet and recommend Excel and MATLAB for error propagation. Use of software allows us to investigate the impact of error estimates: knowing the negligible impact of an error estimate on the final result would justify accepting the estimate without further concern, while discovering the dominance of an estimate should motivate further investigation. The process of determining error limits deepens our understanding of the quality and reliability of our measurement methods and guides us in improving our processes.

Thanks to having an understanding of error propagation, we were able to advance both sets of linked DP-meter examples in this chapter (Figure 1.3). We continue to rely on ordinary least-squares calculations in these examples; in the next chapter we discuss the missing pieces in the linked problems and complete them. Model fitting is a very powerful tool in science and engineering. Chapter 6 presents the ordinary least-squares method of model fitting along with the error propagation calculations that are part of least-squares calculations.

Measurements are stochastic variables and subject to uncertainty. With the tools introduced in this text we see how to determine or estimate uncertainties and combine them in calculated results. We also see how to identify the high-impact uncertainties, leading us to focus our efforts on improving these measurements. With techniques such as these, we are well equipped to advance scientific and engineering knowledge.

## 5.6  Problems

1. On what principles are the error propagation methods of this chapter based? Please answer in your own words.
2. In Example 5.2 we showed that there is more uncertainty in a length determined from adding two measured lengths than in measuring the length all at once. In your own words, why is this true?
3. Covariance between two stochastic variables is zero if they are uncorrelated. Name two uncorrelated stochastic variables; name two correlated stochastic variables. In error propagation, what is the main advantage of combining uncorrelated stochastic variables over correlated stochastic variables?
4. Sarah uses a bathroom scale to measure the weight of some luggage for her trip abroad. She uses two different methods. In the first method, she places the suitcase on the bathroom scale and measures the weight. In the second method she steps onto the scale and gets one reading and then steps onto the scale without the suitcase for a second reading; she then subtracts the two readings. Which method is preferable? Justify your answer including topics from this chapter.
5. We measure the length of an aluminum rod of diameter 1.75 cm two ways. The first way is to use a meter stick (a metal stick precisely etched at intervals of 1 mm). The second way is to use a shorter stick that is

30 cm long and marked with the same precision as the longer stick. The measurement for the longer stick was found to be 1.324 m and with repeated use of the shorter stick the length was found to be 1.333 m (extra digits provided). What are the 95% level of confidence error limits for each measurement? Comment on your answers.

6. What is the smallest volume you can accurately dispense with a graduated cylinder? Assume no more than 25% error in a measurement is acceptable. Explain how you determined your answer in terms of the methods discussed in this book.

7. What is the shortest length you can measure with the ruler in Figure 3.16? Assume no more than 25% error in a measurement is acceptable. Give answers derived from considering both measuring scales, centimeters and inches (not just a unit conversion). Explain how you determined your answer in terms of the methods of this book.

8. The height of an object is determined with an inexpensive meter stick by measuring in two stages – first measuring the lower part to obtain $H_l$ and then measuring the upper part to obtain $H_u$. The standard calibration error associated with the meter stick is $e_{s,cal} = 0.5$ mm. What is the error associated with the total height of the object? If the height had been determined in three stages (adding three lengths), what would the error in the total height be? Extrapolate your answer to the case where the height is determined in $m$ stages.

9. Using a 30 cm (or 12 in) ruler, determine the width of six to ten tables in a classroom or lab (the tables should be a meter wide or so, and they should be similar). Repeat the measurements with a tape measure. Analyze your data as follows:

   a. *Measurement uncertainty of method.* Even without taking any data, we can determine 95% level of confidence error limits on the method using the resources in this chapter. Determine these error limits for the two methods of measuring table width (using the ruler versus using the tape measure).

   b. *Prediction interval of replicates.* For the two measuring methods, determine the 95% prediction interval for the next value of table width. For each, create a plot similar to Figure 2.19. Include individual error bars on each point, determined from your technique-based error limits from part (a).

Discuss your results. How would you summarize the effect of choosing one measurement method over the other?

10. When applying the manufacturer's calibration to Cannon–Fenske viscometer raw data (Equation 5.41), we need the calibration constant $\tilde{\alpha}$ at the appropriate temperature. The function $\tilde{\alpha}(T)$ for a viscometer is a straight-line calibration function provided in the form of two values, one at 40°C and a second one at 100°C (see the caption for Table 5.3). For the viscometers considered in Example 5.8, calculate $\tilde{\alpha}$ at $T_{bath} = 26.7°C$. What is your estimate of the uncertainty in $\tilde{\alpha}$? Justify your answer.

11. A colleague loads a Cannon–Fenske viscometer with a solution and measures the efflux time (the time for the meniscus to pass a set of timing marks) three times. He calculates three values of viscosity, averages them to obtain a mean viscosity, calculates a replicate error on the mean viscosity, and determines 95% level of confidence error limits on his results based on replicate error. Evaluate this process and result and explain any issues you discover.

12. The dimensions of a cylindrical laboratory tank are diameter = 30.5 cm and height = 45.2 cm. The tank is filled with water (20°C) to the following heights: 12.5, 17.0, and 22.9 cm. Calculate the volume of the water for each height; include appropriate error limits and justify any estimates.

13. In Example 5.3, we consider the rate of sensible heat transferred to a liquid water stream in a heat exchanger $\dot{q}$, which is calculated from this equation:

$$\dot{q} = \dot{m}\hat{C}_p\left(T_{out} - T_{in}\right) \qquad \text{(Equation 5.10)}$$

For water as the fluid in the stream, with special limits (SLE) thermocouples used for temperature measurements, and with the data given here, calculate the value of $\dot{q}$ and the appropriate uncertainty limits.

$$\dot{m} = 12.1 \text{ kg/min}$$
$$e_{s,\dot{m}} = 0.05 \text{ kg/min}$$
$$T_{out} = 55.3°C$$
$$T_{in} = 12°C$$

14. Saturated steam condenses at $p_{sat} = 0.95$ atm in a heat exchanger, producing latent heat $\dot{q}$ at the rate given by

$$\dot{q} = \dot{m}\hat{H}_{sat}$$

where $\hat{H}_{sat}$ is the enthalpy of the saturated steam. The flow rate of the steam is $\dot{m} = 5.3$ g/s. What is the rate of latent heat loss from the steam stream as it condenses? Include error limits on your answer.

15. For the data in Table 2.7 on leaves from a maple tree, calculate the ratio of leaf mass to length, provide an average value, and determine the 95% CI and PI for the dataset. Create a plot like Figure 2.19 showing the data and the intervals. Using error propagation and error estimates based on the measurement method, create error bars for the individual data points. Comment on what you can conclude from this figure.

16. For the data in Table 2.8 on leaves from a lilac bush, calculate the ratio of leaf mass to length, provide an average value, and determine the 95% CI and PI for the dataset. Create a plot like Figure 2.19 showing the data and the intervals. Using error propagation and error estimates based on the measurement method, create error bars for the individual data points. Comment on what you can conclude from this figure.

17. For the data in Table 2.9 on leaves from a flowering crab tree, calculate the ratio of leaf mass to length, provide an average value, and determine the 95% confidence interval and prediction interval for the dataset. Create a plot like Figure 2.19 showing the data and the intervals. Using error propagation and error estimates based on the measurement method, create error bars for the individual data points. Comment on what you can conclude from this figure.

18. Average fluid velocity in a tube may be calculated from volumetric flow rate $Q$ according to this equation:

$$\text{Average velocity} \qquad \langle v \rangle = \frac{4 Q \alpha_{0,\langle v \rangle}}{\pi D^2} \qquad (5.48)$$

where $D$ is the tube inner diameter. For the flow rate data in Table 4.2, calculate the average fluid velocities $\langle v \rangle$ and determine appropriate error limits on $\langle v \rangle$. The tube is half-inch nominal, type L copper tubing [8]. Indicate your assumptions and justify your estimates.[12]

19. The Reynolds number for a liquid flowing in a tube is defined as

$$\text{Re} = \frac{\rho \langle v \rangle D \alpha_{0,Re}}{\tilde{\mu}} \qquad (5.49)$$

where $\langle v \rangle$ is average fluid velocity (see Equation 5.48), $\rho$ is fluid density, $D$ is the inner tube diameter, and $\tilde{\mu}$ is fluid viscosity. For the water ($25°C$) flow rate data in Table 5.5, calculate the Reynolds number and determine plausible error limits. Note that there are only four values of flow rate explored in this dataset, so our calculations result in only four values of Reynolds number. Indicate your assumptions and justify your estimates.

[12] From reference [8], for half-inch nominal, type L copper tubing: outer diameter = 0.625 in; inner diameter = 0.545 in; wall thickness = 0.040 in.

20. Fanning friction factor for flow in pipes is calculated as

$$\text{Fanning friction factor} \qquad f = \frac{\Delta p \, D\alpha_{0,f}}{2L\rho \langle v \rangle^2} \qquad (5.50)$$

where $\langle v \rangle$ is average fluid velocity (see Equation 5.48), $D$ is the tube inner diameter, $\rho$ is fluid density, and $\Delta p$ is pressure drop across length $L$. For the water (25°C) data given in Table 5.5, calculate the Fanning friction factors and determine plausible error limits. Note that $\Delta p$ is

Table 5.5. *Data for Problems 19 and 20. Four different Bourdon gauge pairs were used to measure pressure drops across a 6.0-ft length of copper tubing with water (25°C) flowing at a constant volumetric rate. The inner diameter of the tubing is 0.305 in.*

| Index | Gauge ID | $Q$ (gal/min) | $\Delta p$ (psi) | Index | Gauge ID | $Q$ (gal/min) | $\Delta p$ (psi) |
|---|---|---|---|---|---|---|---|
| 1 | 2 | 2.0 | 3.0 | 25 | 7 | 2.0 | 3.0 |
| 2 | 2 | 2.0 | 3.0 | 26 | 7 | 2.0 | 3.0 |
| 3 | 2 | 2.0 | 3.0 | 27 | 7 | 2.0 | 3.0 |
| 4 | 2 | 2.5 | 4.0 | 28 | 7 | 2.5 | 4.0 |
| 5 | 2 | 2.5 | 4.0 | 29 | 7 | 2.5 | 4.0 |
| 6 | 2 | 2.5 | 4.0 | 30 | 7 | 2.5 | 4.0 |
| 7 | 2 | 3.0 | 5.3 | 31 | 7 | 3.0 | 6.0 |
| 8 | 2 | 3.0 | 5.5 | 32 | 7 | 3.0 | 6.0 |
| 9 | 2 | 3.0 | 5.5 | 33 | 7 | 3.0 | 5.5 |
| 10 | 2 | 3.5 | 7.0 | 34 | 7 | 3.5 | 7.5 |
| 11 | 2 | 3.5 | 6.5 | 35 | 7 | 3.5 | 7.5 |
| 12 | 2 | 3.5 | 7.0 | 36 | 7 | 3.5 | 7.5 |
| 13 | 5 | 2.0 | 2.5 | 37 | 8 | 2.0 | 3.0 |
| 14 | 5 | 2.0 | 3.0 | 38 | 8 | 2.0 | 3.0 |
| 15 | 5 | 2.0 | 3.0 | 39 | 8 | 2.0 | 3.0 |
| 16 | 5 | 2.5 | 4.0 | 40 | 8 | 2.5 | 4.0 |
| 17 | 5 | 2.5 | 4.0 | 41 | 8 | 2.5 | 3.5 |
| 18 | 5 | 2.5 | 4.0 | 42 | 8 | 2.5 | 3.5 |
| 19 | 5 | 3.0 | 6.0 | 43 | 8 | 3.0 | 6.0 |
| 20 | 5 | 3.0 | 5.5 | 44 | 8 | 3.0 | 5.5 |
| 21 | 5 | 3.0 | 6.0 | 45 | 8 | 3.0 | 5.5 |
| 22 | 5 | 3.5 | 7.0 | 46 | 8 | 3.5 | 7.5 |
| 23 | 5 | 3.5 | 7.0 | 47 | 8 | 3.5 | 7.5 |
| 24 | 5 | 3.5 | 7.0 | 48 | 8 | 3.5 | 7.5 |

reported as having been obtained with two Bourdon gauges through two independent measurements $p_1, p_2$ that were then subtracted, $\Delta p = p_1 - p_2$.

21. The pressure difference measured by a manometer (see Figure 4.13) is related to manometer fluid density $\rho$ and total height difference $h$ as follows:

$$\Delta p = \rho g h$$

where $g$ is the acceleration due to gravity. A colleague proposes to use water in a manometer to measure $\Delta p$. If the maximum allowable height for the manometer apparatus is 1.6 m, what is the highest $\Delta p$ this device can measure with water as its fluid? Using your own estimates of error, what is the lowest accurate $\Delta p$ this device would be able to measure? Give your answer in both dynes/cm$^2$ and psi $=$ lb$_f$/in$^2$.

22. The Fanning friction factor for pipes is calculated with Equation 5.50. For water (room temperature) flowing in half-inch, type L copper tubing (inner diameter $= 0.545$ in) in turbulent flow, which parameter in the equation controls the accuracy of $f$? In other words, which error dominates? See Problem 5.20 for some context for this problem. A representative data point would be an average velocity of $\langle v \rangle = 4.02$ m/s generating 3.1 psi over a length of 6.0 ft. Make reasonable estimates of the errors you need.

23. In your estimation, which error dominates in the calculation for the rate of sensible heat flow to a liquid water stream in a heat exchanger (Equation 5.10)? How dominant is it? Make reasonable estimates of the errors you need. See Problem 5.13 for some context and representative values for this problem.

24. For 80% glycerin in water ($20°C$, $\rho = 1,211$ kg/m$^3$; $\tilde{\mu} = 0.0601$ Pa-s) slowly flowing in tubes in the laminar flow regime, experimental and theoretical investigations both give the result that the Fanning friction factor $f$ (Equation 5.50) and the Reynolds number Re (Equation 5.49) are related as $f = 16/$Re (laminar regime present for Re $\leq 2100$). How sensitive is experimental verification of this result likely to be with respect to the accuracy of the tube inner diameter $D$? Make reasonable estimates of the errors you need. See Problems 5.19 and 5.20 for some context and representative values for this problem.

$$\text{Re} = \frac{\rho \langle v \rangle D \, \alpha_{0,\,\text{Re}}}{\tilde{\mu}} \qquad \text{(Equation 5.49)}$$

$$f = \frac{\Delta p \, D \, \alpha_{0,\,f}}{2L\rho\langle v \rangle^2} \qquad \text{(Equation 5.50)}$$

where $\langle v \rangle$ is average fluid velocity,

$$\langle v \rangle = \frac{4 Q \alpha_{0,\langle v \rangle}}{\pi D^2} \qquad \text{(Equation 5.48)}$$

25. Equation 2.58 results from an error propagation:

95% prediction interval
for $x_i$, the next        $x_i = \bar{x} \pm t_{0.025,n-1}\, s \sqrt{1 + \frac{1}{n}}$    (Equation 2.58)
value of $x$

Using the error propagation techniques of this chapter, show that Equation 2.58 holds.

26. In Chapter 4 we recommended that the limit of determination be set when the relative error reaches 25%. This implies that LOD $= 8 e_s$. If we can only tolerate 10% error, what is the new value of LOD?

# 6

# Model Fitting

In this chapter we discuss fitting models to data as well as how to assess the uncertainties associated with using the models. We have already been using models: in Chapters 4 and 5, we employed results from model fitting in the context of instrument calibration. We presented a calibration plot in which a straight line had been fit to experimental data consisting of a DP-meter electronic signal $y = (I, \text{mA})_{cal}$ versus differential pressure $x = (\Delta p, \text{psi})_{std}$ (Figure 5.10). The equation for the line, $y = mx + b = 4.326143x + 3.995616$ (Equation 5.38, shown with extra digits), is a compact way to represent the calibration data. The equation for the straight-line fit can be used to apply the DP-meter calibration to future raw data signals from the device, as illustrated in Figure 5.11. In those earlier chapters, the model fit was provided, but we did not discuss how the fit was obtained.

Here, we present the process of determining the coefficients for empirical model fits. Because measurements are stochastic and subject to uncertainty, producing model fits is an optimization. First, we choose a model; Appendix F contains a discussion of the mathematical forms that models typically take. Second, we fit the model to the data. The model-fitting calculation seeks to minimize the net (summed) deviation between what the model predicts and the observed values of the stochastic variable. Third, we interpret the fits, including how to assign uncertainty to model parameters and model predictions. With the methods of this chapter we are able to resolve several of the linked error discussions presented in the text (see Figure 1.3). We conclude the chapter, and the book, with a short discussion of strategies for more complex model identification.

## 6.1 Introduction

Producing a calibration equation is an example of how empirical model fitting is used in science and engineering: data are obtained, the data often show a trend or a systematic variation, and a model is fit to the data to represent that trend. The *model* is an equation that can show a trend similar to that of the data, although for the model equation to actually *fit* the data, we need a method to identify appropriate values for the model's parameters. Figure 6.1 shows examples of model equations fit to data. Obtaining a good fit depends strongly on which model is chosen to represent the data.

The steps for *fitting a model* to a dataset are as follows:

1. *Obtain data*, along with the appropriate error limits for the data points.
2. *Choose a form for the model.* As we discuss in Appendix F, this step may be as simple as noting that the data trend seems to follow a straight line/parabola/power law, although the process of choosing the model form may be considerably more complex, involving an investigation of the

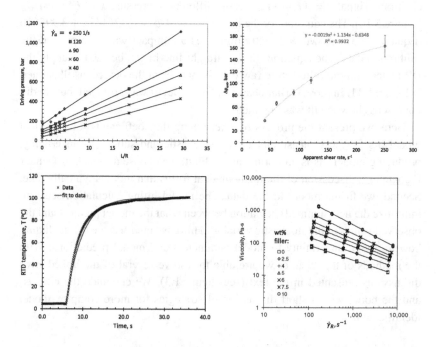

Figure 6.1 Fitting of models to data is an essential tool in most branches of science and engineering. Shown here are four types of data fits that are discussed in this text (see Examples 6.4, 6.9, 6.12, and Appendix F).

physics behind the data (see Example 6.13). The model will contain one or more unknown parameters or coefficients that must be determined by a fitting process.

3. *Fit the model to the data.* This may be accomplished with a computer by trial-and-error or by mathematical algorithm. Excel provides an ordinary least-squares function for model fitting, LINEST. Alternatively, we could use matrix methods and MATLAB or Excel for these calculations. Excel's function SOLVER, which is designed to carry out nonlinear optimization, may also be used [43].

4. *Use the model* to represent the data; *include appropriate propagated uncertainty* in the reported results. The compact nature of a model equation is its advantage – the model may be embedded in follow-on calculations, standing in for the complex dataset. The uncertainty in the original, unprocessed dataset, however, leads to uncertainty in the model parameters, as well as to uncertainty in any calculations employing the model or its parameters. We determine the uncertainty in follow-on calculations by propagating, through all uses of the model, the uncertainty present in the original dataset.

*Empirical* models are those chosen solely for their resemblance to trends in the data. The counterpart to an empirical model is a *theoretical* one; a theoretical model is one that arises from consideration of the physics that produced the data (see Example 6.13). In theoretical models, the models' parameters have physical meaning, while for empirical models this is often not the case. We focus on empirical models. The principal discussion of the chapter is how to fit empirical models to data using an ordinary least-squares algorithm. We also address the uncertainty associated with model parameters and the uncertainty associated with values obtained from model predictions.

## 6.2  Least Squares of Linear Models

The first step in fitting an empirical model to data is to choose the form of the model. The second step is to fit the model to the data – that is, to identify values of the parameters in the model so that the model, with those parameters, represents the data well. For example, if the model is a straight line, $y = mx + b$, we are looking for the values of slope $m$ and $y$-intercept $b$ that give the "best fit" of the model to the data.

Numerous algorithms have been developed to fit a model to data. The basic idea behind these algorithms is to minimize the differences between what the model predicts at various points and the values of the observed data at these points. In this section, we describe one of the most common techniques for model fitting, *ordinary least-squares linear regression*. We first discuss the idea behind "least squares," and then we present how to assess uncertainties related to the model's parameters and predictions.

As we discuss least-squares model fitting, we also introduce the software tools that are an important part of empirical modeling. The calculations for linear regression can be tedious to perform, particularly for a large number of data points, but the algorithm is easy to program into software. We pause here to describe our strategy for presenting the Excel and MATLAB software tools for ordinary least-squares linear regression.

### 6.2.1 Software Tools for Linear Regression

Both Excel and MATLAB are effective at carrying out error calculations, and which tool you choose is a matter of personal preference. Excel contains preprogrammed functions designed to evaluate ordinary least-squares linear regression for a dataset $(x_i, y_i)$, $i = 1, 2, \ldots, n$. This is a "black-box" approach, which hides the detailed calculations and only requires the user to properly call the routines and interpret the results.

The MATLAB *Statistics and Machine Learning Toolbox* also has a wide variety of preprogrammed functions that can calculate the ordinary least-squares linear fit, although this powerful toolbox is not needed for our purposes. As we discuss in Section 6.3.1, because the ordinary least-squares calculation may be expressed as a straightforward matrix algebra problem, basic MATLAB can solve such problems in a few lines of code.

We present both Excel and basic MATLAB approaches to linear regression. The Excel approach is compatible with a topic-by-topic introduction of various aspects of the algorithm. For this reason, as we introduce the least-squares technique and explain how to perform and use calculations, we present the associated Excel functions as they arise in the discussion.

The MATLAB approach requires a slightly different path. To implement linear regression with basic MATLAB, the reader needs familiarity with the matrix version of linear regression; this material is presented in Section 6.3.1. Readers interested primarily in the MATLAB approach are advised to read the introductory material on linear regression presented in the first few pages of Section 6.2.2 (up to Example 6.1), and to follow this with review of Section 6.3.1, which restates that material in matrix format. Following study of the matrix presentation, the reader would return to Example 6.1 and continue to follow the text in order. This strategy gives a MATLAB user the background needed to interpret the MATLAB solutions to the Chapter 6 examples, which are referenced and provided in the text and in Appendix D.

In Appendix D, we provide two important reference tools for the MATLAB reader. The first is a table that maps Excel regression commands to MATLAB commands and code (Table D.1). The second is several home-grown MATLAB functions that provide solutions to the Chapter 6 linear-regression examples. With these resources, the MATLAB reader can link the background information on linear regression to the practical question of how to implement the solutions in MATLAB.

## 6.2.2 Ordinary Least-Squares Linear Regression

Consider the situation where there are $n$ data points $(x_i, y_i)$ for which we seek an equation $y = mx + b$ that is a good fit to the data. At this stage, the values of $m$ and $b$ are unknown to us. To find good values of $m$ and $b$ for the $n$ data points $(x_i, y_i)$, we reason as follows. Unless the line $y = mx + b$ is a perfect fit for all the data points, there will be differences between what the model predicts and the measured $(x_i, y_i)$ values. The best choices for $m$ and $b$ are the choices that result in the smallest *net* deviation between the model-predicted values and the observed values.

To determine the best values of slope and y-intercept, we create an error measure to calculate the overall deviation between the measured data and the model predictions. If we assume that all the experimental $x_i$ values are known with certainty,[1] we can can construct an error measure based on the vertical distances between the measured $y_i$ and the value that the model predicts for $y$ at the same value of $x = x_i$ (Figure 6.2). Let the model-predicted $y$ values be called $\widehat{y}$; for each value of $x_i$ for which we have an experimental data point, we can calculate from the model the corresponding value of $\widehat{y}_i$ as

$$\begin{array}{cc} \text{Predicted } y \text{ values} & \widehat{y}_i = mx_i + b \qquad (6.1) \\ \text{for the } i\text{th data point} & \end{array}$$

The differences between observed and predicted values of $y$ are given by

$$\begin{array}{cc} \text{Deviation between observed } y_i & \\ \text{and model-predicted values } \widehat{y}_i & (y_i - \widehat{y}_i) \qquad (6.2) \\ \text{for the } i\text{th value of } x & \\ (x_i \text{ assumed to be accurate}) & \end{array}$$

---

[1] This is a significant assumption; we return to this assumption in a moment.

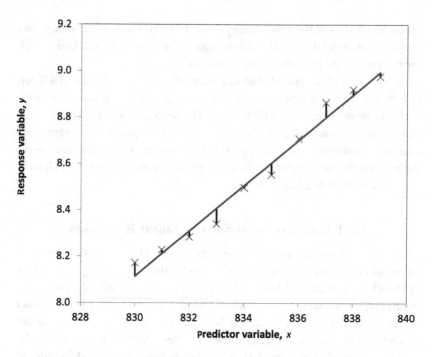

Figure 6.2 If we assume that all the $x_i$ values are known with certainty, we can calculate the values of $y$ predicted by the model by substituting the $x_i$ values of the data points into the equation for the line, $\widehat{y} = mx + b$. The errors between the measured $y_i$ and the predicted $\widehat{y}_i$ are the short vertical segments shown.

We now seek an expression for overall error, which we plan to minimize. If we create the overall error measure by summing the deviations $(y_i - \widehat{y}_i)$ for all data pairs, the positive deviations will cancel out some of the effect of the negative deviations. We care equally about positive deviations (measured value greater than predicted) and negative deviations (predicted value greater than measured); thus, we choose the sum of the *square* of the deviations as the error measure, rather than the sum of the bare deviations. The sum of the squared deviations (squared errors) is abbreviated $SS_E$:

$$\begin{matrix} \text{Sum of squared} \\ \text{deviations between} \\ \text{observed and model, } SS_E \\ \text{(quantity to be minimized)} \end{matrix} \qquad SS_E \equiv \sum_{i=1}^{n} (y_i - \widehat{y}_i)^2 \qquad (6.3)$$

For the straight-line model, this becomes

$$SS_E = \sum_{i=1}^{n} \left[ y_i - (mx_i + b) \right]^2 \qquad (6.4)$$

To find the values of $m$ and $b$ that result in the minimum value of $SS_E$, we use calculus. First, we take the derivatives of $SS_E$ with respect to each parameter $m$ and $b$; second, we set those derivatives to zero, locating the minimum; and finally, we solve, simultaneously, the two resulting expressions for the values of $m$ and $b$ that correspond to minimum $SS_E$ error. This is a straightforward calculation, although it is a bit involved algebraically (related matrix calculations are discussed in Section 6.3). The final results for the values of $m$ and $b$ that give the smallest sum of squared errors for a dataset $(x_i, y_i)$, $i = 1, 2, \ldots, n$, are:

Ordinary least-squares, best fit to slope

$$m = \frac{n \sum_{i=1}^{n} x_i y_i - \left(\sum_{i=1}^{n} x_i\right)\left(\sum_{i=1}^{n} y_i\right)}{n \sum_{i=1}^{n} x_i^2 - \left(\sum_{i=1}^{n} x_i\right)^2} \tag{6.5}$$

$$m = \frac{SS_{xy}}{SS_{xx}}$$

Ordinary least-squares, best fit to intercept

$$b = \frac{\left(\sum_{i=1}^{n} x_i\right)^2 \left(\sum_{i=1}^{n} y_i\right) - \left(\sum_{i=1}^{n} x_i y_i\right)\left(\sum_{i=1}^{n} x_i\right)}{n \sum_{i=1}^{n} x_i^2 - \left(\sum_{i=1}^{n} x_i\right)^2} \tag{6.6}$$

$$b = \bar{y} - m\bar{x}$$

The equivalencies between the two formulas for slope and intercept in Equations 6.5 and 6.6 (definitions of $\bar{x}$, $\bar{y}$, $SS_{xx}$, and $SS_{xy}$ are provided next) may be shown by straightforward algebra. The quantities $\bar{x}$ and $\bar{y}$ above are the mean values of the $x_i$ and $y_i$ in the dataset, $SS_{xx}$ is the sum of the squares of the deviations between the $x_i$ and the mean value of the $x_i$, and $SS_{xy}$ is the sum of the product of the $x$ and $y$ deviations from their respective means. The formulas for these defined quantities and their Excel function calls are given here. Table D.1 in Appendix D gives both the MATLAB and Excel commands for these quantities.

$$\bar{x} \equiv \frac{1}{n} \sum_{i=1}^{n} x_i = \text{AVERAGE(x-range)} \tag{6.7}$$

$$\bar{y} \equiv \frac{1}{n} \sum_{i=1}^{n} y_i = \text{AVERAGE(y-range)} \tag{6.8}$$

$$SS_{xx} \equiv \sum_{i=1}^{n} (x_i - \bar{x})^2 = \text{DEVSQ(x-range)} \qquad (6.9)$$

$$SS_{xy} \equiv \sum_{i=1}^{n} (x_i - \bar{x})(y_i - \bar{y})$$

$$= \text{COUNT(x-range)*COVARIANCE.P(y-range,x-range)} \quad (6.10)$$

Excel is preprogrammed to evaluate the formulas in Equations 6.5 and 6.6 for a dataset $(x_i, y_i)$, $i = 1, 2, \ldots, n$.. The Excel functions for the optimum values of $m$ and $b$ are

Parameters from
ordinary least-squares
best fit to line (Excel)      $m = \text{SLOPE(y-range,x-range)}$ $\qquad (6.11)$
(using Equations 6.5 and 6.6)

$b = \text{INTERCEPT(y-range,x-range)}$ $\quad (6.12)$

Note that in Excel the order of the independent variables in the function call is opposite to the mathematics convention: the $y$ values are first, followed by the $x$ values.

Excel provides a comprehensive least-squares tool, LINEST, which may also be used to calculate least-squares slope and intercept. LINEST is an array function, and its results are accessed by using the INDEX(array, row, column) command to retrieve desired results from the output array. Slope is provided in the LINEST array output in the $(1,1)$ array position (row, column), and intercept is located in the $(1,2)$ position.

Ordinary least-squares
best fit to slope      $m = \text{INDEX(LINEST(y-range,x-range),1,1)}$
from LINEST

$(6.13)$

Ordinary least-squares
best fit to intercept      $b = \text{INDEX(LINEST(y-range,x-range),1,2)}$
from LINEST

$(6.14)$

The values of slope and intercept calculated with the functions SLOPE(), INTERCEPT(), and LINEST are the same values that are calculated by Excel when adding a trendline to a plot of the data points.[2] The function calls in Equations 6.11 and 6.12 are convenient when it is desirable to embed the model parameter values in calculations in an Excel spreadsheet. Alternatively, all calculations associated with a LINEST fit may be calculated at once, improving efficiency (see Appendix C). We will have more to say about LINEST in the discussions that follow.

In Example 6.1, we provide some practice with ordinary least-squares model fitting with Excel by generating the calibration curve fit used in Examples 4.2 and 5.7 (see Figure 1.3 for more context for this example). The MATLAB calculation of intercept, slope, and coefficient of determination $R^2$ is demonstrated for this example in Appendix D. In the MATLAB solution, the data are held in $1 \times n$ vectors $xdata$ and $ydata$ ($n = 15$ data points). Once the $n \times 2$ matrix $A$ (shown here) is created from the $x$ data, the $2 \times 1$ vector $bhat$ containing the intercept and slope is calculated in a single line of MATLAB code.

$$A = [\text{ones}(n, 1), xdata']; \qquad (6.15)$$

$$bhat = A \backslash ydata;$$

See Appendix D and Section 6.3.1 for discussion of the matrix method of calculating ordinary least-squares parameters. The answers for the least-squares model parameters are, of course, the same whether we use Excel or MATLAB.

**Example 6.1: \*LINKED\* Determining a calibration curve for a differential-pressure meter.** *Our laboratory has a differential-pressure (DP) meter that is used to measure fluid pressure differences between two locations in a flow through tubing, as discussed in Examples 5.6 and 5.7. The raw DP-meter signals are in the form of electric-current values ranging from 4.0 to 20.0 mA. To translate these signals to differential pressure in psi, we have obtained DP-meter calibration data by using an air-over-Blue-Fluid manometer as a differential-pressure standard. The calibration data and the errors on $(\Delta p, psi)_{std}$ are given Table 5.2 (Example 5.7). Based on the calibration dataset, what is the equation for the calibration curve?*

**Solution:** In Example 5.7, we matched DP-meter calibration signals to values for $(\Delta p, \text{psi})_{std}$ obtained from a DP standard, the air-over-Blue-Fluid manometer discussed in Example 4.8. In Example 5.7, we plotted the

---

[2] In some least-squares calculations in Excel, issues of computing stability have led Microsoft to make choices that are a bit different than we describe. As a result, there may be small numerical differences in the Excel answers from what one would obtain with the methods discussed.

calibration data as the dependent variable $(I, \text{mA})_{cal}$ versus the independent variable $(\Delta p, \text{psi})_{std}$. We presented the best-fit equation in Equation 5.40 but did not reveal how we arrived at that equation. Our discussion of ordinary least-squares regression now allows us to calculate the best-fit line from the calibration dataset, which then leads to the calibration curve.

The data show a linear trend (Figure 6.3). The $(\Delta p, \text{psi})_{std}$ calibration standards (on the $x$-axis) are very accurate by design; this matches the assumption in the least-squares algorithm of zero $x$-error. In Excel we can add a trendline to the graph using ordinary least squares by right-clicking on the data, and we specify in the dialog box that the equation for the line and the coefficient of determination $R^2$ be shown on the graph. Alternatively, we can calculate the slope and intercept of the data using SLOPE() and INTERCEPT() (Equations 6.11 and 6.12) or with LINEST using Equations 6.13 and 6.14. The three methods yield the same answers, as all three are programmed implementations of ordinary least-squares Equations 6.5 and 6.6.[3]

Least-squares fit
to calibration data
(extra digits shown)

$$(I, \text{mA})_{cal} = m(\Delta p, \text{psi})_{std} + b \qquad (6.16)$$

$$m, \text{mA/psi} = \text{SLOPE(y-range,x-range)} = 4.326143$$
$$b, \text{mA} = \text{INTERCEPT(y-range,x-range)} = 3.995616$$

The *coefficient of determination*, better known as $R^2$, is shown on the graph in Figure 6.3 ($R^2 = 0.9998$). This quantity is informally known as a measure of the quality of the fit of the line to the data. Formally, $R^2$ is the fraction of the variability of the stochastic $y_i$ data that is accounted for by the linear model.

Coefficient of
determination, $R^2$:

$$R^2 \equiv \frac{(\text{error explained by the model})}{(\text{total error})}$$

$$= \frac{(SS_T - SS_E)}{SS_T} \qquad (6.17)$$

$$= \text{RSQ(y-range,x-range)} \qquad (6.18)$$

$$= \text{INDEX(LINEST(y-range,x-range,1,1),3,1)} \qquad (6.19)$$

---

[3] The MATLAB calculations of intercept, slope, and $R^2$ for the DP meter calibration curve are shown in Appendix D, Section D.7.

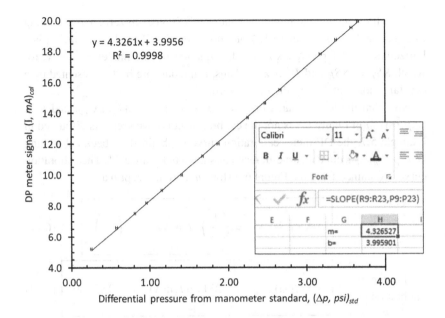

Figure 6.3 The data fall on a straight line. The trendline shown was added in Excel and the equation for the line is displayed on the graph. The inset shows that calculating the slope (command shown) and intercept of the data using Equations 6.11 and 6.12 yields the same answers as shown with the Excel Trendline dialog box. The $x$-error bars were developed in Example 5.7.

where[4] $SS_E$ is the residual sum of squares of errors, the quantity that we minimize in least-squares model fitting (Equation 6.3; this will be small if the model is well represented by the line), and $SS_T$, also written as $SS_{yy}$, is the total sum of squares or the total of the squared $y$-deviations from the mean of the $y_i$ values:

$$\text{Total sum of squares} \quad SS_T = SS_{yy} = \sum_{i=1}^{n}(y_i - \bar{y})^2 \quad (6.20)$$

$$SS_T \text{ or } SS_{yy}$$

$$= \text{DEVSQ(y-range)} \quad (6.21)$$

The "total" sum of squares $SS_T = SS_{yy}$ is the "error" that would be present if we would assume that the $y$ data should be constant and just equal to its average value, $\bar{y}$. When the linear model is a very good fit, there is little deviation

---

[4] To calculate $R^2$ with LINEST, we include two additional optional arguments in Equation 6.19, instructing LINEST to produce the full set of statistical results of the curve fit. See Appendix C for more discussion of LINEST.

between the data and the model; in that case $SS_E$ goes to zero, and according to Equation 6.17, $R^2$ goes to 1. Note, however, that if the best-fit model is a horizontal line (slope equals zero), the equation for the model is $\widehat{y} = \bar{y}$, for which $SS_T = SS_E$ and $R^2$ is zero. Thus, if the data are best represented by a constant (horizontal line), then $R^2$ is zero.

Note that the least-squares fit $y = mx + b$ for $y = (I, \text{mA})_{cal}$ and $x = (\Delta p, \text{psi})_{std}$ in Figure 6.3 is not the calibration curve we seek. As discussed in Example 5.7, the DP-meter calibration curve is obtained by back-calculating $(\Delta p, \text{psi})_{meter}$ from the curve-fit result as shown in Figure 5.11. The calibration curve is obtained from the Figure 6.3 slope $m$ and intercept $b$ as

$$(\Delta p, \text{psi})_{meter} = \left(\frac{1}{m}\right)(I, \text{mA})_{raw} - \left(\frac{b}{m}\right) \qquad (6.22)$$

DP-meter
calibration curve    $$(\Delta p, \text{psi})_{meter} = 0.23115(I, \text{mA})_{raw} - 0.92360 \qquad (6.23)$$

This calibration curve matches the previously presented versions, Equations 4.1 and 5.40.

Equations 6.5 and 6.6 are the best-fit, ordinary least-squares values of $m$ and $b$ for the model $y = mx + b$ and the dataset $(x_i, y_i)$, $i = 1, 2, \ldots, n$. The algorithm is called *least squares* because it minimizes (finds the least value of) the sum of the squared deviations as its method of finding the "best values" of the model's parameters. The term *ordinary* refers to our assumption at the outset that the values of the $x_i$ in the fitting dataset $(x_i, y_i)$ are *certain*. Since the values of the $x_i$ are certain, the deviations between the measured values of the data pairs and the model values are given by vertical distances $(y_i - \widehat{y}_i)$ (Equation 6.2), as shown in Figure 6.2. When the $x$ values under study are related to instrument settings that are quite precise, or when the $x_i$ are carefully determined or set (such as calibration standards are), this is a good assumption. There are circumstances, however, in which the values of the $x_i$ are also uncertain. In this case, the graphical representation of the deviation of the data points from the model is better represented by the shortest distance from each data point to the model line (Figure 6.4). The mathematics of this version of least squares has been worked out, and Excel macros are available to make least-squares calculations with this representation of the deviations [13, 50]. Abandoning the assumption of zero error on $x_i$ also complicates the calculation of errors on $m$ and $b$. Ordinary least squares as described here is by far the most

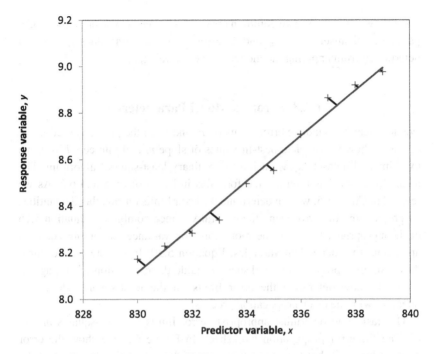

Figure 6.4 When the values of the $x_i$ are assumed to be certain, the deviations between the model values and the measured values of the data pairs are given by $(y_i - \widehat{y}_i)$ (Figure 6.2). If we make no such assumption, the deviation is calculated as the shortest distance to the model line as shown here.

common algorithm in use for determining best-fit parameters for linear model fitting. Because of the assumption that the $x_i$ are certain, our best choice when using ordinary least squares is to designate the variable with the least error as the $x$ variable.

There are many scientific and engineering applications of best-fit models. First, we have mentioned that models are compact and allow the trends in the data to be expressed as a formula. Second, sometimes the parameters of the model (for example, the slope and the intercept) have meaning in their own right. Quite often we use a model function $\widehat{y} = \phi(x)$ to determine values of the function at $x$ values for which we have no experimental data; that is, we use a model to *interpolate* or *extrapolate*. For all these purposes, we need to be able to express the appropriate error limits for quantities calculated from the model. The uncertainties associated with applications of ordinary least-squares results are determined from error propagations based on the calculations used to determine the model parameters (Equations 6.5 and 6.6). In the sections that follow, we address two common error calculations: (1) error on the model parameters themselves $s_b$, $s_m$, and (2) error $s_{\widehat{y}_0}$ on function values $\widehat{y}_0 = \phi(x_0)$ determined from applying the model at a new $x$ value, $x_0$.

### 6.2.3 Error on Model Parameters

We now seek to calculate error limits on $m$ and $b$. In the previous section we discussed how to calculate best-fit values of slope $m$ and intercept $b$ from an experimental dataset $(x_i, y_i)$ using the ordinary least-squares algorithm. The results of this calculation are the formulas in Equations 6.5 and 6.6. As we learned in Chapter 5, we can determine the error limits on calculated quantities by propagating the error using the general variance-combining Equation 5.16 or, if appropriate, by using the more common variance-combining equation applicable to independent variables, Equation 5.20. In Chapter 5, we introduced an error propagation worksheet to guide the calculation of propagated error. We now determine the error limits on the least-squares slope and intercept with the error-propagation process.

Our task is to determine appropriate error limits on least-squares $m$ and $b$ using the error propagation worksheet. In Figure 6.5, we show the error propagation worksheet and the beginnings of the calculation for error on the

| $\phi(x_1, x_2, x_3, x_4, x_5)$: | Formula for $\phi$:[*] $m = \dfrac{n\sum_{i=1}^n x_i y_i - (\sum_{i=1}^n x_i)(\sum_{i=1}^n y_i)}{n\sum_{i=1}^n x_i^2 - (\sum_{i=1}^n x_i)^2}$ *Note: units must work as written. | | Representative value of $\phi$: (include units) $m =$ | 95% C.I. of $\phi$: $(\phi \pm 2e_{s,\phi})$ (include units) $m \pm (t_{0.025,n-2})(s_m)$ |
|---|---|---|---|---|
| **Measured quantities, $x_i$** | | | $\dfrac{\partial \phi}{\partial x_i}$ | $e_{s,x_i}$ | $\left(\dfrac{\partial \phi}{\partial x_i}\right)^2 e_{s,x_i}^2$ |
| $x_i$ | Symbol | Representative value | | | |
| $x_1$ | $y_1$ | units | $\dfrac{\partial m}{\partial y_1}$ | $s_{y,x}$ | $\left(\dfrac{\partial m}{\partial y_1}\right)^2 s_{y,x}^2$ |
| $x_2$ | $y_2$ | units | $\dfrac{\partial m}{\partial y_2}$ | $s_{y,x}$ | $\left(\dfrac{\partial m}{\partial y_2}\right)^2 s_{y,x}^2$ |
| $x_3$ | $y_3$ | units | $\dfrac{\partial m}{\partial y_3}$ | $s_{y,x}$ | $\left(\dfrac{\partial m}{\partial y_3}\right)^2 s_{y,x}^2$ |
| $x_4$ | $y_4$ | units | $\dfrac{\partial m}{\partial y_4}$ | $s_{y,x}$ | $\left(\dfrac{\partial m}{\partial y_4}\right)^2 s_{y,x}^2$ |
| $x_5$ | $\vdots$ | units | $\vdots$ | $\vdots$ | $\vdots$ |

$$e_{s,\phi}^2 = \left(\frac{\partial \phi}{\partial x_1}\right)^2 e_{x_1}^2 + \left(\frac{\partial \phi}{\partial x_2}\right)^2 e_{x_2}^2 + \left(\frac{\partial \phi}{\partial x_3}\right)^2 e_{x_3}^2 + \left(\frac{\partial \phi}{\partial x_4}\right)^2 e_{x_4}^2 + \left(\frac{\partial \phi}{\partial x_5}\right)^2 e_{x_5}^2$$

* All variables $x_i$ must be independent for this equation to hold.

$e_{s,\phi}^2 = s_m^2$

$e_{s,\phi} = s_m$   units

Figure 6.5 We begin the calculation for the error on the least-squares slope by determining the needed information requested in the error propagation worksheet. The errors in $y_i$ have been assumed to be independent of $x_i$, as discussed in the text (homoscedasticity).

slope. The least-squares formula for $m$ in Equation 6.5 is inserted as the formula for the propagation. The variables in this equation are the $n$ data points $(x_i, y_i)$ that were used to determine the fit. In the ordinary least-squares calculation, the $x_i$ are presumed to be accurate, so the error on the $x_i$ is set to zero in the worksheet. Because of this and to save space, the $x_i$ are omitted from the worksheet in Figure 6.5. The errors in the $y_i$ are not zero, and, in general, may vary from point to point. To complete the worksheet we need to determine the error in $y_i$ at each value of $x_i$.

Estimating the uncertainty in $y_i$ at each $x_i$ presents a bit of a problem. Often we have only one data point for each $x_i$ in the fitting dataset. It would seem that we need replicates at each value of $x_i$ to get a good value of the error on each $y_i$. One way out of this dilemma is to assume that the error in $y_i$ is *independent of* $x_i$ – that is, there is a single standard deviation $s_{y,x}$ that characterizes the deviation of the $y_i$ from their expected values on the best-fit line, independent of the value of $x_i$ considered. With this assumption, we could use all $n$ data points in the dataset to calculate the common standard deviation $s_{y,x}$. As $n$ is typically a reasonably large number (usually more than 10), a good estimate of $s_{y,x}$ could be obtained.

In the assessment of errors for model parameters in the ordinary least-squares method, we assume that the distribution of errors on the $y_i$ is *the same for every value of* $x_i$. In statistics, datasets for which this is true are called *homoscedastic*. This assumption is reflected in Figure 6.5 in the second column from the right, where the error in each $y_i$ is the same value, given the symbol $s_{y,x}$, the standard deviation of $y$ at a particular value of $x$.

> In ordinary least squares,
> we assume the data are homoscedastic.

We can test whether this is a good assumption by plotting the differences, or *residuals*, between the model and the data points, as a function of $x$. If the data are homoscedastic, as assumed [19], least-squares residuals should be scattered around zero with no discernable pattern. Residuals for the DP meter calibration are the subject of Problem 6.9.

With the data assumed homoscedastic, the standard deviation of $y$ at a value of $x$, $s_{y,x}$, may be calculated from the fitting dataset using the definition of variance. The definition of variance of one-dimensional variable $u$ was introduced in Chapter 2 (Equation 2.9).

$$\text{Sample variance of one-dimensional dataset, } u_i \qquad s^2 \equiv \left( \frac{\sum_{i=1}^{n}(u_i - \bar{u})^2}{n-1} \right) \qquad (6.24)$$

where there are $n$ observations of $u_i$ in the dataset and $\bar{u}$, the mean of the $u_i$ in the dataset, is the expected value of $u$. Variance is a sort of average of the squared deviations between a quantity and its expected value. It is a *sort of* average because the calculation of variance from the mean uses the degrees of freedom $(n - 1)$ in the denominator instead of the number of data points $n$. The degrees of freedom in the calculation of the variance from the mean for a stochastic variable $u$ is one less than the number of data points because the expected value of $u$ – in that case the mean $\bar{u}$ – is first calculated from the dataset and then used in the variance formula; that calculation uses up a degree of freedom.

For a two-dimensional dataset $(x_i, y_i)$ used to fit the parameters of a line, variance $s_{y,x}^2$ of $y$ at a value of $x$ is once again the average of the squared deviations between the observed quantities and their expected values; the expected value of $y_i$ is given by the model-predicted value at each point $\hat{y}_i$, and the number of degrees of freedom is also adjusted. The number of degrees of freedom in this case is $(n - 2)$, since two parameters, $m$ and $b$, are first determined from the dataset before the variance is calculated, reducing the number of degrees of freedom by 1.[5] For the linear-model fit, the variance of the model-fit residuals, $s_{y,x}^2$, is given by

Sample variance of
$y$ at a given value of $x$
calculated from a dataset $(x_i, y_i)$
for a linear model $y = mx + b$;
$s_{y,x}^2$ assumed independent of $x_i$
$$s_{y,x}^2 = \mathrm{Var}(y, x)$$

$$s_{y,x}^2 = \frac{1}{(n - 2)} \sum_{i=1}^{n} (y_i - \hat{y}_i)^2 \qquad (6.25)$$

Standard deviation of
$y$ at a given value of $x$

$$s_{y,x} = \mathrm{STEYX}(\text{y-range}, \text{x-range}) \qquad (6.26)$$

$$s_{y,x} = \mathrm{INDEX}(\mathrm{LINEST}(\text{y-range}, \text{x-range}, 1, 1), 3, 2)$$

The variance $s_{y,x}^2$ is a measure of the stochastic variability of data points $y_i$ from the expected values of $y$ at each point $x_i$. It captures the average

---

[5] This discussion of the origins of the variance of $y$ at a value of $x$ is a plausibility argument, not a proof. Rigorously, it may be shown that Equation 6.25 is an unbiased predictor of the true variance of $y$ at a value of $x$ [38].

variability of the $y$ values in the dataset, as measured by deviation from the model prediction. We saw a preliminary discussion of $s_{y,x}$ in Example 2.18. The calculation of $y_i$ variability in terms of the standard deviation $s_{y,x} = \sqrt{s_{y,x}^2}$ is programmed into Excel with the function calls just shown; see the Table D.1 entries for $s_{y,x}$, matrix $A$, and $SSE$ for a method of obtaining $s_{y,x}$ with MATLAB.

The standard deviation $s_{y,x}$ allows us to express error limits on $y$ values in the dataset $(y_i, x_i)$, as implied by the quality of the least-squares fit, assuming that (1) the distribution of $y$ at a given $x$ is independent of the value of $x$ (homoscedacity) and (2) the model is a good predictor of the value of $y$. It may be shown that the deviations of $y_i$ from their true values at $x_i$, $y_{i,true}$, are $t$-distributed with $(n-2)$ degrees of freedom [38]. This information allows us to construct 95% confidence interval error limits on the values of the $y_i$ in the dataset, relative to the fit.

$$
\begin{array}{c}
\text{95\% CI on} \\
\text{linearly correlated } y(x) \text{ values} \\
\text{in the fitting dataset}
\end{array}
\qquad
\boxed{y_{i,true} = y_i \pm t_{0.025, n-2} s_{y,x}}
\qquad (6.27)
$$

In Example 6.2, we show the calculation of both $s_{y,x}$ and dataset $y$-error bars for the DP meter calibration of Example 6.1.

**Example 6.2: \*LINKED\* DP-meter signal errors from ordinary least-squares fit.** *The DP meter discussed in Example 6.1 responds to a differential pressure by generating a current signal that is then read from a digital multimeter (DMM) connected to it. The measurement that results from this arrangement, $(I, mA)$, is affected by intrinsic variability (scatter) associated with both the DP meter's and digital multimeter's design and operation. What are appropriate error limits to attribute to signals obtained from the DP meter, $(I, mA)$?*

**Solution:** The observed signal on the DMM attached to the DP meter is given by

$$
\begin{pmatrix}
\text{observed value} \\
\text{of } I\,(mA) \\
\text{(stochastic)}
\end{pmatrix}
=
\begin{pmatrix}
\text{true value} \\
\text{of } I\,(mA)
\end{pmatrix}
+
\begin{pmatrix}
\text{deviation due} \\
\text{to DP meter} \\
\text{functioning} \\
\text{(stochastic)}
\end{pmatrix}
$$

$$
+
\begin{pmatrix}
\text{deviation} \\
\text{due to DMM} \\
\text{calibration} \\
\text{(stochastic)}
\end{pmatrix}
+
\begin{pmatrix}
\text{deviation} \\
\text{due to DMM} \\
\text{reading} \\
\text{(stochastic)}
\end{pmatrix}
\qquad (6.28)
$$

We can evaluate the inherent variability of the setup during the calibration process. To calibrate the setup, we take a set of measurements of known differential pressures; these values serve as calibration standards $(x, y) = [(\Delta p, \text{psi})_{std}, (I, \text{mA})_{cal}]$. These measurements include information about both the signal values and the error limits that accompany measurements with this setup.

As is usual for many devices, the DP meter is designed to give a linear response in electric current signal as a function of differential pressure. Using ordinary least squares, we can fit a line to the set of signals $(I, \text{mA})_{cal}$ versus $(\Delta p, \text{psi})_{std}$. By using this method, we have assumed the $x$ values are accurate (which, since they come from DP standards, they are) and that the $y$ data are homoscedastic – that is, that the errors in $y$ values are the same for every value of $x$. The calculation of the standard deviation $s_{y,x}$ averages the squared $y$ deviations of the data from the best-fit line.

As shown in Equation 6.28, the stochastic contributions to the raw signals may originate with the operation of the DP meter, the calibration of the DMM, and the reading error of the DMM. There is no need to separate the first two stochastic contributions since they influence each point simultaneously, regardless of the value of $x$ (homoscedasticity). The DMM reading error contribution will, however, stochastically vary with $x$. In fact, since a comprehensive set of values of $x$ are sampled in the calibration dataset, it is reasonable to assume that the individual contributions from the reading error will average to zero in the calculation of $s_{yx}$. The error $s_{y,x}$ is thus associated with the combination of calibration error of the digital multimeter and the device variability of the DP meter and represents the uncertainty for each $y$ value of the calibration measurement.

$$
\begin{pmatrix} \text{observed value} \\ \text{of } I(\text{mA}) \\ \text{(stochastic)} \end{pmatrix} = \begin{pmatrix} \text{true value} \\ \text{of } I(\text{mA}) \end{pmatrix} + \begin{pmatrix} \text{deviation due to} \\ \text{DMM and DP meter} \\ \text{functioning} \\ \text{(stochastic)} \end{pmatrix} \quad (6.29)
$$

The quantity $s_{y,x}$, calculated with Equation 6.25, is programmed into the Excel function STEYX(y-range,x-range) and Table D.1 shows how to calculate $s_{y,x}$ with MATLAB. For the Example 6.1 DP-meter calibration dataset given in Table 5.2, we obtain:

$$
y = (I, \text{mA})_{cal}
$$
$$
x = (\Delta p, \text{psi})_{std}
$$
$$
s_{y,x} = \text{STEYX(y-range,x-range)}
$$

DP-meter calibration dataset
standard deviation of $y$
at a value of $x$
(assumed homoscedastic)

$$s_{y,x} = 0.077 \text{ mA} \qquad (6.30)$$

This quantity aggregates the variability observed in the values of $y = (I, \text{mA})_{cal}$ (DP meter signal as recorded by the DMM, mA) observed in the calibration dataset, assuming that the $y$-variability is the same for all values of $x$. Note that this agrees, to one significant figure, with the expected combined standard uncertainty $e_{s,cmbd}^2 = e_{s,reading}^2 + e_{s,cal}^2$ for the DMM when reading current.

The 95% confidence interval error limits on individual $y_i$ values in the calibration dataset are based on the variability captured by $s_{y,x}$. These error limits are written as

DP meter observations
(with error limits)

$$y_{i,true} = y_i \pm t_{0.025, n-2} s_{y,x} \qquad (6.31)$$

Figure 6.16 shows the DP meter calibration data plotted with the error limits determined with Equation 6.31.

With the identification of $s_{y,x}$ as the variability in a fitting dataset's $y_i$ values, we now have all the information we need to carry out the error propagation on a least-squares slope (Figure 6.5). All the $y_i$ are independent variables with error $e_{s,y_i} = s_{y,x}$, and the error-propagation formula at the bottom of the worksheet applies. The result of this calculation gives the variance $s_m^2$ on the slope $m$. An analogous worksheet calculation using the formula for intercept $b$ (Equation 6.6) gives variance on the intercept $s_b^2$ (see Problems 6.11 and 6.12). The results of these two calculations are

Variance of slope
(ordinary least squares)

$$s_m^2 = s_{y,x}^2 \left( \frac{1}{SS_{xx}} \right) \qquad (6.32)$$

Variance of intercept
(ordinary least squares)

$$s_b^2 = s_{y,x}^2 \left( \frac{1}{n} + \frac{\bar{x}^2}{SS_{xx}} \right) \qquad (6.33)$$

Standard deviation may be calculated as the square root of variance. These formulas and software function calls to carry them out are collected in Tables 6.1 and D.1, along with other significant quantities related to least-squares error calculations.

It is helpful to reflect on the results for $s_m^2$ and $s_b^2$ and test if the formulas match our intuition about what should cause uncertainty in the fitting of a line

Table 6.1. *Significant formulas and Excel function calls related to ordinary least-squares error calculations.*

| Formula | Equation | Excel command |
|---|---|---|
| $\bar{x} \equiv \dfrac{1}{n} \displaystyle\sum_{i=1}^{n} x_i$ | 6.7 | AVERAGE(x-range) |
| $SS_{xx} \equiv \displaystyle\sum_{i=1}^{n} (x_i - \bar{x})^2$ | 6.9 | DEVSQ(x-range) |
| $SS_T = SS_{yy} \equiv \displaystyle\sum_{i=1}^{n} (y_i - \bar{y})^2$ | 6.21 | DEVSQ(y-range) |
| $SS_{xy} \equiv \displaystyle\sum_{i=1}^{n} (x_i - \bar{x})(y_i - \bar{y})$ | 6.10 | COUNT(x-range)*<br>COVARIANCE. P(y-range,x-range) |
| $m = \dfrac{SS_{xy}}{SS_{xx}}$ | 6.5 | SLOPE(y-range,x-range) or<br>INDEX(LINEST(y-range,x-range),1,1) |
| $b = \bar{y} - m\bar{x}$ | 6.6 | INTERCEPT(y-range,x-range) or<br>INDEX(LINEST(y-range,x-range),1,2) |
| $s_{y,x} = \sqrt{\dfrac{1}{(n-2)} \displaystyle\sum_{i=1}^{n} (y_i - \widehat{y}_i)^2}$ | 6.25 | STEYX(y-range,x-range) |
| $s_m = \sqrt{s_{y,x}^2 \left(\dfrac{1}{SS_{xx}}\right)}$ | 6.32 | INDEX(LINEST(y-range,x-range,1,1),2,1) |
| $s_b = \sqrt{s_{y,x}^2 \left(\dfrac{1}{n} + \dfrac{\bar{x}^2}{SS_{xx}}\right)}$ | 6.33 | INDEX(LINEST(y-range,x-range,1,1),2,2) |
| $R^2 \equiv \dfrac{(SS_T - SS_E)}{SS_T}$ | 6.17 | RSQ(y-range,x-range) |
| Degrees of freedom | 6.36 | INDEX(LINEST(y-range,x-range,1,1),3,2) |
| $s_{\widehat{y}_0}^2 = s_{y,x}^2 \left(\dfrac{1}{n} + \dfrac{(x_0 - \bar{x})^2}{SS_{xx}}\right)$ | 6.62 | |
| $s_{y_0,n+1}^2 = s_{y,x}^2 \left(1 + \dfrac{1}{n} + \dfrac{(x_0 - \bar{x})^2}{SS_{xx}}\right)$ | 6.73 | |

to data. Both $s_m^2$ and $s_b^2$ are proportional to $s_{y,x}^2$, indicating that the greater the intrinsic scatter in the $y_i$, the greater will be the error in both slope and the intercept, as we would expect. Both expressions also include $SS_{xx}$ in the denominator of one term. $SS_{xx}$ is a measure of the spread of the data along the $x$-direction. Thus, the equations indicate that the more spread out the data, the lower the error in slope and intercept. This is easy to rationalize, since data that are spread out place more strict constraints on the possible slopes and intercepts that could be consistent with the data. Finally, in the expression for the variance of the intercept $s_b^2$, we find an additional $n$-dependence, indicating that if the number of data points in the dataset is reduced, there will be an increase in the uncertainty in intercept. A bit of playing around with a dataset (modifying the points, and the number of points, and seeing the effects on $m$, $b$, $s_m$, and $s_b$) confirms the stronger influence of $n$ on $b$ than on slope $m$.

The variances of ordinary least-squares model parameters are used to construct 95% confidence interval error limits on slope and intercept. The following scaled deviations are $t$-distributed with variances $s_m^2$ and $s_b^2$, and $(n - 2)$ degrees of freedom [38].

$$\begin{array}{c} \text{scaled deviation} \\ \text{of slope} \end{array} = \left( \frac{m - m_{true}}{s_m} \right) \qquad (6.34)$$

$$\begin{array}{c} \text{scaled deviation} \\ \text{of intercept} \end{array} = \left( \frac{b - b_{true}}{s_b} \right) \qquad (6.35)$$

$$\begin{array}{c} \text{Degrees of freedom, } df \\ \text{least squares} \\ \text{for linear model} \end{array} \qquad df = (n - 2) \qquad (6.36)$$

where $m$ and $b$ are the least-squares calculated slope and intercept, and $m_{true}$ and $b_{true}$ are the true slope and intercept, respectively. Following the logic used in Chapter 2 to construct 95% confidence intervals, we can now produce 95% confidence interval error limits on both $m$ and $b$ based on $\pm t_{0.025,df} \times$ (standard deviation).

$$\begin{array}{c} \text{95\% CI on} \\ \text{least-squares slope} \end{array} \qquad \boxed{m_{true} = m \pm t_{0.025,n-2} s_m} \qquad (6.37)$$

$$\begin{array}{c} \text{95\% CI on} \\ \text{least-squares intercept} \end{array} \qquad \boxed{b_{true} = b \pm t_{0.025,n-2} s_b} \qquad (6.38)$$

With 95% confidence, the true values of slope and intercept are bracketed by the upper and lower bounds given here. If $n$ is greater than about 8, $t_{0.025, n-2}$ is about 2, and we obtain the usual $\approx \pm 2e_s$ 95% confidence intervals.

In the examples that follow (from the fields of fluid mechanics and rheology), we show applications of model fitting in which there are physical interpretations of slope and intercept. Included in the examples are instructions on how to use LINEST to perform complete ordinary least-squares calculations, including error calculations. Additional information on using LINEST is available in Appendix C as well as in Table 6.1. The expressions given are also readily evaluated in MATLAB. See Appendix D for code that provides MATLAB solutions to all the examples in this chapter.

**Example 6.3: Uncertainty on slope: Wall shear stress from capillary rheometer data.** *Viscosity is an important physical property of polymers, and capillary rheometers are used to measure the high-shear-rate viscosities of molten polymers. A capillary rheometer allows for the measurement of driving pressure difference $\Delta p$ as a function of volumetric flow rate $Q$ for steady polymer flow through a capillary. Flow pressure drops for a polyethylene at a fixed value of scaled flow rate $\dot{\gamma}_a = 4Q/\pi R^3 = 90 \ s^{-1}$ are given in Table 6.2 as a function of the ratio of capillary length $L$ to radius $R$. The data are analyzed with the following relationship, which is based on a flow-direction momentum balance [39]:*

$$\Delta p = (2\tau_R) \frac{L}{R} + \Delta p_{ent} \tag{6.39}$$

*where $\tau_R$ is the wall shear stress (used in the calculation of viscosity), and $\Delta p_{ent}$ is the entrance pressure loss, which is constant for fixed $4Q/\pi R^3$. We desire to calculate $\tau_R$ and $\Delta p_{ent}$. What value of wall shear stress $\tau_R$ is implied by the data in Table 6.2? Include appropriate 95% confidence interval error limits on your answer.*

**Solution:** The data in Table 6.2 represent driving pressure drops for a material pushed through a variety of capillaries, with all data taken at a fixed value of scaled flow rate $\dot{\gamma}_a = \frac{4Q}{\pi R^3}$, which is called the apparent shear rate [39]. We begin by plotting the data (Figure 6.6). As expected from the supplied theoretical equation (Equation 6.39), the data seem to follow a straight line. The theoretical equation also implies that a plot of driving pressure $\Delta p$ versus $L/R$ has a slope of $2\tau_R$; thus we can calculate the requested wall shear stress, $\tau_R$, from the slope. To obtain the error on wall shear stress, we would need to follow the determination of the slope with a simple error-propagation calculation.

Table 6.2. *For a molten polyethylene, pressure drop* ($\Delta p$, bar) *in pressure-driven flow in a capillary versus $L/R$, for* $\dot{\gamma}_a = 4Q/\pi R^3 = 9.0 \times 10^1 \ s^{-1}$. *Data are from Bagley [1].*

| $\dfrac{L}{R}$ | $\Delta p$, bar |
|---|---|
| 0.00 | 85.2 |
| 1.84 | 117.2 |
| 2.88 | 144.8 |
| 3.74 | 160.7 |
| 4.78 | 182.6 |
| 9.20 | 273.9 |
| 18.42 | 460.8 |
| 29.34 | 673.6 |

$$\text{slope} = 2\tau_R$$

$$\tau_R = \frac{\text{slope}}{2} \tag{6.40}$$

For the dataset in Table 6.2, we use Excel or MATLAB to find the slope and, subsequently, the standard error of the slope. Using Excel's SLOPE() command:

$$\text{Least-squares slope:} \quad m = \text{SLOPE(y-range,x-range)}$$
$$= 20.1605 \text{ bar} \quad \text{(extra digits supplied)} \tag{6.41}$$

Alternatively, we could use LINEST, which returns slope in array position (1,1). These values are accessed by using the INDEX(array, row, column) function and specifying the array (the LINEST call) and the array position of interest.

$$\text{Least-squares slope:} \quad m = \text{INDEX(LINEST(y-range,x-range),1,1)} \tag{6.42}$$
$$= 20.1605 \text{ bar} \quad \text{(extra digits supplied)}$$

To obtain the error on the slope, we also have two choices of method when using Excel. First, we may use Equation 6.32 and Excel function calls for $s_{y,x}$ and $SS_{xx}$.

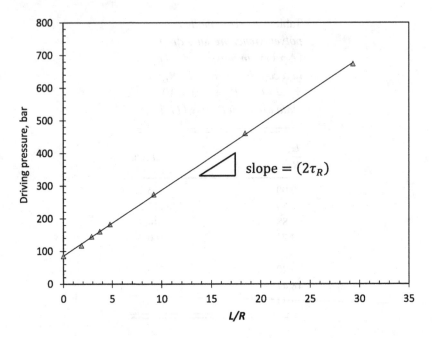

Figure 6.6 Driving pressure drops of a polyethylene flowing through a variety of capillaries, at a fixed value of the apparent shear rate, $\dot{\gamma}_a = \frac{4Q}{\pi R^3} = 9.0 \times 10^1 \text{ s}^{-1}$. Data are from Bagley [1].

$$
\begin{aligned}
s_m &= \frac{s_{y,x}}{\sqrt{SS_{xx}}} \\
&= \frac{\text{STEYX(y-range,x-range)}}{\sqrt{\text{DEVSQ(x-range)}}} \\
&= 0.12477 \text{ bar} \quad \text{(extra digits supplied)}
\end{aligned}
\tag{6.43}
$$

Alternatively, we can use LINEST, which returns the error on the slope in array position (2,1).

$$
\text{Excel} \quad s_m = \text{INDEX(LINEST(y-range,x-range,1,1),2,1)} \tag{6.44}
$$
$$
= 0.12477 \text{ bar} \quad \text{(extra digits supplied)}
$$

In Equation 6.44, to extend the LINEST calculation to provide error measures, we added two logical "TRUEs" or 1s in the LINEST call. The first 1 instructs the intercept to be calculated (using 0 instead would force the intercept to be zero); the second 1 instructs that the complete set of statistical results be calculated, including the standard deviations of slope and intercept. The standard deviation of slope is located in the (2,1) position of the LINEST array output; Appendix C contains a complete listing of the contents of the LINEST array when all error statistics are calculated. Some common LINEST calls are listed in Table 6.1.

A MATLAB solution to this example appears in Appendix D and is based on the matrix formulation of the least-squares problem. In Section 6.3.3, we present the matrix calculations of the errors $s_b = s_{b_0}$ and $s_m = s_{b_1}$. The variances $s_m^2$ and $s_b^2$ are the diagonal elements of the covariance matrix $\text{Cov}(b, m)$. As shown in the solution in Appendix D and in Table D.1, we need only a few lines of code to calculate $\text{Cov}(b, m)$ from $s_{y,x}$ and the previously discussed matrix $A$ (Equation 6.15). The results are the same as when calculated with Excel.

$$\text{MATLAB} \quad \boxed{s_m = 0.01244 \text{ bar}} \quad \text{(extra digits supplied)} \quad (6.45)$$

The 95% confidence interval error limits on slope are constructed as $\pm t_{0.025, n-2} s_m$ (Equation 6.37). For the dataset in this example, $n = 8$, and we calculate $t_{0.025, 6} = \text{T.INV.2T}(0.05, 6) = 2.4469$.

$$\begin{aligned} \text{slope, bar} &= m \pm t_{0.025, 6} s_m \\ &= 20.1605 \pm (2.4469)(0.12477) \\ &= 20.2 \pm 0.3 \end{aligned} \quad (6.46)$$

The final reported value of wall shear stress for $\dot{\gamma}_a = 90 \text{ s}^{-1}$, with error limits, is calculated as follows.

$$\tau_R = \frac{m}{2} = 10.08025 \text{ bar} \quad \text{(extra digits provided)} \quad (6.47)$$

From the error propagation worksheet and Equation 5.18:

$$\begin{aligned} e_{\tau_R}^2 &= \left(\frac{\partial \tau_R}{\partial m}\right)^2 s_m^2 \\ &= \left(\frac{1}{2}\right)^2 s_m^2 \end{aligned}$$

$$e_{\tau_R} = \frac{s_m}{2} = \frac{0.124477}{2} = 0.0624 \text{ bar} \quad (6.48)$$

$$\tau_R = 10.08025 \pm (2.4469)(0.0624)$$

$$\boxed{\tau_R = 10.08 \pm 0.15 \text{ bar}} \quad (6.49)$$

**Example 6.4: Uncertainty on intercept: Bagley plot for capillary entrance-pressure loss.** *Viscosity is an important physical property of polymers, and capillary rheometers are used to measure the high-shear-rate viscosities of molten polymers. To quantify capillary entry and exit effects in capillary rheometer measurements, E. B. Bagley [1] established a now-standard technique to measure the entrance effect in terms of* $\Delta p_{ent}$, *a quantity that is equal to the intercept of a plot of driving pressure versus capillary-length-to-radius ratio,* $L/R$.

$$\Delta p = (2\tau_R)\frac{L}{R} + \Delta p_{ent} \qquad (6.50)$$

*The entrance effect is a function of the rate of polymer deformation in the capillary, and therefore the driving-pressure versus* $L/R$ *plots must represent experiments conducted at the same value of scaled flow rate* $\dot{\gamma}_a = 4Q/\pi R^3$, *a quantity that for non-Newtonian fluids is called the apparent shear rate.*

*Data for a polyethyelene of driving pressure versus* $L/R$ *for five values of* $\dot{\gamma}_a$ *are provided in Table 6.3. Calculate* $\Delta p_{ent}$ *as a function of* $\dot{\gamma}_a$ *from these data. Provide 95% confidence interval error limits for the five values of* $\Delta p_{ent}(\dot{\gamma}_a)$ *determined.*

**Solution:** The first solution step is to visualize the data; they are plotted in Figure 6.7. As expected from Bagley's analysis, for each value of apparent

Table 6.3. *Driving pressure as a function of capillary-length-to-radius ratio* $(L/R)$ *for five values of apparent shear rate* $\dot{\gamma}_a = 4Q/\pi R^3$ *in a capillary rheometer. Q is the volumetric flow rate, and the fluid in the instrument is polyethylene. The data are digitized from Bagley's publication [1]; extra digits provided.*

|         | Driving pressure, bar | | | | |
|---------|---------------------------------|-----------|-----------|------------|------------|
| $L/R$   | $\dot{\gamma}_a = 40\ \text{s}^{-1}$ | $60\ \text{s}^{-1}$ | $90\ \text{s}^{-1}$ | $120\ \text{s}^{-1}$ | $250\ \text{s}^{-1}$ |
| 0.00    | 34.8   | 62.0   | 85.2   | 106.5  | 182.2  |
| 1.84    | 62.9   | 95.9   | 117.2  | 144.2  | 192.6  |
| 2.88    | 77.1   | 115.8  | 144.8  | 171.9  | 257.0  |
| 3.74    | 89.1   | 131.7  | 160.7  | 187.8  | 288.5  |
| 4.78    | 103.2  | 145.8  | 182.6  | 219.3  | 325.8  |
| 9.20    | 161.6  | 217.8  | 273.9  | 324.3  | 471.5  |
| 18.42   | 286.5  | 369.8  | 460.8  | 532.4  | 762.8  |
| 29.34   | 433.5  | 545.8  | 673.6  | 778.2  | 1119.0 |

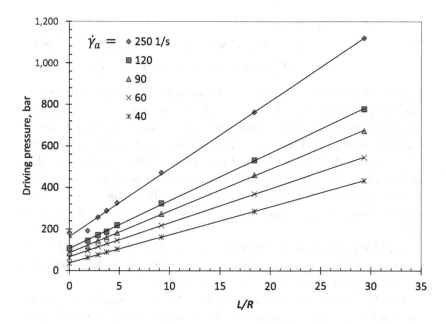

Figure 6.7 Driving pressure versus capillary-length-to-radius ratio, $L/R$, for a polyethylene flowing in a capillary rheometer. Data are from Bagley's original 1957 paper [1]

shear rate $\dot{\gamma}_a$ the data fall along a straight line. Note that one of this example's datasets was used in the previous example to determine a value of wall shear stress, $\tau_R$.

$$\Delta p = (2\tau_R)\left(\frac{L}{R}\right) + \Delta p_{ent} \tag{6.51}$$

$$\text{slope} = 2\tau_R \tag{6.52}$$

$$\tau_R = \frac{\text{slope}}{2} \tag{6.53}$$

To calculate the intercepts for each of the best-fit lines, we use Excel's statistical functions. As with the previous example, we have a choice as to which Excel function calls to use. The INTERCEPT() function is sufficient to calculate the intercepts for the five lines, and we may calculate the standard deviation of the intercept, $s_b$, from Equation 6.33 and the functions AVERAGE(), DEVSQ(), and STEYX().

Our other choice is to use LINEST, which is programmed to calculate both the intercept and its standard deviation directly. For the data supplied, these quantities are obtained from

$$\Delta p_{ent}(\dot{\gamma}_a) = \text{intercept}$$

$$= \text{INDEX(LINEST(y-range,x-range),1,2)} \qquad (6.54)$$

$$s_b = \text{INDEX(LINEST(y-range,x-range,1,1),2,2)} \qquad (6.55)$$

To calculate the intercept (Equation 6.54), we do not need either of the optional LINEST parameters (the default is that the intercept will be calculated; compare with Equation 6.42 for slope) and the result for $b$ is stored in matrix location (1,2). For the standard deviation of the intercept (Equation 6.55), we need both of the optional parameters: "TRUE" or 1 to calculate the intercept and "TRUE" or 1 to calculate the statistical parameters. The quantity $s_b$ is stored in matrix location (2,2). The values of $\Delta p_{ent}(\dot{\gamma}_a)$ and the associated $s_b$ values for Bagley's polyethylene data are given in Table 6.4.

The MATLAB calculations for this example are in Appendix D and follow the same pattern as those in Example 6.3. Extending the calculation from one $(x, y)$ dataset to five is accomplished with a "for loop."

The 95% confidence interval error bars for the $\Delta p_{ent}(\dot{\gamma}_a)$ values are calculated as $\pm t_{0.025,n-2}s_b$ (Equation 6.38), where the degrees of freedom $(n - 2) = 8 - 2 = 6$. The values of $t_{0.025,6}s_b$ for each $\dot{\gamma}_a$ are included in Table 6.4, and the error bars appear on the data points in Figure 6.8. Note that the size of the error bars varies with the apparent shear rate. The plot emphasizes that each $\Delta p$ versus $L/R$ dataset has a different net squared-deviation from its best-fit line (each line in Figure 6.7 has a different amount of scatter). The quality of the fit determines the reliability of the calculated intercept $\Delta p_{ent}$; this is reflected in the error bars on $\Delta p_{ent}$ in Figure 6.8.

Table 6.4. *The results for entrance pressure losses $\Delta p_{ent}$ (extra digits provided) as a function of apparent shear rate $\dot{\gamma}_a$ are given along with their 95% CI error limits [$t_{0.025,6} = T.INV.2T(0.05,6) = 2.4469$]. The material is a polyethylene [1].*

| $\dot{\gamma}_a$ (s$^{-1}$) | $\Delta p_{ent}$ (bar) | $s_b$ (bar) | $t_{0.025,6}s_b$ (bar) |
|---|---|---|---|
| 40  | 37.54  | 0.64 | 1.6  |
| 60  | 66.87  | 1.33 | 3.2  |
| 90  | 85.44  | 1.61 | 3.9  |
| 120 | 106.13 | 2.02 | 4.9  |
| 250 | 164.07 | 7.42 | 18.2 |

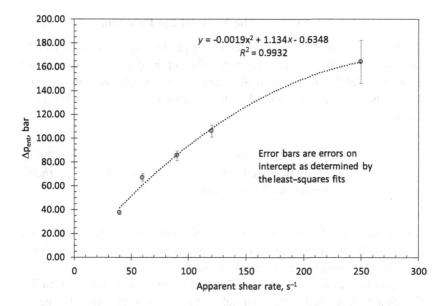

Figure 6.8 By finding best-fit lines to the rheological data of capillary $\Delta p$ versus $L/R$ (Figure 6.7), we are able to use the Bagley method to determine the entrance pressure losses for a polyethylene. In addition, the uncertainty in the fit allows us to calculate the uncertainty on the intercepts, giving us a better idea of the quality of the resulting values of $\Delta p_{ent}$ (error bars shown). In this graph, a second-order polynomial was fit to the $\Delta p_{ent}(\dot{\gamma}_a)$ data using Excel Trendline. We discuss second-order polynomial fits in Section 6.3 and the fitting of these data (in both Excel and MATLAB, including uncertainty in the polynomial fit) in Examples 6.10 and 6.11.

The previous examples show the ease with which uncertainties in least-squares slope and intercept may be determined with the help of modern tools like Excel and MATLAB. If we are interested in quantities that are calculated from the slope or intercept, we can readily carry out the calculations and use $s_m$ and $s_b$ in error propagations to determine the appropriate uncertainties in the quantities calculated.

It is not unusual to use *both* the least-squares slope and intercept in the same calculation (see Example 5.9, addressing the creation of a calibration curve). Since the least-squares slope and intercept are obtained from the same dataset $(x_i, y_i)$, they are *not independent variables*, however. Any error propagation using both $m$ and $b$, including error propagation for the values of $\widehat{y}$ from $\widehat{y} = mx + b$, must take into account the *covariance* between these two parameters. We confront this issue in the next section.

### 6.2.4 Error on the Value of the Function

Often the purpose of fitting a model $y = \phi(x)$ to a dataset is to be able to calculate a value of $y$ for arbitrary new values of the dependent variable. For a linear model that has been fit to a dataset, for example, we can calculate the $y$-value $\widehat{y}_0$ that is associated with any arbitrarily chosen $x$-value $x_0$ as follows:

$$\begin{array}{c}\text{Evaluate a value}\\ \text{of the function } y \qquad \widehat{y}_0 = mx_0 + b \qquad\qquad (6.56)\\ \text{obtained from least squares}\end{array}$$

The calculation in Equation 6.56 is a combination of stochastic variables $x_0$, $m$, and $b$. Therefore we turn to error propagation and the error propagation worksheet to determine $s_{\widehat{y}_0}$, the error on $\widehat{y}_0$.

Figure 6.9 shows the start of filling out the worksheet for error on $\widehat{y}_0$. Equation 6.56 is used for the formula. The variables used to calculate $\widehat{y}_0$ are $x_0$, $m$, and $b$. As discussed previously, in ordinary least squares the $x$ values are assumed to be known exactly, and thus the error on $x_0$ is assumed

| $\phi(x_1, x_2, x_3, x_4, x_5)$: | | Formula for $\phi$:[*]<br>$\widehat{y}_0 = mx_0 + b$<br>*Note: units must work as written. | | Representative value of $\phi$:<br>(include units)<br>$\widehat{y}_0 =$ | 95% C.I. of $\phi$: $(\phi \pm 2e_{s,\phi})$<br>(include units)<br>$\widehat{y}_0 \pm (t_{0.025, n-2})(s_{\widehat{y}_0})$ |
|---|---|---|---|---|---|
| **Measured quantities, $x_i$** | | | $\dfrac{\partial \phi}{\partial x_i}$ | $e_{s,x_i}$ | $\left(\dfrac{\partial \phi}{\partial x_i}\right)^2 e_{s,x_i}^2$ |
| $x_i$ | Symbol | Representative value | | | |
| $x_1$ | $m$ | units | $\dfrac{\partial \widehat{y}_0}{\partial m} = x_0$ | $s_m$ | $(x_0 s_m)^2$ |
| $x_2$ | $x_0$ | units | $\dfrac{\partial \widehat{y}_0}{\partial x_0} = m$ | $0$ | $0$ |
| $x_3$ | $b$ | units | $\dfrac{\partial \widehat{y}_0}{\partial b} = 1$ | $s_b$ | $s_b^2$ |
| $x_4$ | | units | | | |
| $x_5$ | | units | | $+2\left(\dfrac{\partial \widehat{y}_0}{\partial m}\right)\left(\dfrac{\partial \widehat{y}_0}{\partial b}\right)\text{Cov}(m,b)$ | |

$$e_{s_\phi}^2 = \left(\frac{\partial \phi}{\partial x_1}\right)^2 e_{x_1}^2 + \left(\frac{\partial \phi}{\partial x_2}\right)^2 e_{x_2}^2 + \left(\frac{\partial \phi}{\partial x_3}\right)^2 e_{x_3}^2 + \left(\frac{\partial \phi}{\partial x_4}\right)^2 e_{x_4}^2 + \left(\frac{\partial \phi}{\partial x_5}\right)^2 e_{x_5}^2$$

*All variables $x_i$ must be independent for this equation to hold.*

$e_{s,\phi}^2 = s_{\widehat{y}_0}^2$

$e_{s,\phi} = s_{\widehat{y}_0}$   units

Figure 6.9 To calculate the error limits on $\widehat{y}_0$ we again turn to error propagation, employing the error propagation worksheet, but this is not correct for $\widehat{y}_0$. When the propagated variables are not independent, the covariance must be included in the error-propagation calculation.

to be zero.[6] The errors on ordinary least squares $m$ and $b$ were calculated in the previous section ($s_m$ in Equation 6.32 and $s_b$ in Equation 6.33). The rest of the worksheet guides forward the calculation. Unfortunately, the unmodified worksheet *would not be correct* for the calculation of the error on $\hat{y}_0$. The reason for this is that the formula used in the error propagation worksheet is the common, special-case version of error propagation that assumes that all the variables are independent (Equation 5.20). In the calculation of $s_{\hat{y}_0}$, however, the two variables $m$ and $b$ are *not* independent, since they were calculated from the same dataset ($x_i, y_i$). We must proceed with caution.

> Least-squares slope and intercept are *not* independent variables.
> They are calculated from the same dataset, ($x_i, y_i$).

When variables in an error-propagation calculation are not independent, we are required to use the original, more complicated error-propagation equation – that is, Equation 5.16. For the case of a function of three variables $y = \phi(x_1, x_2, x_3)$, the general error-propagation equation for the variance of function $\phi$ is [38]:

$$\sigma_\phi^2 = \left(\frac{\partial \phi}{\partial x_1}\right)^2 \sigma_{x_1}^2 + \left(\frac{\partial \phi}{\partial x_2}\right)^2 \sigma_{x_2}^2 + \left(\frac{\partial \phi}{\partial x_3}\right)^2 \sigma_{x_3}^2 + 2\left(\frac{\partial \phi}{\partial x_1}\right)\left(\frac{\partial \phi}{\partial x_2}\right) \mathrm{Cov}(x_1, x_2)$$

$$+ 2\left(\frac{\partial \phi}{\partial x_1}\right)\left(\frac{\partial \phi}{\partial x_3}\right) \mathrm{Cov}(x_1, x_3) + 2\left(\frac{\partial \phi}{\partial x_2}\right)\left(\frac{\partial \phi}{\partial x_3}\right) \mathrm{Cov}(x_2, x_3) \quad (6.57)$$

where $\sigma_{x_1}^2$, $\sigma_{x_2}^2$, and $\sigma_{x_3}^2$ are the variances of stochastic variables $x_1$, $x_2$, and $x_3$, respectively. The first three terms in Equation 6.57 are the familiar error-propagation terms: they show how the squared errors in the variables are weighted in error propagation by the squares of the partial derivatives of the function with respect to the individual variables. The remaining three terms involve the covariances among the variables. *Covariance* accounts for the interdependency of two variables – that is, that changes in one variable have an effect on another variable. Covariance is defined as (discussed in Chapter 5, Equation 5.7; [38]):

Covariance of a
pair of continuous     $$\mathrm{Cov}(x, z) \equiv \int_{-\infty}^{\infty} (z' - \bar{z})(x' - \bar{x}) f(x', z') dx' dz'$$
stochastic variables

$$(6.58)$$

---

[6] When calibration data are fit to a model, the independent variable is the value given by the standard, which is very accurate by design. If less accurate values of $x_0$ are used, then $e_{s,x_0}$ may not be zero.

where $f(x, z)$ is the joint probability density function for the variables $x$ and $z$ [5, 38]. If two quantities are independent, their covariance is zero.

For the calculation of $\sigma_\phi^2 = s_{\widehat{y}_0}^2$ for $\widehat{y}_0 = mx_0 + b$, Equation 6.57 becomes

$$s_{\widehat{y}_0}^2 = \left(\frac{\partial \widehat{y}_0}{\partial m}\right)^2 \sigma_m^2 + \left(\frac{\partial \widehat{y}_0}{\partial x_0}\right)^2 \sigma_{x_0}^2 + \left(\frac{\partial \widehat{y}_0}{\partial b}\right)^2 \sigma_b^2 + 2\left(\frac{\partial \widehat{y}_0}{\partial m}\right)\left(\frac{\partial \widehat{y}_0}{\partial x_0}\right)\text{Cov}(m, x_0)$$
$$+2\left(\frac{\partial \widehat{y}_0}{\partial m}\right)\left(\frac{\partial \widehat{y}_0}{\partial b}\right)\text{Cov}(m, b) + 2\left(\frac{\partial \widehat{y}_0}{\partial x_0}\right)\left(\frac{\partial \widehat{y}_0}{\partial b}\right)\text{Cov}(x_0, b) \quad (6.59)$$

The two covariance terms involving $x_0$ are zero since the choice of $x_0$ is arbitrary, and therefore its value is independent of both $m$ or $b$. The covariance of the pair $m, b$, however, is not zero – the slope and intercept are calculated from the same dataset $(x_i, y_i)$, and thus $m$ and $b$ are not independent. We need $\text{Cov}(m, b)$ to complete our calculation of $s_{\widehat{y}_0}^2$ in Equation 6.59.

Calculating $\text{Cov}(m, b)$ from its definition in Equation 6.58 is difficult since we need the joint probability density function $f(m, b)$. An alternative strategy [55] is to use an identity that results from the linearity of the definition of covariance ($a$, $b$, $c$, and $d$ are constants and $U$ and $W$ are variables; [38]):

$$\text{Cov}((aU + bW), (cU + dW)) = ac\text{Cov}(U, U) + ad\text{Cov}(U, W)$$
$$+bc\text{Cov}(W, U) + bd\text{Cov}(W, W) \quad (6.60)$$

Writing $\text{Cov}(m, b)$ with $m$ and $b$ written as linear combinations of the $y_i$ (using Equations 6.6 and 6.5; see Appendix E), we arrive at a very simple result for the covariance that we need. Taking this approach, the covariance of slope and intercept may be shown to be equal to (Section E.4.2)

Covariance of ordinary least-squares slope and intercept
$$\text{Cov}(m, b) = -\left(\frac{s_{y,x}^2 \bar{x}}{SS_{xx}}\right) \quad (6.61)$$

The formulas for $\bar{x}$, $SS_{xx}$, and $s_{y,x}^2$ are given in Equations 6.7, 6.9, and 6.25.

When the error-propagation Equation 6.59 is evaluated for $\widehat{y}_0 = mx_0 + b$, including the nonzero covariance term involving the pair $m$ and $b$ (Equation 6.61), we obtain the correct expression for the error in $\widehat{y}_0$:

Variance in the model-predicted value $\widehat{y}_0$ evaluated at $x_0$, Var($\widehat{y}_0$)
$$s_{\widehat{y}_0}^2 = s_{y,x}^2\left(\frac{1}{n} + \frac{(x_0 - \bar{x})^2}{SS_{xx}}\right) \quad (6.62)$$

With some extra effort we can use Excel to do an error propagation when the variables are not independent. To do this we determine and add the covariance terms to the error-propagation spreadsheet; one way of doing this is shown in

| Error propagation worksheet | | | | | | | | |
|---|---|---|---|---|---|---|---|---|
| $\phi(x_1, x_2, x_3) =$ | | | $\phi =$ | | units | $te_{s\phi} =$ | | units |
| | $x_i$ | value | units | $d\phi/dx_i$ | $(d\phi/dx_i)^2$ | $e_{xi}$ | $e_{xi}^2$ | $(d\phi/dx_i)^2 e_{xi}^2$ units |
| $x_1$ | | | | | | | | |
| $x_2$ | | | | | | | | |
| $x_3$ | | | | | | | | |
| | | | | | | | $2\text{Cov}(x_i, x_j)(d\phi/dx_i)(d\phi/dx_j)$: | |
| | | | | | $\text{Cov}(x_1, x_2) =$ | | | |
| | | | | | $\text{Cov}(x_1, x_3) =$ | | | |
| | | | | | $\text{Cov}(x_2, x_3) =$ | | | |
| | | | | | | | $e_{s,\phi}^2$ | |
| | | | | | | | $e_{s,\phi}$ | units |

Figure 6.10 The error propagation worksheet is designed for independent variables, but when implemented in software we can modify it to include covariance terms as shown here in Excel. See Figure 6.14 for an example using this calculation design.

Figure 6.10. The two paths to the correct error propagation for the least-squares model $\widehat{y}_0 = mx_0 + b$ are equivalent: using Equation 6.62 or modifying an error propagation worksheet. Using software to create a modified error propagation worksheet (Figure 6.10, for example) allows us to see the relative magnitude of all the terms in the error propagation, including the influence of the covariance terms (see Example 6.6).

The 95% confidence interval for $\widehat{y}_0$, as for slope and intercept, is $\pm t_{0.025, n-2}$ times its standard deviation.

$$\begin{array}{c} \text{95\% CI for} \\ \text{value of the function, } \widehat{y}_{0,true} \end{array} \qquad \boxed{\widehat{y}_{0,true} = (mx_0 + b) \pm t_{0.025, n-2} s_{\widehat{y}_0}} \qquad (6.63)$$

Note that $n$ is the number of data points in the dataset used in the least-squares fit for $m$ and $b$.

Looking at Equation 6.62, we can see how the 95% confidence interval error limits vary with $x_0$. The standard deviations $s_{\widehat{y}_0}$ (square root of Equation 6.62), and hence the error limits on the predictions of the model, are smallest at $x_0 = \bar{x}$, that is, for values of $x_0$ at the center of the $x$-range of the fitting dataset. The variance and error limits on the model grow modestly for $x_0$ values away from $\bar{x}$ in both directions (higher and lower than $\bar{x}$). The variance $s_{y,x}^2$ is independent of $x$ by assumption (homoscedasticity of the measured dataset). The variances $s_{\widehat{y_0}}^2$ associated with predictions using the model, however, depend on $x$.

The complete set of values of error limits $\widehat{y}_{0,true} = (mx_0 + b) \pm t_{0.025,n-2}s_{\widehat{y_0}}$ for *all* possible values of $x_0$ produces two curves that bound a region within which we expect to find the true values of the function $y = \phi(x)$. This 95% CI region is shown for one of the sets of capillary flow data from Example 6.4 (Figure 6.11). This region, slightly wider at the outer limits and narrower near the middle [near $(\bar{x})$], expresses the 95% confidence limits in the model fit over the range of $x$ used to predict the fit. A related interval shown in Figure 6.11, the 95% prediction interval of the model fit, is discussed in the next section. In Section 6.2.6 we discuss another presentation of these regions, this time for a heat-exchanger dataset (see Figure 6.12). Also included in that section are several examples that employ the methods discussed in this chapter.

### 6.2.5 Prediction Interval on the Next Measured Value

In Chapter 2, we discussed single-variable replicate sampling (using an example concerning fluid density $\rho$, Example 2.5) and introduced the prediction interval. A prediction interval indicates the range within which we expect to find the next data measurement $\rho_{n+1}$ obtained for a dataset under consideration (for fitting dataset $\rho_i$, $i = 1, 2, \ldots, n$, $n$ replicates, sample standard deviation $s$). For a two-dimensional fitting dataset $(x_i, y_i)$, $i = 1, 2, \ldots, n$, we may also be interested in assessing the uncertainty in a next value of $(x_0, y_0)$ at a chosen $x_0$. The error on the next value of $y_0$ is normally and independently distributed. Thus the following scaled deviation is $t$-distributed with $(n - 2)$ degrees of freedom [38]:

$$\text{Scaled deviation of} \quad \frac{(y_0 - \widehat{y}_0)}{s_{y_{0,n+1}}} = t \qquad (6.64)$$
$$\text{next value of } y_0$$

where $s_{y_{0,n+1}}$, which we derive in this section, is the appropriate standard error for the next value of $y_0$. We can construct a 95% prediction interval of the next $y_0$ based on $\pm t_{0.025,n-2}s_{y_{0,n+1}}$.

Prediction interval
of the next measured value
of $y_0$ at $x_0$
(ordinary least-squares fit)

$$y_{0,n+1,true} = (mx_0 + b) \pm t_{0.025,n-2}s_{y_{0,n+1}}$$

$$(6.65)$$

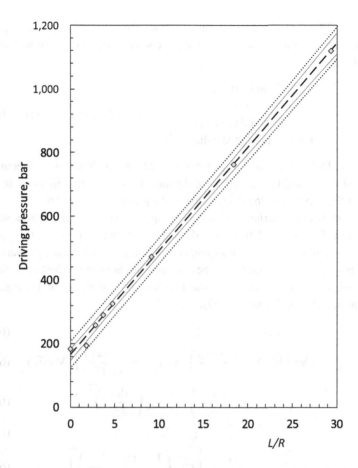

Figure 6.11 Driving pressure versus capillary $L/R$ for a polyethylene flowing in a capillary rheometer [1]; $\dot{\gamma}_a = 4Q/\pi R^3 = 250$ s$^{-1}$. The model fit to the data produces the model line, bracketed by the 95% CI lines, which indicate the confidence region of the model predictions (inner solid lines), and the 95% PI lines, which indicate with 95% confidence where the next values of the data points will lie (outer dotted lines).

The value of $s_{y_{0,n+1}}$ is different for different values of $x_0$. We can construct a prediction-interval region for the entire dataset.

Consider a function $y = \phi(x)$ for which we have a dataset of $n$ data pairs $(x_i, y_i)$. We now intend to take an additional, $(n+1)$th data point for an $x$ value equal to $x_0$. The value predicted by the model for $\widehat{y_0}$ at $x_0$ is

$$\widehat{y_0} \equiv \phi(x_0) = mx_0 + b \qquad (6.66)$$

*which itself is uncertain.* We can construct the error between $y_{0,n+1}$, the observed next value of the function at $x_0$, and the predicted $\widehat{y}_0$ value. We call this $\mathrm{dev}_0$.

$$
\begin{array}{c}
\text{Prediction error at } x_0: \\
\text{deviation between} \\
\text{observed next } y_0 \\
\text{and model-predicted value of } \widehat{y}_0
\end{array}
\qquad \mathrm{dev}_0 = (y_0 - \widehat{y}_0) \quad (6.67)
$$

The standard deviation $s_{y_{0,n+1}}$ we seek, which combines the inherent $y$-variability $s_{y,x}$ and the variability of the model-predicted line $\widehat{y}_0$, is the square root of the variance of the distribution of the prediction error, $\mathrm{dev}_0$.

To calculate the variance of $\mathrm{dev}_0$ we apply the error-propagation equation (Equation 5.16) to the definition of $\mathrm{dev}_0$. The variance of $y_0$ is $s_{y,x}^2$ ($s_{y,x}^2$ is the variance of all values of $y$, by assumption of homoscedacity), and the variance of $\widehat{y}_0$ is given by Equation 6.62. The two variables are independent (covariance is equal to zero). We combine the two variances using error propagation (Equation 5.16 and Equation 6.67).

$$
\mathrm{dev}_0 = (y_0 - \widehat{y}_0) \tag{6.68}
$$

$$
s_{y_{0,n+1}}^2 = \mathrm{Var}(\mathrm{dev}_0) = \left(\frac{\partial \mathrm{dev}_0}{\partial y_0}\right)^2 \mathrm{Var}(y_0) + \left(\frac{\partial \mathrm{dev}_0}{\partial \widehat{y}_0}\right)^2 \mathrm{Var}(\widehat{y}_0) \tag{6.69}
$$

$$
= \left(\frac{\partial \mathrm{dev}_0}{\partial y_0}\right)^2 s_{y,x}^2 + \left(\frac{\partial \mathrm{dev}_0}{\partial \widehat{y}_0}\right)^2 s_{\widehat{y}_0}^2 \tag{6.70}
$$

$$
= s_{y,x}^2 + s_{\widehat{y}_0}^2 \tag{6.71}
$$

$$
= s_{y,x}^2 + \left[ s_{y,x}^2 \left( \frac{1}{n} + \frac{(x_0 - \bar{x})^2}{SS_{xx}} \right) \right] \tag{6.72}
$$

Variance in the next value of $y_0$ at $x_0$ $(y_{0,n+1})$ (ordinary least-squares fit)

$$
s_{y_{0,n+1}}^2 = s_{y,x}^2 \left( 1 + \frac{1}{n} + \frac{(x_0 - \bar{x})^2}{SS_{xx}} \right) \tag{6.73}
$$

With this result we can calculate the 95% prediction interval of the next value of $y_0$ (Equation 6.65) for all possible values of $x_0$, producing a region. The prediction limits indicate where a next data point would be expected to lie, while the tighter confidence limits of the function (Equation 6.62) indicate the region where we expect the model-predicted values to lie. Both of these regions are shown in Figure 6.11 for rheological flow data and are calculated for data on overall heat transfer coefficient in Example 6.5.

### 6.2.6 Working with Uncertainty in Models

To see the least-squares tools of this chapter in action, we now present four examples, including the resolutions of several examples we have pursued previously. We first turn to an example involving heat-transfer data, in which we determine the confidence-interval region of a model fit and the prediction interval region of the next value of the stochastic variable.

**Example 6.5: Uncertainty in measurement of overall heat transfer coefficient of double-pipe heat exchangers.** *Measurements of the overall heat-transfer coefficient U as a function of steam mass flow rate for 10 similar double-pipe heat exchangers were aggregated and plotted and found to follow a linear model (Table 6.5, Figure 6.12 [40]). What is the equation for the best-fit model for the data? What is the uncertainty associated with the fit? What are the uncertainties on future measurements of heat-transfer coefficient for these heat exchangers?*

**Solution:** There are 86 measurements provided for the overall heat-transfer coefficient $U$ $(\text{kW/m}^2 \ K)$ for various values of steam mass flow rate $\dot{M}$ between about 1.0 g/s and 6.0 g/s. Although there is considerable scatter in the dataset, the data display a linear trend when plotted (Figure 6.12; $R^2 = 0.9073$). To calculate the uncertainty associated with the fit, we need the 95% confidence intervals (CI) on the model (Section 6.2.4, Equation 6.62); to determine the uncertainty on a future measurement of heat-transfer coefficient, we need the 95% prediction interval (PI) (Section 6.2.5, Equation 6.73).

Each of these intervals is calculated with respect to the model. In Excel we use LINEST or SLOPE() and INTERCEPT() to determine the equation of the best-fit line. We subsequently calculate the confidence and prediction intervals from the dataset $(x_i, y_i)$ and Equations 6.63 and 6.65, which are repeated here.

95% CI of model fit: $\quad \widehat{y}_{0,true} = (mx_0 + b) \pm t_{0.025,n-2} s_{\widehat{y}_0} \quad$ (Equation 6.63)

95% PI of next value: $\quad y_{0,n+1,true} = (mx_0 + b) \pm t_{0.025,n-2} s_{y_{0,n+1}}$

$$\text{(Equation 6.65)}$$

where

$$s_{\widehat{y}_0}^2 = s_{y,x}^2 \left( \frac{1}{n} + \frac{(x_0 - \bar{x})^2}{SS_{xx}} \right) \qquad \text{(Equation 6.62)}$$

$$s_{y_{0,n+1}}^2 = s_{y,x}^2 \left( 1 + \frac{1}{n} + \frac{(x_0 - \bar{x})^2}{SS_{xx}} \right) \qquad \text{(Equation 6.73)}$$

Table 6.5. *Experimental data for overall heat-transfer coefficient U versus steam mass flow rate Ṁ for 10 double-pipe heat exchangers [40]. Extra digits are provided.*

| Index | Ṁ (g/s) | U (kW/m²K) | Index | Ṁ (g/s) | U (kW/m²K) | Index | Ṁ (g/s) | U (kW/m²K) |
|---|---|---|---|---|---|---|---|---|
| 1  | 1.30 | 0.42 | 30 | 2.27 | 1.89 | 59 | 4.72 | 3.98 |
| 2  | 1.52 | 1.39 | 31 | 3.54 | 3.24 | 60 | 3.91 | 3.69 |
| 3  | 1.30 | 0.42 | 32 | 3.32 | 3.11 | 61 | 4.76 | 4.15 |
| 4  | 1.80 | 1.12 | 33 | 3.81 | 3.56 | 62 | 3.91 | 3.69 |
| 5  | 1.80 | 1.37 | 34 | 3.54 | 3.24 | 63 | 4.49 | 3.84 |
| 6  | 1.84 | 1.14 | 35 | 1.76 | 1.67 | 64 | 4.00 | 3.38 |
| 7  | 4.25 | 3.24 | 36 | 2.14 | 2.13 | 65 | 2.87 | 2.41 |
| 8  | 4.69 | 3.65 | 37 | 1.47 | 1.37 | 66 | 2.66 | 2.45 |
| 9  | 2.80 | 2.04 | 38 | 2.28 | 2.29 | 67 | 2.76 | 2.54 |
| 10 | 3.83 | 2.82 | 39 | 3.09 | 2.94 | 68 | 2.40 | 2.26 |
| 11 | 2.48 | 1.92 | 40 | 2.06 | 1.95 | 69 | 2.91 | 2.41 |
| 12 | 2.80 | 2.27 | 41 | 4.44 | 3.66 | 70 | 2.66 | 2.45 |
| 13 | 3.05 | 2.08 | 42 | 4.25 | 3.66 | 71 | 2.60 | 2.23 |
| 14 | 2.10 | 1.60 | 43 | 3.43 | 2.83 | 72 | 3.30 | 2.83 |
| 15 | 1.81 | 1.65 | 44 | 4.44 | 3.66 | 73 | 4.05 | 3.31 |
| 16 | 2.81 | 2.98 | 45 | 2.23 | 1.95 | 74 | 4.88 | 4.45 |
| 17 | 2.40 | 2.00 | 46 | 4.25 | 3.66 | 75 | 5.25 | 4.64 |
| 18 | 2.70 | 2.54 | 47 | 3.45 | 2.98 | 76 | 5.67 | 5.46 |
| 19 | 3.40 | 2.90 | 48 | 2.79 | 2.98 | 77 | 1.95 | 1.40 |
| 20 | 2.80 | 2.98 | 49 | 3.54 | 2.86 | 78 | 3.00 | 2.61 |
| 21 | 3.70 | 3.16 | 50 | 3.45 | 2.98 | 79 | 3.17 | 2.67 |
| 22 | 4.10 | 3.72 | 51 | 2.23 | 1.95 | 80 | 3.65 | 3.06 |
| 23 | 4.50 | 4.02 | 52 | 2.79 | 2.98 | 81 | 2.04 | 1.43 |
| 24 | 3.62 | 3.02 | 53 | 2.84 | 3.31 | 82 | 3.56 | 2.37 |
| 25 | 3.51 | 3.24 | 54 | 3.86 | 3.51 | 83 | 1.20 | 1.02 |
| 26 | 4.07 | 3.54 | 55 | 2.83 | 2.51 | 84 | 2.05 | 1.71 |
| 27 | 4.50 | 4.02 | 56 | 4.62 | 4.03 | 85 | 2.09 | 1.67 |
| 28 | 3.51 | 3.24 | 57 | 2.83 | 2.51 | 86 | 2.34 | 1.80 |
| 29 | 3.81 | 3.56 | 58 | 3.82 | 4.07 |    |      |      |

Note that there is no need to calculate the quantities $s_m$ or $s_b$ to complete this example, as we have propagated the errors on the parameters $m$ and $b$, along with the appropriate covariance, to obtain Equations 6.62 and 6.73.

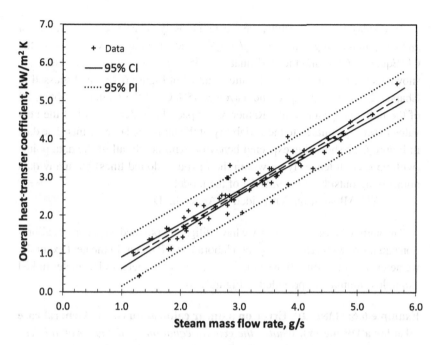

Figure 6.12 Experimental data for overall heat-transfer coefficient $U$ versus steam mass flow rate for double-pipe heat exchangers [40]. Shown are the linear model fit (dashed line), the 95% confidence interval of the model (inner, solid lines), and the 95% prediction interval of the next value of $U$ (outer, dotted lines).

Software function calls (collected in Tables 6.1 and D.1) make the calculations straightforward, as outlined here (extra digits shown).

$$\widehat{y}_0 = mx_0 + b$$

$$\text{x-range} = \dot{M}, \text{g/s}$$

$$\text{y-range} = U, \text{kW/m}^2 K$$

$$n = 86$$

$$m = \text{SLOPE(y-range,x-range)} = 0.917502 \ (\text{kW/m}^2\text{K})/(\text{g/s})$$

$$b = \text{INTERCEPT(y-range,x-range)} = -0.153737 \ \text{kW/m}^2\text{K}$$

$$t_{0.025,84} = \text{T.INV.2T(0.05,84)} = 1.98861$$

$$s_{y,x} = \text{STEYX(y-range,x-range)} = 0.2969174 \ \text{kW/m}^2\text{K}$$

$$SS_{xx} = \text{DEVSQ(x-range)} = 85.833033 \ (\text{g/s})^2$$

$$\bar{x} = \text{AVERAGE(x-range)} = 3.151395 \ \text{g/s}$$

$$\text{95\% CI} = \pm t_{0.025,84}s_{\widehat{y}_0}, \text{ for all } x_0$$

$$\text{95\% PI} = \pm t_{0.025,84}s_{y_0,n+1}, \text{ for all } x_0$$

To display the final results, we create a closely spaced vector of values of $x_0$ that span the range $1.0 \leq (\dot{M}, \text{g/s}) \leq 6.0$ and calculate the variances for the CI (Equation 6.63) and the PI (Equation 6.65) for all $x_0$ values. The results are shown as the CI (inner) and PI (outer) curves in Figure 6.12. As discussed in Chapter 2, we do not expect the narrower 95% CI of the model to capture all of the individual data points. Rather, we expect the wider 95% PI of the next value to set the limits as to the variability of the individual data points. The data in Figure 6.12 show this expected behavior, with nearly all of the data points (with 95% confidence) lying within the PI (outer dotted lines) but many data points lying outside of the 95% CI of the model.

A MATLAB solution is provided in Appendix D.

Throughout several chapters we have diligently pursued an accurate calibration equation (with error limits) for a laboratory DP meter. In the next example, we determine the calibration error, $e_{s,cal}$, for this DP meter (a DP-meter linked examples summary is provided in Figure 1.3).

**Example 6.6: \*LINKED\* Error on a linear calibration curve: General case (also for a DP meter).** *What is the general equation for the error on a linear calibration curve? Also, the calibration curve for a DP meter was discussed in Example 6.1 (Equation 6.23) and throughout the text (Figure 1.3). What is the standard calibration error $e_{s,cal}$ associated with the installed performance of this DP meter?*

**Solution:** We need to determine $e_{s,cal}$ from calibration data that exhibit a linear relationship. This is a common situation. First we derive the general equation for the error on a linear calibration curve [19]. Subsequently, we apply these results to the DP meter we have been discussing.

The calibration process consists of selecting physical values $x_i$ associated with calibration standards and collecting the corresponding signals $y_i$ from the device being calibrated. These data pairs form the calibration dataset $(x_i, y_i)$. A straight line is fit to the calibration dataset using ordinary least-squares regression.

$$\text{Least-squares model fit to calibration data} \qquad y = mx + b \qquad (6.74)$$

We have associated the values of the standard with the $x$-axis of the fit since the least-squares algorithm assumes that the values on the $x$-axis are certain.

To convert future device readings $y = y_0$ to physical values $x = \widehat{x_0}$, we rearrange the calibration data model-fit equation (Equation 6.74) to obtain the calibration curve, which is explicit in the new, predicted value $\widehat{x_0}$.

| $\phi(x_1, x_2, x_3, x_4, x_5)$: | **Formula for $\phi$:**$\hat{x}_0 = \left(\dfrac{1}{m}\right) y_0 - \left(\dfrac{b}{m}\right)$*Note: units must work as written.* | | **Representative value of $\phi$:** (include units) | **95% C.I. of $\phi$: $(\phi \pm 2e_{s,\phi})$** (include units) |
|---|---|---|---|---|
| **Measured quantities, $x_i$** | | $\dfrac{\partial \phi}{\partial x_i}$ | $e_{s,x_i}$ | $\left(\dfrac{\partial \phi}{\partial x_i}\right)^2 e_{s,x_i}^2$ |
| $x_i$ / Symbol | Representative value | | | |
| $x_1$ / $m$ | units | $\dfrac{\partial \hat{x}_0}{\partial m} = (y_0 - b)\left(-\dfrac{1}{m^2}\right)$ | $s_m = \sqrt{s_{y,x}^2/SS_{xx}}$ | Carry out algebra and sum with covariance term |
| $x_2$ / $y_0$ | units | $\dfrac{\partial \hat{x}_0}{\partial y_0} = \left(\dfrac{1}{m}\right)$ | $s_{y,x}$ | |
| $x_3$ / $b$ | units | $\dfrac{\partial \hat{x}_0}{\partial b} = \left(-\dfrac{1}{m}\right)$ | $s_b = s_{y,x}\sqrt{\dfrac{1}{n} + \dfrac{\bar{x}^2}{SS_{xx}}}$ | |
| $x_4$ | units | $Cov(m, b) = -\left(\dfrac{s_{y,x}^2 \bar{x}}{SS_{xx}}\right)$ | | |
| $x_5$ | units | | | |

$$e_{s,\phi}^2 = \left(\frac{\partial \phi}{\partial x_1}\right)^2 e_{x_1}^2 + \left(\frac{\partial \phi}{\partial x_2}\right)^2 e_{x_2}^2 + \left(\frac{\partial \phi}{\partial x_3}\right)^2 e_{x_3}^2 + \left(\frac{\partial \phi}{\partial x_4}\right)^2 e_{x_4}^2 + \left(\frac{\partial \phi}{\partial x_5}\right)^2 e_{x_5}^2$$

$$+2\left(\frac{\partial \hat{x}_0}{\partial m}\right)\left(\frac{\partial \hat{x}_0}{\partial b}\right) Cov(m, b)$$

$e_{s,\phi}^2 = s_{\hat{x}_0}^2$

$e_{s,\phi} = s_{\hat{x}_0}$ units

Figure 6.13 The error on a value obtained from a calibration curve may be obtained from Equation 6.77, which is derived by following the error propagation shown in this error propagation worksheet.

$$\text{Calibration curve} \qquad \hat{x}_0 = \left(\frac{1}{m}\right) y_0 - \left(\frac{b}{m}\right) \qquad (6.75)$$

(previously introduced as Equation 6.22; see also Figure 5.11). The calibration error is related to $s_{\hat{x}_0}^2$, the variance of $\hat{x}_0$.

To calculate $s_{\hat{x}_0}$ we perform an error propagation on Equation 6.75 as shown in the error propagation worksheet in Figure 6.13 (this calculation was started for the DP meter in Example 5.9, Figure 5.15). As discussed earlier in this chapter, since the least-squares-fit parameters $m$ and $b$ are obtained from the same dataset $(x_i, y_i)$, they are not independent variables, and we must include their covariance in any error propagation involving them both. The covariance of $m$ and $b$ is given by (see Equation 6.61 and Appendix E)

$$Cov(m, b) = -\left(\frac{s_{y,x}^2 \bar{x}}{SS_{xx}}\right) \qquad (6.76)$$

We also need the errors in least-squares slope and intercept, $s_m$ and $s_b$, and these have been derived (Equations 6.32 and 6.33); the formulas appear in Figure 6.13. The error associated with the new reading $y_0$ is shown in

Figure 6.13 as $s_{y,x}$, an assignment discussed in Section 6.2.3. We elaborate on this assignment next, as it impacts device calibration curves.

The general calibration-curve equation, Equation 6.75, expresses the process in Figure 5.11 in which a new reading $y_0$ from a device is back-propagated through the least-squares calibration fit to give an associated value $\widehat{x}_0$. During the collection of the calibration dataset $(x_i, y_i)$ the readings $y_i$ have been influenced by all the intrinsic variations of the device that produced the reading – that is, the device that was calibrated (see discussion in Example 6.2). A measure of this stochastic variability is $s_{y,x}^2$, the variance of $y$ at a value of $x$ for the calibration dataset. When we fit the model $y = mx + b$ to the calibration dataset, we assumed the data were *homoscedastic* – that is, that the variability of $y$ is the same throughout the entire range of $x$. This assumption implies that a new observed reading $y_0$ would likewise have the same variance as the calibration dataset itself exhibited, $s_{y,x}^2$. Thus, in the current error propagation to determine error on $\widehat{x}_0$ (Equation 6.75), we use $s_{y,x}$ from the calibration dataset for the error on a new $y_0$ measurement from the same device.

Following the error-propagation calculations implied in the worksheet in Figure 6.13, including the covariance term and carrying out some algebra, we obtain the final variance on a linear calibration curve, $s_{\widehat{x}_0}^2$ (see Problem 6.19).

| Variance of value from linear calibration curve, $\widehat{x}_0$ (single value of $y_0$) | $s_{\widehat{x}_0}^2 = \dfrac{s_{y,x}^2}{m^2}\left(1 + \dfrac{1}{n} + \dfrac{(y_0 - \bar{y})^2}{m^2 SS_{xx}}\right)$ |
|---|---|

$$(6.77)$$

The stochastic variable $\widehat{x}_0$ is $t$-distributed with $(n - 2)$ degrees of freedom. Thus, a 95% confidence interval on $\widehat{x}_0$ may be calculated from the following interval:

| 95% confidence interval on values of $\widehat{x}_0$ determined from a calibration curve | $\widehat{x}_{0,true} = \left[\left(\dfrac{1}{m}\right)y_0 - \left(\dfrac{b}{m}\right)\right] \pm t_{0.025,n-2}s_{\widehat{x}_0}$ |
|---|---|

$$(6.78)$$

Finally, we wish to calculate standard calibration error $e_{s,cal}$ for the device being calibrated. The calibration error limits given by Equation 6.78 are equal to $2e_{s,cal}$ based on the methods of Chapter 4.

Calibration error limits
general case
(95% level of confidence)

$$2e_{s,cal} = t_{0.025,n-2}s_{\hat{x}_0} \qquad (6.79)$$

Thus, in general, the calibration error is determined for a calibration dataset by calculating $m$, $b$, and $s_{\hat{x}_0}^2$, with the final standard error assignment made with Equation 6.79.

We have all the information we need to calculate the standard calibration error of the DP meter. For illustration, we carry out two calculations of the DP-meter calibration error. First, we use the error propagation worksheet employing a representative value $y_0 = (I, mA)_{raw} = 10$ mA. Second, we create a vector of all possible values of $y_0$ to determine $e_{s,cal}$ over the entire range of possible values of $y_0$.

The worksheet and Excel spreadsheet calculations for $y_0 = (I, mA)_{raw} = 10$ mA are shown in Figure 6.14. From the rightmost column of the worksheet we see that the dominant error contribution to $s_{\hat{x}_0}$ is the error from the DP meter reading, $(I, mA)_{raw}$. The covariance term is far from negligible, being equal in magnitude to the contribution associated with the intercept $b$ and is also equal to three times the contribution associated with the slope $m$.

To calculate the calibration error $e_{s,cal}$ over the whole range of possible values of $y_0 = (I, mA)$, we create a closely spaced vector that spans all possible values of current (4–20 mA) and use Equations 6.77 and 6.79 and spreadsheet software (Figure 6.15) to calculate $e_{s,cal}$ for each value of DP-meter signal in that vector. The calibration data with the error limits on $(\Delta p, psi)$ indicated are plotted in Figure 6.16. The calculations show that the standard calibration error $e_{s,cal}$ is an almost constant over the entire range of the instrument and averages 0.021 psi.

Calibration error
on DP meter

$$e_{s,cal,DP} = 0.021 \text{ psi} \qquad (6.80)$$

One additional general comment: the calibration variance in Equation 6.77 applies when the future value $y_0$ comes from a single observation. Sometimes when a measurement is performed, the value of $y_0$ is determined as an average of $k$ replicate values. When this is the case, the error on $y_0$ would be modified to be a standard error of $k$ replicates of $y_0$.

$$s_{y_0}^2 = \frac{s_{y,x}^2}{k} \qquad (6.81)$$

| $\phi(x_1, x_2, x_3, x_4, x_5)$: | Formula for $\phi$:* $\Delta p = \left(\dfrac{1}{m}\right)I - \left(\dfrac{b}{m}\right)$ *Note: units must work as written. | Representative value of $\phi$: (include units) | 95% C.I. of $\phi$: ($\phi \pm 2e_{s,\phi}$) (include units) |
|---|---|---|---|
| | | 1.39 psi | 1.39 ± 0.04 psi |

| Measured quantities, $x_i$ | | | $\dfrac{\partial \phi}{\partial x_i}$ | $e_{s,x_i}$ | $\left(\dfrac{\partial \phi}{\partial x_i}\right)^2 e^2_{s,x_i}$ |
|---|---|---|---|---|---|
| $x_i$ | Symbol | Representative value | | | |
| $x_1$ | $m$ | 4.326143 mA/psi (units) | $\dfrac{\partial \Delta p}{\partial m} = (I-b)\left(-\dfrac{1}{m^2}\right)$ | $s_m = 0.018133$ | $3.4 \times 10^{-5}$ psi² |
| $x_2$ | $(I, mA)_{raw}$ | 10 mA (units) | $\dfrac{\partial \Delta p}{\partial I} = \left(\dfrac{1}{m}\right)$ | $s_{y,x} = 0.076928$ | $32 \times 10^{-5}$ psi² |
| $x_3$ | $b$ | 3.995616 mA (units) | $\dfrac{\partial \Delta p}{\partial b} = \left(-\dfrac{1}{m}\right)$ | $s_b = 0.041271$ | $9.1 \times 10^{-5}$ psi² |
| $x_4$ | | (units) | | | |
| $x_5$ | | (units) | *Cov$(m,b) = -6.6 \times 10^{-4}$ => ——— | | $-9.7 \times 10^{-5}$ psi² |

$$e^2_{s_\phi} = \left(\frac{\partial \phi}{\partial x_1}\right)^2 e^2_{x_1} + \left(\frac{\partial \phi}{\partial x_2}\right)^2 e^2_{x_2} + \left(\frac{\partial \phi}{\partial x_3}\right)^2 e^2_{x_3} + \left(\frac{\partial \phi}{\partial x_4}\right)^2 e^2_{x_4} + \left(\frac{\partial \phi}{\partial x_5}\right)^2 e^2_{x_5}$$

$e^2_{s,\phi} = 3.4 \times 10^{-4}$ psi² (units)
$e_{s,\phi} = 0.019$ psi

**Error propagation Worksheet**

| $\phi(x_1, x_2, x_3, x_4, x_5)$ | | $\phi =$ | $\Delta p =$ | 1.39 psi | $te_{s\phi} =$ | | 0.04 psi | |
|---|---|---|---|---|---|---|---|---|

| | $x_i$ | value | units | $d\phi/dx_i$ | $(d\phi/dx_i)^2$ | $e_{xi}$ | $e_{xi}^2$ | $(d\phi/dx_i)^2 e_{xi}^2$ | units |
|---|---|---|---|---|---|---|---|---|---|
| $x_1$ | m | 4.326143 | mA/psi | -3.2E-01 | 1.0E-01 | 1.81E-02 | 3.3E-04 | 3.4E-05 | psi² |
| $x_2$ | $(I,mA)_{raw}$ | 10.0 | mA | 2.3E-01 | 5.3E-02 | 7.69E-02 | 5.9E-03 | 3.2E-04 | psi² |
| $x_3$ | b | 3.995616 | mA | -2.3E-01 | 5.3E-02 | 4.13E-02 | 1.7E-03 | 9.1E-05 | psi² |
| | | | | | Cov(m,b)= | -6.6E-04 | | -9.7E-05 | psi² |
| | | | | | | | $e_{s,\phi}^2$ | 3.4E-04 | psi² |
| | | | | | | | $e_{s,\phi}$ | 0.019 | psi |

Figure 6.14  For a single value of the measurement, $y_0 = (I, mA)_{raw} = 10$ mA, we use the modified error propagation worksheet to carry out the calculation of the error on the reported measurement value $\hat{x}_0$. The worksheet is modified to include covariance since $m$ and $b$ are not independent variables.

If this value is used for the error in $y_0$ in the error propagation in Figure 6.14, the variance $s^2_{\hat{x}_0}$ becomes [19]

Variance of value
from calibration, $\hat{x}_0$
($y_0$ obtained from $k$ replicates)

$$s^2_{\hat{x}_0} = \frac{s^2_{y,x}}{m^2}\left(\frac{1}{k} + \frac{1}{n} + \frac{(y_0 - \bar{y})^2}{m^2 SS_{xx}}\right)$$

(6.82)

Figure 6.16 shows the final calibration curve for the DP meter along with 95% level of confidence error limits; it also includes the limit of determination,

| $\Delta p_{meter}$ (xhat$_p$) | DP meter readings ($y_p$) | $s^2_{xhatp}$ | $S_{xhatp}$ | $t_{0.025,13}$*$t_{xhatp}$ | $e_{s,cal}$ | Error limit, low | Error limit, high | % Error on $\Delta p_{meter}$ |
|---|---|---|---|---|---|---|---|---|
| (psi) | (mA) | (psi²) | (psi) | (psi) | (psi) | (psi) | (psi) | |
| 0.00 | 4.0 | 4.1E-04 | 2.0E-02 | 0.04 | 0.022 | -0.043 | 0.045 | |
| 0.05 | 4.2 | 4.0E-04 | 2.0E-02 | 0.04 | 0.022 | 0.004 | 0.091 | 92% |
| 0.09 | 4.4 | 4.0E-04 | 2.0E-02 | 0.04 | 0.022 | 0.050 | 0.137 | 46% |
| 0.14 | 4.6 | 4.0E-04 | 2.0E-02 | 0.04 | 0.022 | 0.097 | 0.183 | 31% |
| 0.19 | 4.8 | 3.95E-04 | 1.99E-02 | 0.04 | 0.021 | 0.143 | 0.229 | 23% |
| 0.23 | 5.0 | 3.92E-04 | 1.98E-02 | 0.04 | 0.021 | 0.189 | 0.275 | 18% |
| 0.28 | 5.2 | 3.89E-04 | 1.97E-02 | 0.04 | 0.021 | 0.236 | 0.321 | 15% |

$\bullet$ $\bullet$ $\bullet$

| | | | | | | | | |
|---|---|---|---|---|---|---|---|---|
| 0.74 | 7.2 | 3.65E-04 | 1.91E-02 | 0.04 | 0.021 | 0.699 | 0.782 | 6% |
| 0.79 | 7.4 | 3.63E-04 | 1.91E-02 | 0.04 | 0.021 | 0.746 | 0.828 | 5% |
| 0.83 | 7.6 | 3.61E-04 | 1.90E-02 | 0.04 | 0.021 | 0.792 | 0.874 | 5% |
| 0.88 | 7.8 | 3.59E-04 | 1.90E-02 | 0.04 | 0.020 | 0.838 | 0.920 | 5% |
| 0.93 | 8.0 | 3.57E-04 | 1.89E-02 | 0.04 | 0.020 | 0.885 | 0.966 | 4% |
| 0.97 | 8.2 | 3.56E-04 | 1.89E-02 | 0.04 | 0.020 | 0.931 | 1.013 | 4% |
| 1.02 | 8.4 | 3.54E-04 | 1.88E-02 | 0.04 | 0.020 | 0.977 | 1.059 | 4% |
| 1.06 | 8.6 | 3.53E-04 | 1.88E-02 | 0.04 | 0.020 | 1.024 | 1.105 | 4% |
| 1.11 | 8.8 | 3.51E-04 | 1.87E-02 | 0.04 | 0.020 | 1.070 | 1.151 | 4% |
| 1.16 | 9.0 | 3.50E-04 | 1.87E-02 | 0.04 | 0.020 | 1.116 | 1.197 | 3% |
| 1.20 | 9.2 | 3.48E-04 | 1.87E-02 | 0.04 | 0.020 | 1.163 | 1.243 | 3% |
| 1.25 | 9.4 | 3.47E-04 | 1.86E-02 | 0.04 | 0.020 | 1.209 | 1.289 | 3% |
| 1.30 | 9.6 | 3.46E-04 | 1.86E-02 | 0.04 | 0.020 | 1.255 | 1.336 | 3% |
| 1.34 | 9.8 | 3.45E-04 | 1.86E-02 | 0.04 | 0.020 | 1.302 | 1.382 | 3% |
| 1.39 | 10.0 | 3.44E-04 | 1.85E-02 | 0.04 | 0.020 | 1.348 | 1.428 | 3% |

$\bullet$ $\bullet$ $\bullet$

| | | | | | | | | |
|---|---|---|---|---|---|---|---|---|
| 3.51 | 19.2 | 3.78E-04 | 1.94E-02 | 0.04 | 0.021 | 3.473 | 3.557 | 1% |
| 3.56 | 19.4 | 3.80E-04 | 1.95E-02 | 0.04 | 0.021 | 3.519 | 3.603 | 1% |
| 3.61 | 19.6 | 3.83E-04 | 1.96E-02 | 0.04 | 0.021 | 3.565 | 3.649 | 1% |
| 3.65 | 19.8 | 3.86E-04 | 1.96E-02 | 0.04 | 0.021 | 3.611 | 3.696 | 1% |
| 3.70 | 20.0 | 3.88E-04 | 1.97E-02 | 0.04 | 0.021 | 3.657 | 3.742 | 1% |

Figure 6.15 To calculate the calibration error over the whole range of DP meter readings ($\Delta p$, psi), we use Excel to create a closely spaced vector of all the possible values of electric current (second column from left) and use Equations 6.77 and 6.78 to calculate the error limits (second and third columns from the right). Also shown is the percent error on ($\Delta p$, psi), calculated as $(2e_{s,cal})/(\Delta p, \text{psi})$ as a function of $y_0 = (I, \text{mA})$. The limit of determination (LOD) is taken as the value of $(\Delta p, \text{psi})_{meter}$ at which the percent error reaches 25%; the LOD for this DP meter is $\approx 0.2$ psi. Values that are below the LOD are shaded in gray. See Example 6.7 for more on LOD.

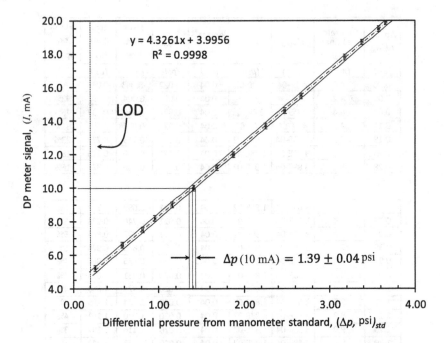

Figure 6.16 The final calibration curve for the DP meter discussed in the linked examples (Figure 1.3). Meter values of electric current in mA (y-axis) may be converted to the corresponding value of differential pressure in psi (x-axis). Uncertainty limits on the DP values are indicated with dashed lines (95% CI on $\widehat{x}_0$). The LOD, below which $\Delta p$ readings on this meter are unreliable, is indicated (see Example 6.7). The individual data points shown are the members of the calibration dataset along with their associated uncertainties in the form of $x$ and $y$ error bars. The $x$ error bars were discussed in Example 5.7. The $y$ data error bars are $\pm t_{0.025,n-2} s_{y,x}$ ($n$ is the number of points in the calibration dataset), as discussed in Section 6.2.3.

LOD (see the next example, Example 6.7 [19]). The calibration curve is used as follows: (1) take a reading $y_0 = (I, \text{mA})_{raw}$ with the DP meter [in the figure $y_0 = (I, \text{mA})_{raw} = 10\,\text{mA}$ is shown as an example] and (2) read off the x-axis of the calibration curve the corresponding $\widehat{x}_0 = (\Delta p, \text{psi})_{meter}$. Alternatively, we can calculate $\widehat{x}_0 = (\Delta p, \text{psi})_{meter}$ from the formula of the calibration curve, Equation 6.75, which for the DP meter is

$$\widehat{x}_0 = \left(\frac{1}{4.326143}\right) y_0 - \left(\frac{3.995616}{4.326143}\right)$$

$$\boxed{\begin{array}{l} \text{DP meter} \\ \text{calibration} \quad (\Delta p, \text{psi})_{meter} \pm 0.04 = 0.23115(I, \text{mA})_{raw} - 0.92360 \\ \text{curve} \end{array}} \quad (6.83)$$

We write the calibration error limits on the $(\Delta p, \text{psi})$ values obtained as $\pm 2e_{s,cal} = \pm 0.04$ psi. For $y_0 = (I, \text{mA})_{raw} = 10$ mA, we obtain $\hat{x}_0 = (\Delta p, \text{psi})_{meter} = 1.39 \pm 0.04$ psi (Figure 6.16).

In Section 4.3 we introduced the concept of limit of determination (LOD). The LOD is the lowest signal that we accept from a sensor or device, and we recommend that this limit be defined with respect to a maximum accepted percentage error on the signal (approximately 25%). We have established how to determine error limits on a calibration curve; thus, we are now in a position to determine the LOD for the user-calibrated differential pressure meter we have been discussing.

**Example 6.7: \*LINKED\*  Determine LOD from a DP-meter calibration curve.** *What is the LOD for the DP meter calibrated in Example 6.6?*

**Solution:** LOD is discussed in Section 4.3. The concept addresses an additional error concern present at low signal values, in which the signal drops to close to, or below, the uncertainty in the signal. For the DP meter, we wish to assess error on $(\Delta p, \text{psi})_{meter}$ at low signal strength.

$$100\% \text{ error on } (\Delta p, \text{psi})_{meter} \qquad (\Delta p, \text{psi})_{meter} = \text{error limits}$$
$$= 2e_{s,cal} \qquad (6.84)$$

The LOD is an evaluation of the quality of an individual data point. The question is, is this data point sufficiently accurate to retain? The rule of thumb we use is that the error on the point should be no more than 25% of the value of the point itself. We check this limit for the DP meter by calculating the ratio of the calibration error limits to the value of $(\Delta p, \text{psi})_{meter}$ provided by the calibration curve.

For the DP meter calibration data, we use Excel to calculate percent error = $2e_{s,cal}/(\Delta p, \text{psi})_{meter}$ for representative low-signal values (shown here and in Figure 6.15).

| $(I, \text{mA})_{raw}$ | $(\Delta p, \text{psi})_{meter}$ | $\left( \frac{2e_{s,cal}}{(\Delta p, \text{psi})_{meter}} \right)$ |
| --- | --- | --- |
| 4.2 | 0.05 | 92% |
| 4.4 | 0.09 | 46% |
| 4.6 | 0.14 | 31% |
| 4.8 | 0.19 | 23% |

For the lowest differential pressures, the error is a large fraction of the signal. The value of $\Delta p$ at which error drops below 25% of the signal is $(\Delta p, \text{psi})_{meter} \approx 0.2$ psi. We fix the LOD at this value.

LOD
for the DP meter          $\boxed{\text{LOD}_{\Delta p} = 0.2 \text{ psi}}$          (6.85)
[25% error on $(\Delta p, \text{psi})$]:

The determination of the LOD means that data values lower than the LOD will be discarded as being insufficiently accurate, by our standards (maximum relative error 25%). The LOD level for the DP meter is shown graphically in Figure 6.16 as a vertical line at $(\Delta p, \text{psi})_{meter} = 0.2$ psi. This LOD corresponds to a lowest acceptable limit on observed current of 4.7 mA.

$$(I, \text{mA})_{cal} = 4.326143(\Delta p, \text{psi})_{std} + 3.995616 \qquad \text{(Equation 5.38)}$$
$$(I, \text{mA})_{limit} = 4.326143(0.2 \text{ psi}) + 3.995616$$
$$= 4.7$$

Example 6.7 completes the first set of linked examples outlined in Section 1.5.2 (Figure 1.3), those directly concerned with calibration of the DP meter. The next example completes the second set of linked examples, those related to flow pressure-drop data obtained with this DP meter.

**Example 6.8: \*LINKED\* Using a calibration curve: DP meter, concluded.**
*Water driven by a pump is made to flow through copper tubing at a variety of flow rates depending on the position of a manual valve (Figure 4.4). Along a 6.0-ft section of copper tubing, the pressure drops are measured with a DP meter as a function of water flow rate, Q(gal/min). The raw differential-pressure signals from the DP meter are in the form of electric current values ranging from 4.0 to 20.0 mA (Table 4.2). These signals may be translated to differential pressures by using the calibration curve for the device, which is given by Equation 4.1 (or Equation 6.83, with error limits), as determined in Examples 6.1 and 6.6. For the flow data provided, what are the associated differential pressures in psi? What are the appropriate error limits (95% level of confidence) associated with each differential pressure? Plot the final results for pressure drop versus flow rate, including the appropriate error bars.*
    **Solution:** We began this problem in Example 4.2. The first part of the question is straightforward and was readily completed there, since the calibration curve was provided (Equation 4.1). The requested values of flow-loop $\Delta p(Q)$ are given in Table 4.3.
    The second part of the question, related to the error limits on the $\Delta p(Q)$ values from the water-flow data in Table 4.3, has been more difficult to address. Our usual way of determining error limits on a quantity is to consider three contributions – replicate error, reading error, and calibration error – and combine these in quadrature. We can follow this process for the flow $\Delta p$ data.

Each of the $\Delta p(Q)$ data points is a single measurement value, and thus we have no direct measure of replicate error. The reading error would be determined by the precision of the display that gives the electric current signal. Since the calibration error for these data was determined by the user – that is, from data on the same setup as was used for these flow data – the impact of reading precision, if any, has already been incorporated into the calibration error. Thus, neither replicate error nor reading error contributes independently to the error on these data; rather, it is pure calibration error.

In this chapter we have shown how to assess calibration error for the DP meter. In Example 6.6 we determined the calibration error for the DP meter as $e_{s,cal} = 0.021$ psi (Equation 6.80). Having considered all three sources of error, we see that the best estimate we can make of error on the pressure-drop measurements $e_{s,cmbd}$ is to use the calibration error.

$$e^2_{s,cmbd} = e^2_{s,replicate} + e^2_{s,reading} + e^2_{s,cal}$$
$$= e^2_{s,cal} = (0.021 \text{ psi})^2 \tag{6.86}$$

The final differential-pressure results as a function of flow rate, including error bars $\pm 2e_s = \pm 0.042$ psi, appear in Figure 6.17. Note that all the data in Table 4.2 are above the LOD $= 4.7$ mA, so there is no need to discard any because of a high percentage error.

The final results of $\Delta p(Q)$ shown in Figure 6.17 follow a gently curving polynomial trend. Polynomial curve fits of the type shown are discussed in the next section.

We note that the data in Figure 6.17 have greater scatter from the trend than is accounted for by the DP-meter calibration error limits we used as error bars. Several effects could be responsible for this outcome.

1. The scatter may be due to stochastic pressure effects of the flow process. Any scatter in the $\Delta p$ produced by the flow – due to air bubbles, pump variation, or outside vibration, for example – would not be reflected in the calibration error bars, but instead reflect stochastic variations due to the generation of the differential pressures by the experiment itself (rather than variability intrinsic to the instrument). The experiment's stochastic effects could be included in error limits by considering deviations of the data from the observed polynomial trend (see Problem 6.25).[7]

---

[7] This is accomplished by using the standard deviation $s_{y,x}$, the standard deviation of $y$ values at a given $x$ value, for a polynomial model fit to the data. That method treats the entire dataset as "replicates," in the sense that the deviation from the model is assumed to be independent of the $x$ value (data assumed homoscedastic).

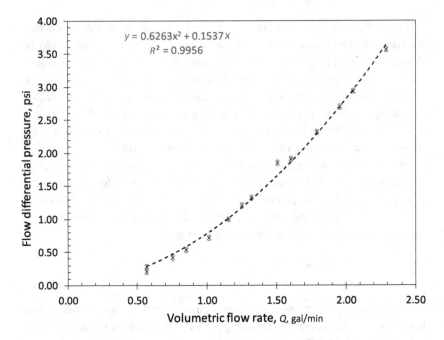

Figure 6.17 The $\Delta p(Q)$ data from Example 4.2 plotted with error bars $\pm 2e_s =$ $\pm 0.04$ psi (Equation 6.86). Also shown is a least-squares regression curve fit to a polynomial of order 2 (with the intercept set to zero). Polynomial least-squares fits are discussed in Section 6.3; see also Problem 6.26.

2. The scatter may be due to stochastic pressure effects of the rate of volumetric flow. Another factor that may be influencing the scatter in the data in Figure 6.17 is the presence of uncertainty in $x$ values – that is, uncertainty in flow rate. The meter or method used to determine $Q$ could be investigated to assess the expected error in that variable.
3. The scatter may be due to physics that has been overlooked. Importantly, we have not explored the physics of $\Delta p(Q)$ to see what relationship is *expected* between $\Delta p$ and $Q$. The work to understand these data is not finished. The task of determining the $e_{s,cal}$ for the DP meter, however, is an essential step that we have completed.

This ends our consideration of these data. If understanding the variation of the data continues to be an interest, additional efforts to investigate possible error sources would need to be conducted.

This concludes our discussion of ordinary least squares applied to the linear model $y = \phi(x) = mx + b$. The least-squares method is quite powerful, and with the computer tools that are available, it is straightforward to carry out such fits and their associated error calculations. The limitations of the ordinary least-squares method include the assumption of no error on the $x$ values and that the data are homoscedastic. With appropriate care, including examination of residuals ([19]; see also Problem 6.7), the least-squares technique can help us to analyze and communicate the significance of our measurements.

The next section introduces calculations that extend the least-squares method to some nonlinear functions. This extension is based on the matrix formulation of least squares.

## 6.3 Least Squares of Polynomial Models

In the previous section we discussed how to obtain best-fit values of the model parameters slope $m$ and intercept $b$ when we have data that show linear behavior. The technique we use is ordinary least-squares regression. The errors on the model parameters, $s_m$ and $s_b$, are calculated by taking the square roots of the variances $s_m^2$ and $s_b^2$ given in Equations 6.32 and 6.33. Excel's LINEST (see the commands in Table 6.1 and Appendix C) is programmed to calculate all four quantities, $m$, $b$, $s_m$, and $s_b$ for a dataset $(x_i, y_i)$. We also pointed to methods of obtaining these four quantities with MATLAB.

It is very common for data to follow a straight line (polynomial of order 1), but as we discussed earlier in this chapter (and in Appendix F), sometimes we seek fits of other types of models to data, including higher-order polynomials and exponentials. In this section we present a brief discussion of how to use ordinary least-squares regression to fit several types of nonlinear models to experimental data. The matrix formulation presented here underlies the MATLAB calculations of ordinary least squares.

### 6.3.1 First-Order Polynomial Fits

The ordinary least-squares method we have discussed so far was specific to using a simple line as the model of the data. To extend our discussion of curve fitting to models other than the straight line, we begin by casting the straight-line fitting problem as a linear algebra problem [7]. From the linear algebra approach, it is straightforward to extend the least-squares method to other models. Our MATLAB solutions of first-order polynomial fits (straight line) are based on the linear algebra approach.

For the straight-line model with intercept $b_0$ and slope $b_1$, the function is written as

Linear model
2 parameters, $b_0$, $b_1$      $y = b_0 + b_1 x$         (6.87)
(polynomial, order 1)

We seek to fit a model to a dataset of $n$ data points $(x_i, y_i)$. We begin by constructing three matrices $\mathbf{A}$, $\mathbf{y}$, and $\mathbf{b}$ from the points in the fitting dataset.[8]

$$\begin{matrix} n \times 2 \text{ matrix} \\ (x \text{ values of the dataset}) \end{matrix} \qquad \mathbf{A} \equiv \begin{pmatrix} 1 & x_1 \\ 1 & x_2 \\ \vdots & \vdots \\ 1 & x_n \end{pmatrix} \qquad (6.88)$$

$$\begin{matrix} n \times 1 \text{ matrix} \\ (y \text{ values of the dataset}) \end{matrix} \qquad \mathbf{y} \equiv \begin{pmatrix} y_1 \\ y_2 \\ \vdots \\ y_n \end{pmatrix} \qquad (6.89)$$

$$2 \times 1 \text{ matrix of coefficients} \qquad \mathbf{b} \equiv \begin{pmatrix} b_0 \\ b_1 \end{pmatrix} \qquad (6.90)$$

If the data sit perfectly on the line produced by the model, then each data pair $(x_i, y_i)$ would satisfy the equation $y_i = b_0 + b_1 x_i$. In other words, if the line $y_i = b_0 + b_1 x_i$ and the data were a perfect fit, each of the $n$ equations would use the same values of slope and intercept. We could write this compactly as the following matrix equation:

$$\begin{matrix} \text{All data pairs} \\ \text{sit on the line} \end{matrix} \qquad \begin{pmatrix} 1 & x_1 \\ 1 & x_2 \\ \vdots & \vdots \\ 1 & x_n \end{pmatrix} \begin{pmatrix} b_0 \\ b_1 \end{pmatrix} = \begin{pmatrix} y_1 \\ y_2 \\ \vdots \\ y_n \end{pmatrix} \qquad (6.91)$$

$$(n \times 2) \qquad (2 \times 1) \; [=] \quad (n \times 1)$$

$$\mathbf{A}\mathbf{b} = \mathbf{y} \qquad (6.92)$$

With experimental data, which are stochastic, the usual case is that the data do not sit exactly on a common line, so the $n$ equations $y_i = b_0 + b_1 x_i$ would not hold for every point. Instead, for each data pair there would be a residual amount $r_i$ that is the difference between the $y$ value represented by the line and the $i$th data point value $y_i$. This relationship may be written in matrix form as

---

[8] If it is appropriate and desirable to force the $y$-intercept to be zero, we accomplish this in the fitting process by omitting the column of 1s in $\mathbf{A}$ [7]; see also Problem 6.26.

$$\begin{pmatrix} 1 & x_1 \\ 1 & x_2 \\ \vdots & \vdots \\ 1 & x_n \end{pmatrix} \begin{pmatrix} b_0 \\ b_1 \end{pmatrix} - \begin{pmatrix} y_1 \\ y_2 \\ \vdots \\ y_n \end{pmatrix} \equiv \begin{pmatrix} r_1 \\ r_2 \\ \vdots \\ r_n \end{pmatrix} \tag{6.93}$$

With residuals present

$$\mathbf{A}\mathbf{b} - \mathbf{y} = \mathbf{r} \tag{6.94}$$

where $\mathbf{r}$ is a new $(n \times 1)$ matrix of residuals for each data point. The least-squares fit for slope and intercept is obtained by finding values of $b_0$ and $b_1$ that minimize the magnitude of the $n \times 1$ vector $\mathbf{r}$.

It can be shown [7] that the minimum squared summed residual vector $\mathbf{r}$ for the dataset $(x_i, y_i)$ is associated with parameter values $\widehat{\mathbf{b}}$, where $\widehat{\mathbf{b}}$ is obtained from solving the following matrix equation, which is called the *Normal equation*:

Optimized values of model parameters

$$\widehat{\mathbf{b}} = \begin{pmatrix} b_0 \\ b_1 \end{pmatrix} \tag{6.95}$$

Normal equation (minimum, least-squares summed residuals)

$$\boxed{\left( \mathbf{A}^T \mathbf{A} \right) \widehat{\mathbf{b}} = \mathbf{A}^T \mathbf{y}} \tag{6.96}$$

$\mathbf{A}^T$ is the transpose of the matrix $\mathbf{A}$ (rows and columns are switched). Solving the Normal equation for $\widehat{\mathbf{b}}$ gives us the ordinary least-squares values of the model coefficients (slope and intercept) for the dataset. The Normal equation is a linear matrix equation, which we can solve by left-multiplying both sides of Equation 6.96 with the matrix inverse $\mathbf{C} \equiv \left( \mathbf{A}^T \mathbf{A} \right)^{-1}$:

Normal equation:

$$\left( \mathbf{A}^T \mathbf{A} \right) \widehat{\mathbf{b}} = \mathbf{A}^T \mathbf{y} \tag{6.97}$$

$$\left( \mathbf{A}^T \mathbf{A} \right)^{-1} \left( \mathbf{A}^T \mathbf{A} \right) \widehat{\mathbf{b}} = \left( \mathbf{A}^T \mathbf{A} \right)^{-1} \mathbf{A}^T \mathbf{y} \tag{6.98}$$

Solution to linear least squares:

$$\boxed{\widehat{\mathbf{b}} = \begin{pmatrix} b_0 \\ b_1 \end{pmatrix} = \left( \mathbf{A}^T \mathbf{A} \right)^{-1} \mathbf{A}^T \mathbf{y}}$$

$$\tag{6.99}$$

$$\widehat{\mathbf{b}} = \mathbf{C} \mathbf{A}^T \mathbf{y} \tag{6.100}$$

This calculation is straightforward to compute using MATLAB or Excel's matrix operations. LINEST is carrying out this type of calculation.[9] The least-squares calculation in Equation 6.99 is sufficiently common that MATLAB has a dedicated syntax to calculate $\widehat{b}$ directly. The syntax is modeled after Equation 6.92.

$$\begin{array}{cc} \text{MATLAB solution} \\ \text{to least squares} \end{array} \qquad \boxed{\widehat{\mathbf{b}} = bhat = \mathbf{A}\backslash\mathbf{y}} \qquad (6.101)$$

The MATLAB least squares solutions provided in Appendix D use this command.

We obtain the same answers with Equation 6.99 (or Equation 6.101) as we did previously with Equation 6.5 ($m = b_1$) and Equation 6.6 ($b = b_0$).

### 6.3.2 Second-Order Polynomial Curve Fits

The discussion of the Normal equation and its solution using linear algebra allow us to consider more complex models for least-squares curve fitting. Consider the problem of a polynomial function of order 2 (parabola), which we would like to fit to a dataset of $n$ data pairs ($x_i, y_i$).

$$\text{Polynomial of order } k = 2: \quad y = b_0 + b_1 x + b_2 x^2 \qquad (6.102)$$

Following the linear algebra approach that led to the Normal equation, we define, for a polynomial of order $k = 2$, three matrices that cast the problem in analogous form.[10]

$$n \times 3 \text{ matrix} \qquad \mathbf{A} \equiv \begin{pmatrix} 1 & x_1 & x_1^2 \\ 1 & x_2 & x_2^2 \\ \vdots & \vdots & \vdots \\ 1 & x_n & x_n^2 \end{pmatrix} \qquad (6.103)$$

$$n \times 1 \text{ matrix} \qquad \mathbf{y} \equiv \begin{pmatrix} y_1 \\ y_2 \\ \vdots \\ y_n \end{pmatrix} \qquad (6.104)$$

---

[9] Excel could use the Normal equation to calculate $\widehat{b}$, and did so in earlier versions, but to obtain greater stability with some datasets it now uses a different calculation method. See the Microsoft online support pages for details.

[10] As before, we may force parameter $b_0$ to zero by omitting the column of 1s.

$$3 \times 1 \text{ matrix of coefficients} \qquad \mathbf{b} \equiv \begin{pmatrix} b_0 \\ b_1 \\ b_2 \end{pmatrix} \qquad (6.105)$$

As before, if all data pairs sit on the line, then $\mathbf{Ab} = \mathbf{y}$. With residuals present, we again have

$$\mathbf{Ab} - \mathbf{y} = \mathbf{r} \qquad (6.106)$$

where $\mathbf{r}$ is the $n \times 1$ array of residuals. When the residuals are minimized using the least-squares formulation, the result for the best-fit vector of model coefficients $\widehat{\mathbf{b}}$ is again found by solving the Normal equation (Equation 6.96). In matrix form, the solution to the Normal equation is the same for polynomials of all orders; the only difference for the polynomial of order $k = 2$ compared to the linear model (order $k = 1$) is that the solution matrix $\widehat{\mathbf{b}}$ is now $3 \times 1$ and contains three ($k + 1 = p$) model parameters.

$$\begin{array}{c} \text{Solution to} \\ \text{polynomial, order } k = 2, \\ \text{least squares:} \end{array} \qquad \boxed{\widehat{\mathbf{b}} = \begin{pmatrix} b_0 \\ b_1 \\ b_2 \end{pmatrix} = (\mathbf{A}^T\mathbf{A})^{-1}\mathbf{A}^T\mathbf{y}} \qquad (6.107)$$

$$\widehat{\mathbf{b}} = \mathbf{CA}^T\mathbf{y} \qquad (6.108)$$

$$\text{MATLAB:} \quad \widehat{\mathbf{b}} = bhat = \mathbf{A}\backslash\mathbf{y} \qquad (6.109)$$

MATLAB, Excel, or other mathematics matrix software can readily evaluate Equation 6.107 for second-order polynomial fits to a dataset $(x_i, y_i)$ (see Example 6.9).

Higher-order polynomial fits are constructed analogously. To fit a cubic or higher-order model using the Normal equation method, we graft to the $\mathbf{A}$ matrix in Equation 6.103 an additional rightside column containing the higher powers of the dataset's $x_i$ and add additional rows to the bottom of the vector $\mathbf{b}$ for the additional model parameters. The best-fit parameters are calculated with Equation 6.107.

In Example 6.9, we use linear algebra to carry out a polynomial curve-fit calculation on rheological entrance pressure loss ($\Delta p_{ent}$) data. In that example, we use MATLAB, Excel matrix tools, as well as LINEST, which, with modification of the function call, is able to carry out the entire polynomial curve-fit calculation.

**Example 6.9: Polynomial fit to rheological entrance-pressure data by matrix methods.** *Using the matrix formulation, calculate the fit of a second-order polynomial model to the entrance-pressure-loss versus* *apparent-shear-rate results of Example 6.4 (Table 6.4, Figure 6.8). Also show the calculation using LINEST.*

**Solution:** We follow the method outlined in this section. First we construct the needed data matrices $\mathbf{A}$ and $\mathbf{y}$. The data are found in Table 6.4 ($n = 5$).

$$\mathbf{A} = \begin{pmatrix} 1 & x_1 & x_1^2 \\ 1 & x_2 & x_2^2 \\ \vdots & \vdots & \vdots \\ 1 & x_n & x_n^2 \end{pmatrix} = \begin{pmatrix} 1 & 40 & 1,600 \\ 1 & 60 & 3,600 \\ 1 & 90 & 8,100 \\ 1 & 120 & 14,400 \\ 1 & 250 & 62,500 \end{pmatrix} \tag{6.110}$$

$$\mathbf{y} = \begin{pmatrix} y_1 \\ y_2 \\ \vdots \\ y_n \end{pmatrix} = \begin{pmatrix} 37.54 \\ 66.87 \\ 85.44 \\ 106.13 \\ 164.07 \end{pmatrix} \tag{6.111}$$

$$\mathbf{Ab} = \mathbf{y} \tag{6.112}$$

where $\mathbf{b}$ is given by Equation 6.105.

The second step is to calculate the solution to the Normal equation, Equation 6.96. The solution is

$$\widehat{\mathbf{b}} = \begin{pmatrix} b_0 \\ b_1 \\ b_2 \end{pmatrix} = \left( \mathbf{A}^T \mathbf{A} \right)^{-1} \mathbf{A}^T \mathbf{y} \tag{6.113}$$

The evaluation of this expression may be carried out using array operations.

The MATLAB solution to Equation 6.113 is one command.

$$\widehat{\mathbf{b}} = bhat = \mathbf{A} \backslash \mathbf{y} \tag{6.114}$$

$$= \begin{pmatrix} b_0 \\ b_1 \\ b_2 \end{pmatrix} = \begin{pmatrix} -0.6390 \\ 1.134 \\ -0.001905 \end{pmatrix} \tag{6.115}$$

A MATLAB function including all steps to arrive at this solution is provided in Appendix D.

Excel has a full range of matrix operations that may be used to solve Equation 6.113. The Excel function calls use the range of cells holding the matrices as their arguments. Matrix commands are invoked in an Excel spreadsheet in a three-step sequence: (1) select a range of the appropriate size for the results; (2) type the matrix-related command into the command line; and (3) simultaneously press the keys "control," "shift," and "return." We use three matrix functions: matrix transpose TRANSPOSE(), matrix multiplication MMULT(), and matrix inversion MINVERSE().

$$\mathbf{A}^T = \text{TRANSPOSE(A-range)}$$

$$= \begin{pmatrix} 1 & 1 & 1 & 1 & 1 \\ 40 & 60 & 90 & 120 & 250 \\ 1{,}600 & 3{,}600 & 8{,}100 & 14{,}400 & 62{,}500 \end{pmatrix} \quad (6.116)$$

$$\mathbf{A}^T \mathbf{A} = \text{MMULT}(\mathbf{A}^T\text{-range, A-range})$$

$$= \begin{pmatrix} 5.000 \times 10^0 & 5.600 \times 10^2 & 9.020 \times 10^4 \\ 5.600 \times 10^2 & 9.020 \times 10^4 & 1.836 \times 10^7 \\ 9.020 \times 10^4 & 1.836 \times 10^7 & 4.195 \times 10^9 \end{pmatrix} \quad (6.117)$$

$$\mathbf{C} = \left[\mathbf{A}^T\mathbf{A}\right]^{-1} = \text{MINVERSE(array-range)}$$

$$= \begin{pmatrix} 3.529 \times 10^0 & -5.935 \times 10^{-2} & 1.839 \times 10^{-4} \\ -5.935 \times 10^{-2} & 1.100 \times 10^{-3} & -3.538 \times 10^{-6} \\ 1.839 \times 10^{-4} & -3.538 \times 10^{-6} & 1.177 \times 10^{-8} \end{pmatrix}$$

$$(6.118)$$

We obtain the final result for $\widehat{\mathbf{b}}$ with successive applications of the Excel matrix command MMULT().

$$\widehat{\mathbf{b}} = \left(\mathbf{A}^T\mathbf{A}\right)^{-1}\mathbf{A}^T\mathbf{y} \quad (6.119)$$

$$= \mathbf{C}\mathbf{A}^T\mathbf{y}$$

$$= \begin{pmatrix} b_0 \\ b_1 \\ b_2 \end{pmatrix} = \begin{pmatrix} -0.6390 \\ 1.134 \\ -0.001905 \end{pmatrix} \quad (6.120)$$

The result for the fit of a polynomial of order $k = 2$ ($p = 3$ parameters) to the Table 6.4 rheological $\Delta p_{ent}(\dot{\gamma}_a)$ results (extra digits shown) is the same whether MATLAB or Excel matrix commands are used:

$$b_0 = -0.6390 \text{ bar}$$

$$b_1 = 1.134 \text{ bar s}$$

$$b_2 = -0.001905 \text{ bar s}^2$$

$$(\Delta p_{ent}, \text{bar}) = b_0 + b_1 \left(\dot{\gamma}_a, \text{s}^{-1}\right) + b_2 \left(\dot{\gamma}_a, \text{s}^{-1}\right)^2$$

The solution agrees with that displayed on the graph in Figure 6.8, which was obtained by using the Excel Trendline option.[11]

---

[11] The reader may ask, why would one go to the trouble of using any of these methods when the Excel TRENDLINE option gives us the polynomial fit? The answer is addressed by our next set of subjects: we need matrix or LINEST methods to calculate the uncertainties on the polynomial model parameters and other related uncertainties.

· Alternatively, polynomial-fit calculations are supported by LINEST, greatly simplifying the effort needed to produce such fits in Excel. To produce a fit to a polynomial of order 2, the modification represented by right-appending the $n \times 1$ column $[x_1^2, x_2^2, \ldots x_n^2]$ to matrix $A$ (Equation 6.103) is brought about by replacing "x-range" in the LINEST function call with "x-range^{1,2}." The 1 and the 2 serve as powers to which the data points are raised. This command transforms the second argument of LINEST to an $n \times 2$ array composed of the rightmost two columns of matrix $A$. The model parameters in Equation 6.120 are thus directly calculated from LINEST as follows:[12]

$$\hat{y} = b_0 + b_1 x + b_2 x^2 \qquad (6.121)$$

$$b_0 = \text{INDEX(LINEST(y-range,x-range^\{1,2\},1,1),1,3)} \qquad (6.122)$$

$$= -0.6390 \text{ bar}$$

$$b_1 = \text{INDEX(LINEST(y-range,x-range^\{1,2\},1,1),1,2)} \qquad (6.123)$$

$$= 1.134 \text{ bar s}$$

$$b_2 = \text{INDEX(LINEST(y-range,x-range^\{1,2\},1,1),1,1)} \qquad (6.124)$$

$$= -0.001905 \text{ bar s}^2$$

The result for the fit of a polynomial of order $k = 2$ ($p = 3$ parameters) to the Table 6.4 rheological $\Delta p_{ent}(\dot{\gamma}_a)$ results (extra digits shown) is the same whether MATLAB, Excel matrix commands, or LINEST are used:

$$b_0 = -0.6390 \text{ bar}$$
$$b_1 = 1.134 \text{ bar s}$$
$$b_2 = -0.001905 \text{ bar s}^2$$
$$(\Delta p_{ent}, \text{bar}) = b_0 + b_1 \left(\dot{\gamma}_a, \text{s}^{-1}\right) + b_2 \left(\dot{\gamma}_a, \text{s}^{-1}\right)^2$$

The solution agrees with that displayed on the graph in Figure 6.8, which was obtained by using the Excel Trendline option.

## 6.3.3 Error on Polynomial Fit Parameters

With polynomial fits, as with fits to a straight-line model, we need to know what error limits to associate with the values of the parameters obtained

---

[12] To force the $y$-intercept to zero with LINEST, set the first logical argument to "No" or zero. This works for the linear case or the polynomial case:

To set $b = 0$: LINEST(y-range,x-range,0,1)

To set $b_0 = 0$: LINEST(y-range,x-range^{1,2},0,1)

See Appendix C.

$(b_0, b_1, b_2)$. The calculation of the errors requires the appropriate use of the full error-propagation equation, Equation 5.16, including comprehensive inclusion of the covariances among the model parameters. Here we provide the equations for error calculations for polynomial fits. We do not include the derivations for this complex error-propagation calculation; rather, we summarize the results. The reader may pursue the details in the literature [38].

For fits to second- and higher-order polynomials (of order $p$), uncertainties on the parameters $b_0$, $b_1$, and $b_2$, or in general $b_i$ for $i = 0, 1, \ldots, k$ ($k$ is the order of the polynomial), and on the fit $\widehat{y}$ (expected value of the function $y$ at each $x$), and for the next value $y_{0,n+1}$ are obtainable by linear-algebra calculations. The needed matrix operations are based on the square matrix, $\mathbf{C}$, which we have already encountered in the solution to the Normal equation (Equation 6.99) and the covariance matrix, which is readily obtained from $\mathbf{C}$.

$$
\begin{array}{c}
\mathbf{C} \text{ matrix} \\
(p \times p) \\
\text{polynomial, order } k
\end{array}
\qquad
\boxed{\mathbf{C} \equiv \left[ \mathbf{A}^T \mathbf{A} \right]^{-1}}
\qquad (6.125)
$$

$$
\begin{array}{c}
\text{Covariance matrix,} \\
ij\text{-component} = \mathrm{Cov}(b_i, b_j) \\
(p \times p)
\end{array}
\qquad
\boxed{s^2_{y,x} \, \mathbf{C}}
\qquad (6.126)
$$

$\mathbf{C}$ is a $p \times p$ symmetric matrix, where $p = k + 1$ is the number of model parameters and the order $k$ is the number of independent variables in the model (variables over which the regression is performed: $x$ and $x^2$ for the second-order polynomial; $x$, $x^2$, and $x^3$ for the third-order polynomial; etc.). For the second-order polynomial $k = 2$ and $p = 3$. The previously defined matrix $\mathbf{A}$ is of size $n \times p$, where $n$ is the number of data points in the fitting dataset. The data are assumed to be homoscedastic, so $s^2_{y,x}$ is the variance associated with $y_i$ values of the dataset at each value of $x_i$, independent of $x$. The diagonal elements of $\mathbf{C}$ are related to the variances of the model parameters, $s_{b_i}$, for $i = 0, 1, \ldots, k$; the off-diagonal elements are related to the covariances of pairs of $b_i$. For $k = 2$:

$$
\mathrm{Cov}(b_i, b_j) = C_{ij} s^2_{y,x}
\qquad (6.127)
$$

$$
\begin{pmatrix}
s^2_{b_0} & \mathrm{Cov}(b_0, b_1) & \mathrm{Cov}(b_0, b_2) \\
\mathrm{Cov}(b_1, b_0) & s^2_{b_1} & \mathrm{Cov}(b_1, b_2) \\
\mathrm{Cov}(b_2, b_0) & \mathrm{Cov}(b_2, b_1) & s^2_{b_2}
\end{pmatrix}
=
\begin{pmatrix}
C_{00} s^2_{y,x} & C_{01} s^2_{y,x} & C_{02} s^2_{y,x} \\
C_{10} s^2_{y,x} & C_{11} s^2_{y,x} & C_{12} s^2_{y,x} \\
C_{20} s^2_{y,x} & C_{21} s^2_{y,x} & C_{22} s^2_{y,x}
\end{pmatrix}
$$

$$(6.128)$$

Diagonal elements
of **C** are related to
variances of $b_i$
$i = 0, 1, \ldots k$

$$s_{b_i}^2 = C_{ii} \, s_{y,x}^2 \qquad (6.129)$$

Off-diagonal elements
of **C** are related to
covariances of pairs of $b_i$

$$\text{Cov}(b_i, b_j) = C_{ij} \, s_{y,x}^2 \qquad (6.130)$$

The 95% confidence intervals on the model parameters $b_i$ are given by [38]:

95% CI on
polynomial curve fit
parameters $b_i$
$i = 0, 1, \ldots k$

$$\boxed{b_{i,true} = b_i \pm t_{0.025, n-p} \sqrt{C_{i,i} \, s_{y,x}^2}} \qquad (6.131)$$

For a second-order polynomial,

95% CI on
second-order polynomial
curve fit parameters

$$\boxed{\begin{aligned} b_{0,true} &= b_0 \pm t_{0.025, n-3} \sqrt{C_{0,0} \, s_{y,x}^2} \\ b_{1,true} &= b_1 \pm t_{0.025, n-3} \sqrt{C_{1,1} \, s_{y,x}^2} \\ b_{2,true} &= b_2 \pm t_{0.025, n-3} \sqrt{C_{2,2} \, s_{y,x}^2} \end{aligned}} \qquad (6.132)$$

In Example 6.10 (in the next section), we use MATLAB and LINEST to calculate the errors on the parameters of the model fit to the data of Example 6.9 (rheological entrance pressure loss versus $\dot{\gamma}_a$).

### 6.3.4 Confidence and Prediction Intervals on Polynomial Fits

As with model fitting for straight lines, with fits to higher-order polynomials we express the uncertainty in the model-predicted value of $y$ with a 95% confidence region for the best-fit value of the model (see Figure 6.12 for a linear model). A next value of the variable $y$ is determined with the related 95% prediction interval. To calculate the CI and PI regions for a fit of a second-order polynomial, the process begins with choosing an $x$ value, $x_0$. We ultimately choose many closely spaced $x_0$ values, and for each value we carry out the 95% confidence and prediction interval calculations, as described in this section.

The calculation of the 95% CI for polynomial curve fits begins with constructing the following $p \times 1$ matrix $\mathbf{x}_0$ (for polynomial of order $k = 2$, $p = k + 1 = 3$), using a single chosen $x$ value, $x_0$.

$$p \times 1 \text{ matrix} \qquad \mathbf{x}_0 = \begin{pmatrix} 1 \\ x_0 \\ x_0^2 \end{pmatrix} \qquad (6.133)$$

The 95% confidence interval of the best-fit line is given by [38]

95% CI on
best-fit polynomial line $\quad y_{0,true} = \widehat{y}_0 \pm t_{0.025,n-p} \sqrt{s_{y,x}^2 \, \mathbf{x}_0^T \left[ \mathbf{A}^T \mathbf{A} \right]^{-1} \mathbf{x}_0}$
of order $k$, $p = k+1$

$$(6.134)$$

$$\boxed{ y_{0,true} = \widehat{y}_0 \pm t_{0.025,n-p} \sqrt{s_{y,x}^2 \, \mathbf{x}_0^T \mathbf{C} \, \mathbf{x}_0} }$$

$$(6.135)$$

Note that the matrix operations beneath the square root in Equation 6.134 involve the multiplication of matrices of the following ranks: $1 \times p$, $p \times p$, and $p \times 1$. The result is thus a $1 \times 1$ matrix containing a single scalar value, which is used to complete the calculation of the 95% confidence interval of the model fit for the chosen $x$ value, $x_0$. The calculation can be performed for closely spaced values of $x_0$ to produce a 95% CI region. Such a region is calculated in Example 6.11. This region, with 95% confidence, will bracket the model-predicted "true" value of $y$ at a chosen $x$.

For a second-order polynomial curve fit, the region that encloses the likely next, $(n+1)$th, value of the function for a chosen $x_0$ (the 95% prediction interval) is given by

95% PI on
next value of $y_0 \quad y_{0,n+1,true} = \widehat{y}_0 \pm t_{0.025,n-p} \sqrt{s_{y,x}^2 \left( 1 + \mathbf{x}_0^T \left[ \mathbf{A}^T \mathbf{A} \right]^{-1} \mathbf{x}_0 \right)}$
at $x_0$

$$(6.136)$$

$$\boxed{ y_{0,n+1,true} = \widehat{y}_0 \pm t_{0.025,n-p} \sqrt{s_{y,x}^2 \left( 1 + \mathbf{x}_0^T \mathbf{C} \mathbf{x}_0 \right)} }$$

$$(6.137)$$

This region, with 95% confidence, will bracket the next observed values of $y_0$ at a chosen $x_0$ for the same conditions under which the fitted data were taken (also demonstrated in Example 6.11).

We can carry out polynomial-fit error calculations in MATLAB and Excel (using the array functions); some of the polynomial-fit error calculations are also supported by the LINEST array function. In the two examples that follow, we calculate the error on the polynomial curve-fit parameters and the 95% confidence and prediction interval regions for the rheological entrance-pressure-loss versus apparent-shear-rate data described in Examples 6.4 and 6.9.

**Example 6.10: Error in parameters of polynomial fit to rheological entrance-pressure data.** *In Example 6.9, we fit a second-order polynomial to entrance-pressure loss versus apparent shear rate results for a polyethylene. What are the uncertainty limits on the parameters determined by the least-squares fit performed in that example?*

**Solution:** In Example 6.4 we performed straight-line fits according to the Bagley method for entrance-pressure loss. The resulting data, shown in Figure 6.8, are five values of $\Delta p_{ent}$ for a polymer melt as a function of apparent shear rate $\dot\gamma_a$, and the data are gently curving.

In Example 6.9, we performed a second-order fit with MATLAB using the command "bhat = $\mathbf{A}\backslash\mathbf{y}$." We determine the uncertainties on the $b_i$ by calculating the covariance matrix $\mathrm{Cov}(b_i, b_j)$ (Equation 6.127), which has the variances $b_i^2$ along the diagonal (Equation 6.128). The MATLAB code is found in Appendix D, and the results are

$$\begin{pmatrix} s_{b_0} \\ s_{b_1} \\ s_{b_2} \end{pmatrix} = stedev\ b = \begin{pmatrix} 1.038 \times 10^1 \\ 1.832 \times 10^{-1} \\ 5.992 \times 10^{-4} \end{pmatrix} \qquad (6.138)$$

We perform the same second-order fit calculation with LINEST by modifying the $x$-range in the second argument of the LINEST function to be a $5 \times 2$ matrix of two columns, where one column contains the $\dot\gamma_{a_i}$ values and the second column contains the values of $\dot\gamma_{a_i}^2$. This modification is carried out as follows to obtain the model parameters:

$$y = b_0 + b_1 x + b_2 x^2 \qquad (6.139)$$

$$b_0 = \mathrm{INDEX}(\mathrm{LINEST}(\text{y-range},\text{x-range}^\wedge\{1,2\},1,1),1,3)$$

$$= -6.390 \times 10^{-1}\ \mathrm{bar}$$

$$b_1 = \mathrm{INDEX}(\mathrm{LINEST}(\text{y-range},\text{x-range}^\wedge\{1,2\},1,1),1,2)$$

$$= 1.134\ \mathrm{bar\ s}$$

$$b_2 = \mathrm{INDEX}(\mathrm{LINEST}(\text{y-range},\text{x-range}^\wedge\{1,2\},1,1),1,1)$$

$$= -1.905 \times 10^{-3}\ \mathrm{bar\ s}^2$$

The variances of the model parameters are also calculated by LINEST and stored in designated locations in the output array. We access the

LINEST-calculated statistical parameters with the commands shown here. For the data of this example, the results are (extra digits shown)

$$s_{b_0} = \text{INDEX(LINEST(y-range,x-range}^\wedge\{1,2\},1,1),2,3)$$
$$= 1.038 \times 10^1 \text{ bar} \tag{6.140}$$

$$s_{b_1} = \text{INDEX(LINEST(y-range,x-range}^\wedge\{1,2\},1,1),2,2)$$
$$= 1.832 \times 10^{-1} \text{ bar} \tag{6.141}$$

$$s_{b_2} = \text{INDEX(LINEST(y-range,x-range}^\wedge\{1,2\},1,1),2,1)$$
$$= 5.992 \times 10^{-4} \text{ bar s}^2 \tag{6.142}$$

$$R^2 = \text{RSQ(y-range,x-range)}$$
$$= 9.932 \times 10^{-1} \tag{6.143}$$

$$s_{y,x} = \text{INDEX(LINEST(y-range,x-range}^\wedge\{1,2\},1,1),3,2)$$
$$= 5.524 \text{ bar} \tag{6.144}$$

For each of the three ($p = k + 1 = 3$) model parameters, the 95% confidence intervals on the parameters $b_i$, $i = 0, 1, 2$, are constructed by including $\pm t_{0.025,n-p}s_{b_i}$ error limits (Equation 6.132). There are five ($n = 5$) data points in the polynomial fit, and thus $(5 - 3) = 2$ degrees of freedom and $t_{0.025,2} = 4.30265$. Carrying out the calculations for the data supplied in Example 6.4, we obtain the final 95% CI on the model parameters, expressed with the appropriate number of significant digits.

$$(\Delta p_{ent}, \text{bar}) = b_0 + b_1(\dot{\gamma}_a, s^{-1}) + b_2\left(\dot{\gamma}_a, s^{-1}\right)^2$$
$$b_0 = (-0.6) \pm (4.30265)(10.45) = -1 \pm 45 \text{ bar}$$
$$b_1 = 1.1 \pm 0.8 \text{ bar s}$$
$$b_2 = -0.002 \pm 0.003 \text{ bar s}^2$$

The low number of degrees of freedom (2) and the high variability of the parameters (for example, $b_0$, for which $s_{b_0} = 10.45$ bar) lead to high uncertainty in the parabolic model parameters. To improve certainty, we need both more and better data.

Note that the coefficient of determination associated with the model fit is close to 1 ($R^2 = 0.9932$), which implies that the model form (quadratic) explains the variation of the data. The errors in the individual data points are known to be high, however (see the error bars in Figure 6.8 and the value of $s_{y,x} = 5.5$ bar, Equation 6.144). Fitting to an arbitrary curve does not diminish the uncertainty we know is present.

**Example 6.11: Error limits on the polynomial fit to rheological entrance-pressure data.** *What is the uncertainty on the order-2 polynomial model fit produced in Examples 6.9 and 6.10 (95% CI on the model fit)? What is the 95% PI of the next value of a measurement of entrance pressure loss?*

**Solution:** We wish to calculate the 95% confidence and prediction limits for the second-order polynomial fit to the rheological entrance pressure losses from Example 6.9. The relationships that we need are provided in Equations 6.135 and 6.137.

$$y_{0,true} = \widehat{y}_0 \pm t_{0.025,n-p}\sqrt{s_{y,x}^2 \, \mathbf{x}_0^T \mathbf{C} \, \mathbf{x}_0} \qquad \text{(Equation 6.135)}$$

$$y_{0,n+1,true} = \widehat{y}_0 \pm t_{0.025,n-p}\sqrt{s_{y,x}^2 \left(1 + \mathbf{x}_0^T \mathbf{C} \mathbf{x}_0\right)} \qquad \text{(Equation 6.137)}$$

For the entrance-pressure loss data, using both MATLAB and Excel's matrix functions, we have already calculated the $3 \times 3$ matrix $\mathbf{C} = \left(\mathbf{A}^T\mathbf{A}\right)^{-1}$ (Equation 6.118), and we have calculated $s_{y,x}$, the standard deviation of $y$ at a value of $x$ (Equation 6.144). The $t$-statistic we need is $t_{0.025,n-p}$, where the number of data points is $n = 5$ and the fit is a polynomial of order $k = 2$ and $p = k + 1 = 3$; $t_{0.025,2} = \text{T.INV.2T}(0.05, 2) = [-\text{tinv}(0.025,2)] = 4.30265$. The remaining piece is to calculate values of the vector $\mathbf{x}_0$, which are $p \times 1 = 3 \times 1$ vectors set up as

$$\mathbf{x}_0 = \begin{pmatrix} 1 \\ x_0 \\ x_0^2 \end{pmatrix} \qquad (6.145)$$

The value $x_0$ is an arbitrary value of $x$ within the $x$-range of the data. Since the entrance-pressure-loss data span from about $\dot{\gamma}_a = 10 \text{ s}^{-1}$ to $250 \text{ s}^{-1}$, we produce a large number of $\mathbf{x}_0$-vectors in that range (we choose 126) and calculate enough values of the 95% CI to be able to plot smooth CI limits. The 95% CI region is determined with Equation 6.135, which is calculated for 126 choices of $x_0$.

In Excel we used a mixture of matrix calculations and conventional dot-product calculations to complete the determination of the intervals (see Figure 6.18). We first produce $\mathbf{X}_0$, a $126 \times 3$ matrix created by stacking all 126 of the arbitrarily selected $1 \times 3$ matrices $\mathbf{x}_0^T$. Next, we calculate "hold," hold $\equiv \mathbf{X}_0\mathbf{C}$, by matrix multiplication. Subsequently, we carry out a series of conventional dot-product calculations (from equation 6.135, $t_{0.025,2}s_{y,x}\sqrt{(\text{hold})_{i^{th}row} \cdot \mathbf{x}_{0,i}}$ is the CI, for $i = 1, 2, \ldots 126$). The calculation under the square root is a vector inner product of two 3-component vectors. Shown in Figure 6.18 is a calculation excerpt of values of $x_{0,i}$ from $10 \text{ s}^{-1}$ to $64 \text{ s}^{-1}$. The 95% confidence limits on the model are shown in Figure 6.19 (inner dashed region).

126 × 3 matrix $X_0$,
with each row = $x_{0,i}$ ≡ hold    hold · $x_{0,i}$

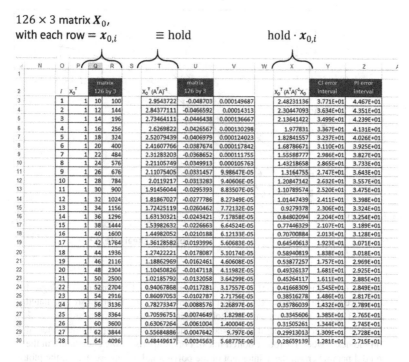

| N | I | $x_0^T$ | matrix 126 by 3 | | $x_0^T(A^TA)^{-1}$ | matrix 126 by 3 | | $x_0^T(A^TA)^{-1}x_0$ | CI error interval | PI error interval |
|---|---|---|---|---|---|---|---|---|---|---|
| 1 | 1 | 10 | 100 | | 2.9543722 | -0.048703 | 0.000149687 | 2.48231136 | 3.771E+01 | 4.467E+01 |
| 2 | 1 | 12 | 144 | | 2.84377111 | -0.0466592 | 0.00014313 | 2.30447093 | 3.634E+01 | 4.351E+01 |
| 3 | 1 | 14 | 196 | | 2.73464111 | -0.0446438 | 0.000136667 | 2.13641422 | 3.499E+01 | 4.239E+01 |
| 4 | 1 | 16 | 256 | | 2.6269822 | -0.0426567 | 0.000130298 | 1.977831 | 3.367E+01 | 4.131E+01 |
| 5 | 1 | 18 | 324 | | 2.52079439 | -0.0406979 | 0.000124023 | 1.82841557 | 3.237E+01 | 4.026E+01 |
| 6 | 1 | 20 | 400 | | 2.41607766 | -0.0387674 | 0.000117842 | 1.68786671 | 3.110E+01 | 3.925E+01 |
| 7 | 1 | 22 | 484 | | 2.31283203 | -0.0368652 | 0.000111755 | 1.55588777 | 2.986E+01 | 3.827E+01 |
| 8 | 1 | 24 | 576 | | 2.21105749 | -0.0349913 | 0.000105763 | 1.43218658 | 2.865E+01 | 3.733E+01 |
| 9 | 1 | 26 | 676 | | 2.11075405 | -0.0331457 | 9.98647E-05 | 1.3164755 | 2.747E+01 | 3.643E+01 |
| 10 | 1 | 28 | 784 | | 2.0119217 | -0.0313283 | 9.40606E-05 | 1.20847142 | 2.632E+01 | 3.557E+01 |
| 11 | 1 | 30 | 900 | | 1.91456044 | -0.0295393 | 8.83507E-05 | 1.10789574 | 2.520E+01 | 3.475E+01 |
| 12 | 1 | 32 | 1024 | | 1.81867027 | -0.0277786 | 8.27349E-05 | 1.01447439 | 2.411E+01 | 3.398E+01 |
| 13 | 1 | 34 | 1156 | | 1.72425119 | -0.0260462 | 7.72132E-05 | 0.9279378 | 2.306E+01 | 3.324E+01 |
| 14 | 1 | 36 | 1296 | | 1.63130321 | -0.0243421 | 7.17858E-05 | 0.84802094 | 2.204E+01 | 3.254E+01 |
| 15 | 1 | 38 | 1444 | | 1.53982632 | -0.0226663 | 6.64524E-05 | 0.77446329 | 2.107E+01 | 3.189E+01 |
| 16 | 1 | 40 | 1600 | | 1.44982052 | -0.0210188 | 6.12133E-05 | 0.70700884 | 2.013E+01 | 3.128E+01 |
| 17 | 1 | 42 | 1764 | | 1.36128582 | -0.0193996 | 5.60683E-05 | 0.64540613 | 1.923E+01 | 3.071E+01 |
| 18 | 1 | 44 | 1936 | | 1.27422221 | -0.0178087 | 5.10174E-05 | 0.58940819 | 1.838E+01 | 3.018E+01 |
| 19 | 1 | 46 | 2116 | | 1.18862969 | -0.0162461 | 4.60608E-05 | 0.53877257 | 1.757E+01 | 2.969E+01 |
| 20 | 1 | 48 | 2304 | | 1.10450826 | -0.0147118 | 4.11982E-05 | 0.49326137 | 1.681E+01 | 2.925E+01 |
| 21 | 1 | 50 | 2500 | | 1.02185792 | -0.0132058 | 3.64299E-05 | 0.45264117 | 1.611E+01 | 2.885E+01 |
| 22 | 1 | 52 | 2704 | | 0.94067868 | -0.0117281 | 3.17557E-05 | 0.41668309 | 1.545E+01 | 2.849E+01 |
| 23 | 1 | 54 | 2916 | | 0.86097053 | -0.0102787 | 2.71756E-05 | 0.38516278 | 1.486E+01 | 2.817E+01 |
| 24 | 1 | 56 | 3136 | | 0.78273347 | -0.0088576 | 2.26897E-05 | 0.35786039 | 1.432E+01 | 2.789E+01 |
| 25 | 1 | 58 | 3364 | | 0.70596751 | -0.0074649 | 1.8298E-05 | 0.3345606 | 1.385E+01 | 2.765E+01 |
| 26 | 1 | 60 | 3600 | | 0.63067264 | -0.0061004 | 1.40004E-05 | 0.31505261 | 1.344E+01 | 2.745E+01 |
| 27 | 1 | 62 | 3844 | | 0.55684886 | -0.0047642 | 9.797E-06 | 0.29913013 | 1.309E+01 | 2.728E+01 |
| 28 | 1 | 64 | 4096 | | 0.48449617 | -0.0034563 | 5.68775E-06 | 0.28659139 | 1.281E+01 | 2.715E+01 |

Figure 6.18 With a mixture of Excel matrix calculations and conventional dot-product calculations, as described in the text, we determined the 95% confidence interval region for the model fit.

The MATLAB calculations follow a similar logic, with a "for loop" serving to step through the "hold" = $X_0C$ calculation. The code is given in Appendix D.

The 95% PI region was calculated analogously with both Excel and MATLAB, using the relationship in Equation 6.137. This region is also plotted in Figure 6.19 (dotted, outer lines).

The 95% CI of the fit, the inner region, represents the quality of the model's prediction of the data's trend. The outer region, the 95% PI of the next value of $y$, represents the region in which we should expect a subsequent $-(n+1)$th or sixth $-$ measurement of $\Delta p_{ent}$ to fall. Both regions are based only on the observed variability of the data compared to the trend. If we have additional knowledge of the reliability of the data points (error bars on the individual data points, for example, determined from the measurement process), this should be considered when making decisions based on these data. The second-order polynomial trendline we chose to use as our model is an empirical model, not one that is tied to any physics linked to the data. Note that the encouraging

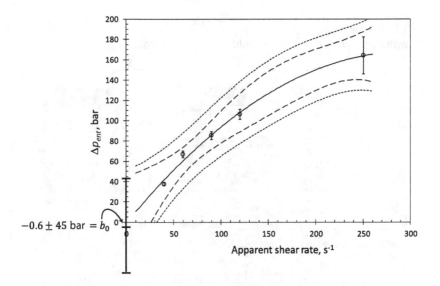

Figure 6.19 The best fit of a second-order polynomial to data for rheological entrance pressure loss as a function of apparent shear rate. The data are calculated from Bagley's measurements on polyethylene [1]. Also shown are the 95% confidence interval of the fit (inner dashed lines) and the 95% prediction interval on the next value of $y$ (outer dotted lines). For comparison, the uncertainty in the model parameter $b_0$, the $y$-intercept of the polynomial, is represented on the plot as error bars on a point $(0, b_0)$. Error bars on individual points were determined during the linear fits of $\Delta p$ versus $L/R$ (Example 6.4) used to produce $\Delta p_{ent}$. The quantity $s_{y,x}$, a measure of the scatter of the data from the polynomial curve, is $s_{y,x} = 5.5$ bar.

close-to-1 value of $R^2 = 0.9932$ does not at all reflect the actual uncertainty that is associated with both the model and the probable next values of $\Delta p_{ent}$ for these data.

## 6.4  Extensions of Least Squares to Other Models

Thus far, we have presented nonlinear model–fitting in terms of polynomials of various orders, and we have shown that the solution methods are readily adapted to higher-order polynomials. To adapt MATLAB to perform fits of exponential, logarithmic, and power-law models, the appropriate matrices $A$, $\widehat{b}$, and $y$ are shown in Table 6.6. Once the matrices are constructed, the solution for $\widehat{b}$ proceeds as in the polynomial case. When using LINEST, the commands are given in Table 6.7. The matrix techniques used are a subset of the general

Table 6.6. *MATLAB arrays to perform least-squares fits for various empirical models discussed in Appendix F.*

| | |
|---|---|
| General equation | $\mathbf{A\,b = y}$ |
| Linear fit | $y = b_1 x + b_0$ |

$$\begin{pmatrix} 1 & x_1 \\ 1 & x_2 \\ \vdots & \vdots \\ 1 & x_n \end{pmatrix} \begin{pmatrix} b_0 \\ b_1 \end{pmatrix} = \begin{pmatrix} y_1 \\ y_2 \\ \vdots \\ y_n \end{pmatrix}$$

Linear fit (zero intercept)　　$y = b_1 x$

$$\begin{pmatrix} x_1 \\ x_2 \\ \vdots \\ x_n \end{pmatrix} \begin{pmatrix} b_1 \end{pmatrix} = \begin{pmatrix} y_1 \\ y_2 \\ \vdots \\ y_n \end{pmatrix}$$

Exponential fit　　$y = ae^{bx}$
$\ln(y) = \ln(a) + bx$

$$\begin{pmatrix} 1 & x_1 \\ 1 & x_2 \\ \vdots & \vdots \\ 1 & x_n \end{pmatrix} \begin{pmatrix} \ln(a) \\ b \end{pmatrix} = \begin{pmatrix} \ln(y_1) \\ \ln(y_2) \\ \vdots \\ \ln(y_n) \end{pmatrix}$$

Logarithmic fit　　$y = a + b \ln x$

$$\begin{pmatrix} 1 & \ln(x_1) \\ 1 & \ln(x_2) \\ \vdots & \vdots \\ 1 & \ln(x_n) \end{pmatrix} \begin{pmatrix} a \\ b \end{pmatrix} = \begin{pmatrix} y_1 \\ y_2 \\ \vdots \\ y_n \end{pmatrix}$$

Power-law fit　　$y = ax^b$
$\ln(y) = \ln(a) + b \ln(x)$

$$\begin{pmatrix} 1 & \ln(x_1) \\ 1 & \ln(x_2) \\ \vdots & \vdots \\ 1 & \ln(x_n) \end{pmatrix} \begin{pmatrix} \ln(a) \\ b \end{pmatrix} = \begin{pmatrix} \ln(y_1) \\ \ln(y_2) \\ \vdots \\ \ln(y_n) \end{pmatrix}$$

Table 6.7. *LINEST commands to perform least-squares fits for various empirical models discussed in Appendix F.*

| | |
|---|---|
| Linear fit | $y = mx + b$ |
| | $m =$ INDEX(LINEST(y-range,x-range,1,1),1,1) |
| | $b =$ INDEX(LINEST(y-range,x-range,1,1),1,2) |
| Linear fit | $y = mx$ |
| (zero intercept) | $m =$ INDEX(LINEST(y-range,x-range,0,1),1,1) |
| Exponential fit | $y = ae^{bx}$ |
| | $a =$ EXP(INDEX(LINEST(LN(y-range),(x-range),,),1,2)) |
| | $b =$ INDEX(LINEST(LN(y-range),(x-range),1,1) |
| Logarithmic fit | $y = a + b \ln x$ |
| | $a =$ INDEX(LINEST(y-range,LN(x-range)),1,2) |
| | $b =$ INDEX(LINEST(y-range,LN(x-range)),1,1) |
| Power-law fit | $y = ax^b$ |
| | $a =$ EXP(INDEX(LINEST(LN(y-range),LN(x-range),,),1,2)) |
| | $b =$ INDEX(LINEST(LN(y-range),LN(x-range),,),1,1) |

technique of multivariable regression. For more on *multivariable regression,* see the literature [38].

The final example of this chapter considers a fit to a more complicated model, the first-order response (see Appendix F, Equation F.17). To make use of linear fitting methods, the data are transformed in several ways. Also important in that example is to consider the LOD. When fitting models to data, it is important to recognize that sometimes measured values become very small and error-prone. We can lose many hours to data misinterpretation if we fail to recognize when measured values have become meaningless.

**Example 6.12: Fitting the first-order response model to temperature data.** *Water temperature is measured as a function of time as a resistance temperature detector (RTD) is moved from one water bath to another. The $T(t)$ data are shown in Figure 6.20. What is an empirical model that gives a good fit to the data?*

**Solution:** We begin by plotting and examining the data. From the catalog of models discussed in Appendix F, the data most closely resemble the first-order response model (Equation F.17).

$$\text{First-order response model} \qquad \boxed{(T - T_0) = (T_\infty - T_0)\left(1 - e^{-(t-t_0)/\lambda}\right)} \qquad (6.146)$$

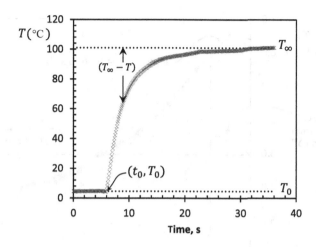

Figure 6.20 The time trace of temperature as an RTD is moved from one water bath to another ($T_0 = 4.6°C$, $T_\infty = 100.9°C$, $t_0 = 5.9$ s). We desire to find an empirical model that captures the trend once the temperature begins to change.

where $t_0$ is the starting time for the experiment, $T_0$ is the initial water temperature, and $T_\infty$ is the temperature at long times. The rate of change of temperature as a function of time is governed by the parameter $\lambda$. In the dataset we are considering (Figure 6.20), the location of the start of the experiment, $(t_0, T_0) = (5.9\,s, 4.6°C)$, may be determined by inspecting the plot. The value of $T_\infty = 100.9°C$, the temperature at long times, likewise may be determined by inspection. The remaining parameter to determine is the time constant $\lambda$.

As discussed in Appendix F Section F.2 (Equation F.18), we can rearrange Equation 6.146 and determine $\lambda$ by least-squares fitting a line to the transformed data.

$$\ln\left(\frac{T_\infty - T}{T_\infty - T_0}\right) = -\frac{1}{\lambda}(t - t_0) \tag{6.147}$$

The current data, transformed, are shown in Figure 6.21. The slope of the plot, $-1/\lambda$, will allow us to determine $\lambda$.

To interpret Figure 6.21, we must account for the low accuracy of the long-time data, as transformed. As the signal in Figure 6.20 approaches its ultimate value of $T_\infty = 100.9°C$, the quantity $(T_\infty - T)$ approaches zero and becomes quite susceptible to error fluctuations. We only wish to consider data that have a reasonably small amount of error; we initially choose at most 25% error (see

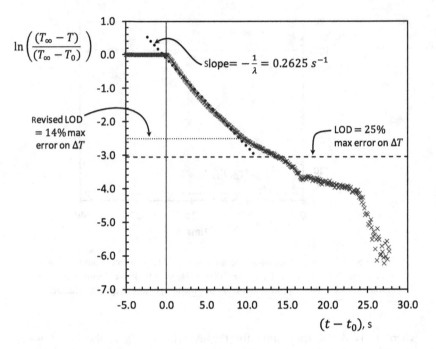

Figure 6.21 Plotting data to allow the first-order-response parameter λ to be calculated. The quantity displayed on the y-axis in this plot becomes very small as $(T_\infty - T) = \Delta T$ approaches zero. As $\Delta T$ becomes small, we discard data for which the error on $\Delta T$ exceeds a chosen threshold (the LOD).

Section 4.3). We need to determine the LOD for $(T_\infty - T)$ and disregard data below the LOD.

To determine the LOD, first we need to know about the calibration of the RTD, the device we used to measure temperature. From the literature, we find that the calibration error for a class-B RTD, in the temperature range of this experiment, is $2e_{s,cal} = 0.8°C$ [53].

Second, for our data we calculate $e_{s,\Delta T}$ for $\Delta T = (T_\infty - T)$ from an error propagation.

$$\text{Error propagation} \qquad \phi = \Delta T = (T_\infty - T)$$

$$e_{s,\phi}^2 = \left(\frac{\partial \phi}{\partial T_\infty}\right)^2 e_{s,RTD}^2 + \left(\frac{\partial \phi}{\partial T}\right)^2 e_{s,RTD}^2$$

$$= 2e_{s,RTD}^2$$

$$e_{s,\Delta T} = 0.565685°C$$

Third, we determine the LOD for $\Delta T$.

$$\begin{array}{cc} \text{LOD of } (T_\infty - T) & \boxed{8e_{s,\Delta T} = 4.5^\circ\text{C}} \\ (25\% \text{ error maximum}) & \end{array} \qquad (6.148)$$

The 25% maximum error LOD cutoff for the temperature data of this example is indicated in Figure 6.21 with a horizontal dashed line:

$$\ln\left(\frac{T_\infty - T}{T_\infty - T_0}\right) = \ln\left(\frac{4.5}{100.9 - 4.6}\right)$$
$$= -3.06$$

Omitting data below the LOD eliminates all the least-linear data (the data below a $y$-axis value of about $-3$), and for the very good reason that the error on $\Delta T$ has risen above 25%. A significant portion of the remaining data make a nonlinear contribution (below about $-2.5$ on the $y$-axis). Eliminating from the fit the data below that level implies discarding data with error above about 14%. That choice is reflected in the fit in Figure 6.21.

It is essential to recognize that at long times, $\Delta T$ is very small; hence we are not able to make accurate measurements in this region. The accurate portions of the data yield a value of slope $= -0.2625$ s$^{-1}$, which corresponds to $\lambda = 3.8$ s. We have made judgment calls on which data to retain, and we are transparent about our reasoning, revealing we use an LOD based on 14% maximum error.

The final fit is shown in Figure 6.22. This fit is valid for 6 s $\leq t \leq$ 36 s. This model gives a reasonable representation of the data over this range.

## 6.5  Least Squares with Physics-Based Models

The modeling in this book has been empirical modeling – that is, modeling of convenience. Often we do not have any firm physical reasons for the models we choose; rather, we choose what seems plausible based on how the data appear. The models thus obtained are convenient ways of representing data.

Alternatively, we can use physics to identify models. Models based on physics may be linear, polynomial, logarithmic, exponential, etc., as well as of more complicated structures. For models derived from physics or from a model based on physics, the parameters of the model may have meaning. We saw this in Examples 6.3 and 6.4, which allowed us to calculate the shear stress and entrance pressure loss in a capillary-flow experiment based on Bagley's

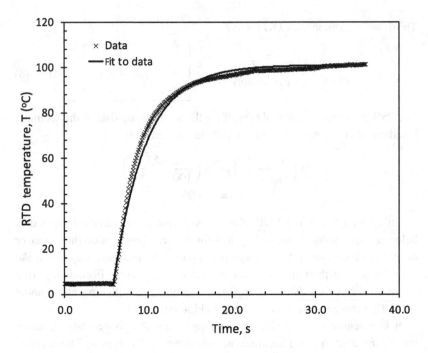

Figure 6.22 With the four parameters of the first-order response appropriately identified, the model gives a reasonable representation of the temperature data for the RTD moved from one water bath to another.

modeling of the physics of flow through a contraction. The techniques of this chapter, used on empirical models, may equally well be used to identify model parameters for physics-based models, as we outline in Example 6.13.

**Example 6.13: Measuring a forced-convection heat-transfer coefficient of a sphere.** *Using the apparatus shown schematically in Figure 6.23, we can measure a forced-convection heat-transfer coefficient $\tilde{h}$ between a submerged sphere (radius R, initially at uniform temperature $T_0$) to a stirred liquid (at bulk temperature $T_\infty$). The physics of the heat transfer can be modeled using the appropriate transient microscopic energy balance [6, 44]. The sphere is initially at uniform temperature $T_0$; at time $t = 0$ it is plunged into a stirred bath held at $T_\infty$. The temperature at the center of the sphere is monitored as a function of time by a thermocouple embedded in the sphere, with its tip located at the sphere center. For data from this experiment, how can we accurately determine $\tilde{h}$ and error limits on $\tilde{h}$?*

**Solution:** We do not provide a complete solution here, but rather discuss how the tools of this text can allow a solution.

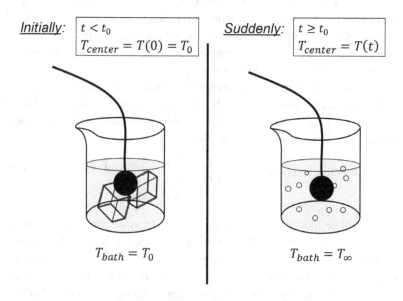

Figure 6.23 We are able to determine the forced-convection heat-transfer coefficient $\tilde{h}$ through temperature measurements at the center of a solid sphere subjected to the process shown here. The sphere is initially equilibrated at $T_0$. Suddenly, it is plunged into a stirred bath at $T_\infty$ (boiling water, for instance). The transient heat conduction may be modeled with an energy balance. In the text we describe how this experiment can lead to a value for $\tilde{h}$, including error limits.

We seek to determine $\tilde{h}$, the forced-convection heat-transfer coefficient between the sphere and the stirred liquid. The energy-balance modeling of a sphere undergoing the conditions described provides a prediction of temperature in the sphere as a function of time and position $r$ (spherical coordinates [6]). The data we have are of temperature of the center of the sphere $r = 0$ as a function of time $T(r,t) = T(0,t)$. The physical model makes a prediction that the center-point temperature of a sphere of radius $R$ as a function of scaled time (Fo) will depend on the Biot number [6, 44].

$$\text{Fourier number:} \quad \text{Fo} = \frac{\bar{\alpha}t}{R^2} \qquad (6.149)$$
$$\text{(scaled time)}$$

$$\text{Biot number:} \quad \text{Bi} = \frac{\tilde{h}R}{\tilde{k}} \qquad (6.150)$$

where $\bar{\alpha} = \tilde{k}/\rho\hat{C}_p$ is the thermal diffusivity of the material used to produce the sphere, $\tilde{k}$ is sphere thermal conductivity, $\rho$ is sphere density, $\hat{C}_p$ is sphere heat capacity, and $\tilde{h}$ is the quantity we seek to extract from the data. The Biot number represents the ratio of the internal (conductive) resistance to heat transfer in the sphere (proportional to $1/\tilde{k}$) to the external (convective) resistance to heat transfer at the boundary (proportional to $1/R\tilde{h}$).

The predicted temperature response at short times is complex: the solution to the microscopic energy balance is an infinite sum of complex terms [6, 44]). However, at long times the model-predicted temperature at the center of the sphere approaches a simple exponential decay.

$$\text{Long-time response} \quad Y \equiv \frac{(T - T_\infty)}{(T_0 - T_\infty)} = b_0 e^{-(b_1 \text{Fo})} \tag{6.151}$$

$$\ln Y = (\ln b_0) - (b_1)\text{Fo} \tag{6.152}$$

The complete modeling solution tells us the decay of scaled temperature $Y$ (slope $b_1$) as a function of the Biot number. We determine $\tilde{h}$ in the laboratory by measuring $T(t)$ for the experiment shown in Figure 6.23, determining the slope $b_1$ of $\ln Y$ versus Fo at longer times. We match the measured decay to the slopes predicted by the modeling solution, deducing the Biot number. From the Biot number thus determined from our measurements, we calculate $\tilde{h}$ from Equation 6.150.

An important aspect of matching model predictions to the data is to use only valid data to determine the long-time slope of $\ln Y$ versus Fo. As we did in Example 6.12, we must determine the LOD of $Y$ and truncate data when the error grows above an appropriate error tolerance level (LOD).

We proceed as follows. To identify $\tilde{h}$ we prepare a plot of $\ln Y = (T - T_\infty)/(T_0 - T_\infty)$ versus Fo $= \bar{\alpha}t/R^2$, where the values of $T(t)$ are the experimental data for the temperature at the center of the sphere as a function of time when a sphere initially at $T_0$ is suddenly submerged in the fluid at bulk temperature $T_\infty$ (the scenario we have modeled with a microscopic energy balance [6, 44]). The data for which error on $(T - T_\infty)$ exceeds 25% of the value (or perhaps 15%? – see the discussion in Example 6.12) are disregarded as unreliable. The slope of the remaining data is determined with a least-squares fit. The experimental slope from the fit is compared with the energy-balance model results, either directly using the solution to the model's partial differential equation [44] or through comparison with the appropriate Heissler chart [14], which is a graphical representation of the mapping of long-time slope with the Biot number for this problem. By these techniques the Biot number is determined only within certain error limits, where the limits depend on which of those two pathways of mapping slope to Bi is chosen; the methods of this text may be used to estimate the uncertainty in Bi (using, for example, Equation 4.3). The final error on $\tilde{h}$ is obtained through error propagation with Equation 6.150.

## 6.6 Summary

The goal of this text has been to provide tools to allow the reader to quantify uncertainty in both measurements and in calculations from measurements. These tools are in the form of worksheets, recommended processes, and rules of thumb. Understanding the reliability of our results, and the assumptions behind them, is essential to good decision making.

Another purpose of this discussion has been to engage the reader in critical analysis of the interpretation of numbers and measurements – to demystify the processes by which meaning is ascribed to data. We encourage the reader to go further than the content in this text, to understand the reasons behind methods we use. An example of the need for reflection is associated with the use of coefficient of determination $R^2$, which is very well-known statistic that is widely quoted as evidence of achieving a good fit or having accurate data. And yet, as was discussed earlier in this chapter, the coefficient of determination has a very limited meaning: it describes the extent to which the trend of the data is captured by the model. Issues such as goodness of fit and degree of data variability are captured by different statistical tools, such as $s_{y,x}$ and 95% confidence and prediction intervals. Understanding our fits and models requires understanding some of the "how" and "why" behind the tools that we use.

We have some additional recommendations.

One practice we recommend is to complete and save error worksheets for devices and processes. The filled-in and appropriately annotated worksheets can be updated as we learn more about the validity and applicability of our devices. And they can be shared with colleagues. Having this information at hand can be a great advantage, allowing us to keep track of the hard work that we have done in understanding the context of our results. We make the same recommendation for error propagation worksheets, particularly for recurring calculations.

We offer a warning about automatic error bars provided by instrument software. When an instrument identifies error bars for your data, in the best-case scenario they reflect the calibration error of the instrument. If the instrument calibration error is the dominant error in the measurement that you are making, those error bars would be appropriate to use for your data (providing a 95% level of confidence). If, however, the data you are reporting have their own substantial intrinsic scatter that is larger than the calibration error (see Figure 6.17), then the automatic calibration error bars are not correct. And, if these kinds of calibration and replication effects are both present, and of comparable magnitudes, the programing in the instrument will not have considered both error sources and added them in quadrature.

In addition to error bars provided by scientific instruments, analysis software, including Excel, offers to insert automatic error bars. The assumptions going into determining those error bars need to be evaluated, including the asserted level of confidence. The software may be making an arbitrary choice (such as to add 5% error bars), or it could be using other assumptions, depending on the programming. Your data are your data. We encourage you to examine the premises behind any assignment of error limits on data that you will use and report. MATLAB, Excel, and other plotting programs allow us to insert custom error bars, once appropriate error limits are determined.

Finally, a word about error analysis in the world outside of the scientific literature. The use of error analysis leads to an important ability to critically examine numbers and arguments that appear in public conversations into which scientific results are being introduced. Numbers from measurements are uncertain; this is beyond dispute. There are good methods to ascertain *how* uncertain the data are. When one is given a number, either in a medical context, during governmental rule-making, or in some other setting, the techniques described in this book can be used to estimate the degree to which those numbers may be relied upon. As we affirm in this text, estimates and judgment calls will be part of that decision making. Keeping track of the assumptions made, sharing these transparently, and reassessing as needed is the best path to reliable understanding of our physical world.

We hope that you found this text useful and that the ease with which the tools can be implemented with software will encourage widespread and appropriate use of error analysis techniques, especially in this era in which significant decisions are justified by numbers based on scientific and engineering measurements and their interpretation.

## 6.7 Problems

1. What is the purpose of fitting a model to experimental data? Give three specific examples, not directly from the text, of when it would be desirable to fit a model to data. Put your answer in your own words.

Table 6.8. *Densities of aqueous sugar solutions, 20°C; from* CRC Handbook, *88$^{th}$ edition [26].*

| Weight fraction | g/cm$^3$ |
|---|---|
| 0.10 | 1.0381 |
| 0.20 | 1.0810 |
| 0.26 | 1.1082 |
| 0.28 | 1.1175 |
| 0.30 | 1.1270 |
| 0.40 | 1.1765 |
| 0.44 | 1.1972 |
| 0.45 | 1.2078 |
| 0.46 | 1.2079 |
| 0.50 | 1.2295 |
| 0.60 | 1.2864 |
| 0.65 | 1.3078 |
| 0.70 | 1.3472 |

2. What assumptions about the data are explicitly made when using the ordinary least-squares method of finding model parameters? How can a user check whether the assumptions are met? What are the consequences if the assumptions are not met? Put your answers in your own words.

3. Show that the two formulas for the least-squares slope in Equation 6.5 are equivalent.

4. Show that the two formulas for the least-squares intercept in Equation 6.6 are equivalent.

5. Using the data in Table 6.8 for the density of various aqueous sugar solutions [26], fit a straight-line model and use it to predict the density of a 35 wt% solution.

6. We plan to measure the density of an aqueous sugar solution and use this information to predict the solution's concentration. Obtain an equation for a solution's concentration as a function of density. Calculate the concentration of a solution with measured density of 1.248 g/cm$^3$. The data for the density of various aqueous sugar solutions are given in Table 6.8.

7. The effect of a systematic error in reading a rotameter was presented in the discussion of randomizing data acquisition in Chapter 2 (Section 2.3.4; see Figures 2.20 and 2.22). For the three sets of rotameter

Table 6.9.  *The complete dataset used to construct Examples 2.17*
*and 2.18 [46], data taken at 19.1°C; see Problems Problem 6, 10, 17, and 21.*
*Two methods were used to read the rotameter at each mass flow rate – the*
*correct method and an incorrect method. When two readers divided the data*
*acquisition task, they first divided the data systematically, resulting in the*
*correct method being used for the low flow rates and the incorrect method*
*being used for the high flow rates (see Figure 2.21). An alternative choice*
*would have been to assign the flow rates randomly.*

| Mass flow rate (kg/s) | One reader, correct method (all) (R%) | One reader, wrong method (all) (R%) | Two readers, randomly divided method (R%) | Two readers, systematically divided method (R%) |
|---|---|---|---|---|
| 0.04315 | 14.67 | 17.67 | 17.67 | 14.67 |
| 0.05130 | 17.33 | 20.00 | 17.33 | 17.33 |
| 0.05515 | 18.50 | 21.50 | 18.50 | 18.50 |
| 0.06512 | 21.50 | 24.50 | 24.50 | 21.50 |
| 0.07429 | 24.33 | 27.00 | 27.00 | 24.33 |
| 0.08545 | 27.80 | 30.50 | 27.80 | 27.80 |
| 0.09585 | 31.00 | 33.50 | 31.00 | 31.00 |
| 0.10262 | 33.00 | 35.50 | 33.00 | 33.00 |
| 0.11449 | 36.67 | 39.00 | 39.00 | 36.67 |
| 0.13201 | 42.00 | 45.00 | 45.00 | 42.00 |
| 0.14325 | 45.50 | 48.50 | 45.50 | 45.50 |
| 0.15100 | 48.33 | 51.00 | 48.33 | 51.00 |
| 0.16882 | 53.80 | 56.33 | 53.80 | 56.33 |
| 0.18326 | 58.00 | 60.67 | 60.67 | 60.67 |
| 0.19643 | 62.33 | 65.00 | 62.33 | 65.00 |
| 0.20447 | 62.50 | 67.00 | 62.50 | 67.00 |
| 0.21361 | 67.00 | 69.67 | 67.00 | 69.67 |
| 0.22880 | 72.00 | 74.67 | 74.67 | 74.67 |
| 0.23625 | 74.33 | 77.00 | 77.00 | 77.00 |
| 0.24323 | 76.00 | 79.00 | 79.00 | 79.00 |
| 0.26620 | 83.00 | 85.50 | 83.00 | 85.50 |
| 0.27831 | 87.00 | 90.67 | 90.67 | 90.67 |

calibration data from this discussion (shown in Table 6.9, divided up
systematically, divided up randomly, and all taken correctly), calculate
the least-squares best-fit lines and plot the residuals. In few sentences,
describe what you see in the residuals and discuss the implications.

Table 6.10. *Viscosity in centipoise (cp; mPa-s) for three aqueous sugar solutions as a function of temperature; from the International Critical Tables, as presented in* Perry's Handbook of Chemical Engineering *[49]*.

| | Viscosity, cp | | |
|---|---|---|---|
| $T(°C)$ | 20 wt% | 40 wt% | $60wt\%$ |
| 0 | 3.818 | 14.82 | |
| 5 | 3.166 | 11.6 | |
| 10 | 2.662 | 9.83 | 113.9 |
| 15 | 2.275 | 7.496 | 74.9 |
| 20 | 1.967 | 6.223 | 56.7 |
| 25 | 1.71 | 5.206 | 44.02 |
| 30 | 1.51 | 4.398 | 34.01 |
| 35 | 1.336 | 3.776 | 26.62 |
| 40 | 1.197 | 3.261 | 21.3 |
| 45 | 1.074 | 2.858 | 17.24 |
| 50 | 0.974 | 2.506 | 14.06 |
| 55 | 0.887 | 2.227 | 11.71 |
| 60 | 0.811 | 1.989 | 9.87 |
| 65 | 0.745 | 1.785 | 8.37 |
| 70 | 0.688 | 1.614 | 7.18 |
| 75 | 0.637 | 1.467 | 6.22 |
| 80 | 0.592 | 1.339 | 5.42 |
| 85 | 0.552 | 1.226 | 4.75 |
| 90 | | 1.127 | 4.17 |
| 95 | | 1.041 | 3.73 |

8. Calculate and plot the model, and model residuals, for the rheological entrance pressure loss versus apparent shear rate data associated with Figure 6.6 (Table 6.2). Do the residuals form any pattern that is of concern? Patterns within residuals can indicate systematic errors.

9. Calculate and plot the model, and model residuals, for the DP meter signal $(I, mA)_{raw}$ versus $(\Delta p, psi)_{std}$ calibration data given in Table 5.2 and plotted in Figure 5.10. Do the residuals form any pattern that is of concern? Patterns within residuals can indicate systematic errors.

10. Table 6.9 contains rotameter calibration data both with and without an unintended systematic error. Using the data that are most correct,

calculate the fit to the data of a straight-line model, calculate the standard deviation of $y$ at a value of $x$ ($s_{y,x}$), and create and display error bars for the individual calibration points. Comment on your results.

11. Using the equation that gives the least-squares slope (Equation 6.5), carry out an error propagation to arrive at the error on the slope, $s_m$ (answer in Equation 6.32).

12. Using the equation that gives the least-squares $y$-intercept (Equation 6.6), carry out an error propagation to arrive at the error on the intercept, $s_b$ (answer in Equation 6.33).

13. For the maple leaf mass and length data in Table 2.7, create a plot of leaf length versus mass. Calculate a best-fit line and the standard deviations of the least-squares slope and intercept. What are the 95% confidence intervals for the slope and intercept for these data? Using the error limits on slope, create a plot that shows how a fit line would vary if its slope ranged between the values associated with the error limits on the slope. Do the same for the effect of uncertainty on the intercept. Write a paragraph summarizing what you see.

14. In the worksheet given in Figure 6.9, we show the structure of an error propagation that leads to the error on $\hat{y}_0$, the model-predicted value of $y$ at $x_0$ for a least-squares curve fit. Starting with the general error propagation equation (Equation 6.57), show that $s_{\hat{y}_0}^2$ is given by Equation 6.62.

15. What is the difference in meaning between the two standard deviations $s_{\hat{y}_0}$ (Equation 6.62) and $s_{y_0,n+1}$ (Equation 6.73)? When might you use each one? Put your answers in your own words.

16. For the DP meter signal $(I, \text{mA})_{raw}$ versus $(\Delta p, \text{psi})_{std}$ calibration data given in Table 5.2 and plotted in Figure 5.10, plot the data and create a least-squares fit with 95% CI and 95% PI regions. What is the meaning of each of these regions? What do you see in the results?

17. For the randomly acquired rotameter data discussed in Example 2.18 (given in Table 6.9), plot the data and create a least-squares fit with 95% CI and 95% PI regions. What is the meaning of each of these regions? What do you see in the results?

18. Use the rheological entrance pressure loss versus apparent shear rate data associated with Figure 6.6 (Table 6.2) to make a figure like Figure 6.12, including the model and 95% CI and 95% PI regions. What is the meaning of each of these regions?

19. Derive Equation 6.77 for the error on a calibration curve (no replicates).

20. Derive Equation 6.82 for the error on a calibration curve (with replicates).

21. For the set of correct rotameter calibration data in Table 6.9, create a calibration curve (mass flow rate as a function of rotameter reading; see Example 6.6) that may be used with future data. What is the calibration error?

22. Using matrix methods (see Section 6.3), calculate a linear best-fit line for the DP meter raw calibration data given in the linked Examples 4.8 and 6.1 (Table 4.4). The answers for the best-fit slope and intercept should match those given in Equation 6.16.

23. Using array methods (see Section 6.3), carry out the calculation of the errors $s_{b_0}$, $s_{b_1}$, and $s_{b_2}$ for the Bagley data discussed in Example 6.10. Show that the results match what is obtained with LINEST. The data are rheological entrance pressure loss versus apparent shear rate [1] of a polyethylene.

24. In several examples, we considered rheological entrance pressure loss versus apparent shear rate for Bagley's data [1] on a polyethylene (a non-Newtonian fluid). In your own words, what is the meaning of the set of error bars calculated in Example 6.4 and shown in Table 6.4? Calculate a second set of error bars based on $s_{y,x}$ and the polynomial fit. Do the two sets of error match? If not, which is "right"? Discuss the data and your answer.

25. Figure 6.17 shows flow pressure drop $\Delta p$ versus volumetric flow rate $Q$ for data obtained with a flow loop (water driven by a pump, 6-ft section of tubing between pressure taps). The data show a gentle parabolic trend, to which a model is fit using polynomial least squares ($y$-intercept set to zero). What are the two model parameters $b_1$ and $b_2$? What is the uncertainty on these model parameters?

26. Figure 6.17 shows flow pressure drop $\Delta p$ versus volumetric flow rate $Q$ for data obtained with a flow loop (water driven by a pump, 6-ft section of tubing between pressure taps). The data show a gentle parabolic trend, to which a model is fit using polynomial least squares ($y$-intercept set to zero). Data error bars established through examination of the devices used to take the data ($\pm 2e_{s,\Delta p}$) are too small to account for deviations from the trendline (see the discussion in Example 6.8). Use LINEST or MATLAB to calculate error bars based on deviations from the model fit. What are the 95% CI and the 95% PI for the polynomial model? What are the meanings of the ranges bounded by these intervals? We observe $s_{y,x} > e_{s,\Delta p}$; what may be causing the additional uncertainty?

27. For the data on density of aqueous sugar solutions as a function of weight fraction $x$ in Table 6.8 (data at 20°C), find a model $\rho(x)$ that provides a

good fit (predictive) to the data. Using the model, what is the density of a solution of weight fraction 0.42? Include uncertainty limits on the value of density found.

28. For the data on density of aqueous sugar solutions as a function of weight fraction $x$ in Table 6.8 (data at 20°C), find a model $\rho(x)$ that provides a good fit (predictive) to the data. Calculate and plot the 95% CI of the model and the 95% PI of the next value of the function. Comment on your results.

29. Using the aqueous sugar solution viscosity data in Table 6.10, determine empirical models for viscosity versus temperature for the three solutions presented there (20 wt%, 40 wt%, and 60 wt% sugar in water). For each solution, calculate and plot the 95% CI of the model and the 95% PI of the next value. Do either of these regimes indicate plausible error limits for a next value of $\mu(T)$ measured in your lab? Explain your answer. (*Hint: try transformations of y or x or both.* )

30. Using the aqueous sugar solution viscosity data in Table 6.10, determine empirical models for viscosity versus temperature for the three solutions (20 wt%, 40 wt%, and 60 wt% sugar in water). What are the viscosities of the three solutions at 48°C? Include uncertainty limits on the values of viscosity found. (*Hint: Try transformations of y or x, or both. To "untransform" the error limits, use the error propagation worksheet.*)

# Appendix A Worksheets for Error Analysis

# Replicate Error Worksheet

Faith A. Morrison

*Uncertainty Analysis for Engineers and Scientists: A Practical Guide*

**(Cambridge University Press, 2021)**

This worksheet guides the user through the calculation of the standard error and 95% confidence interval on a quantity that has been measured $n$ times (replicated). The replicate-error-related standard error $e_s$ may subsequently be used in propagation-of-error calculations of derived quantities.

## Replicated Variable, $Y$:

Units: _____

| Measured values $Y_1, Y_2, ..., Y_n$ | Sample Mean, $\bar{Y}$ | Sample Variance, $s^2$ | Sample Standard Deviation, $s$ | Standard Error, $e_s = \frac{s}{\sqrt{n}}$ | | 95% Confidence Interval based on $n$ replicates (Student's $t$ distribution) |
|---|---|---|---|---|---|---|
| | | | | | | (include units) |
| | | | | | | $\pm$ |
| $Y_1$ | | | | | $n = 1$ | n/a |
| $Y_2$ | | | | | $n = 2$ | $\pm 12.7 e_s$ |
| $Y_3$ | | | | | $n = 3$ | $\pm 4.30 e_s$ |
| $Y_4$ | | | | | $n = 4$ | $\pm 3.18 e_s$ |
| $Y_5$ | | | | | $n = 5$ | $\pm 2.78 e_s$ |
| $Y_6$ | | | | | $n = 6$ | $\pm 2.57 e_s$ |
| $Y_7$ | | | | | $n \geq 7$ | $\pm 2 e_s$ |
| | | | | | $\infty$ | $\pm 1.96 e_s$ |

$$\bar{Y} \equiv \frac{1}{n} \sum_{i=1}^{n} Y_i$$

$$s^2 \equiv \frac{1}{(n-1)^2} \sum_{i=1}^{n} (Y_i - \bar{Y})^2$$

# Reading Error Worksheet

Faith A. Morrison
*Uncertainty Analysis for Engineers and Scientists:  A Practical Guide*
(Cambridge University Press, 2021)

This worksheet guides the user through the determination of the standard reading error and 95% confidence limits for the reading of a scale or from a digital readout. The standard reading error $e_{s,reading}$ may be used in propagation of error calculations of derived quantities.

| Device name: | | | |
|---|---|---|---|
| Measured Quantity: (give symbol) | | | |
| Representative value: | (include units) | | Quantity, or **Not Applicable** |
| issue | contribution to error | | |
| *Sensitivity* (from manufacturer or rule of thumb) | How much signal does it take to cause the reading to change? | 1 | |
| *Resolution:* limitation on marked scale or digital readout | Half smallest division or decimal place | 2 | |
| *Fluctuations* with time of observation | (max-min)/2 | 3 | |
| | Maximum of 1, 2, & 3: | $e_R =$ | |
| Standard reading error: | $e_{s,reading} = e_R/\sqrt{3}$ | $e_s =$ | (units) |
| | 95% Confidence Interval based on reading error: | $\pm 2e_s =$ | (units) |

*Note:* If a quantity is supplied by, for example, a manufacturer, with no indication of the uncertainty, we do not use this worksheet. Instead, see the Calibration Error worksheet.

*Rule of thumb for sensitivity:* 1 (optimistic) or 15 (pessimistic) times the last retained digit. The optimistic choice assumes any minor change is sensed; the pessimistic choice assumes that the manufacturer has displayed two uncertain digits.

Reading error, $e_R$:

# Calibration Error Worksheet

Faith A. Morrison

*Uncertainty Analysis for Engineers and Scientists: A Practical Guide*

(Cambridge University Press, 2021)

The error $e_s$ is defined as the "best-case" standard error for a quantity as determined for a brand-new unit by a manufacturer or for a particular device by someone with authority to certify the value. For example, the technical specifications of a device may indicate that it is accurate to a value $\pm 2e_s$. Alternatively, a value of a constant (the viscometer constant $\alpha$, for example) may be provided by the manufacturer with no specific uncertainty. In this case, the rule of thumb method of "least significant digit" is acceptable for evaluating the uncertainty. Finally, a user may take steps to calibrate a meter on site; this determination of error (likely to be greater than the "best case" error) has the advantage of reflecting issues associated with the particular unit in question.

| Device name: | | | |
|---|---|---|---|
| **Measured quantity:** | **Symbol:** | **Representative value:** (include units) | |
| | | | |
| | | *Estimate* of $e_s$: (or **Not Applicable**) | |
| Rule of Thumb Method: Least significant digit on provided value | Least significant digit varies by at least $\pm 1 = \pm 2e_s$ | | |
| Rigorous Method: Manufacturer maximum error allowable | $2\,e_s \approx$ | | |
| Method 3: User calibration | $2e_s \approx$ | | |
| | | | |
| | Maximum of Methods 1 – 3 | $e_s =$ | **95% CI, Calibration error only:** quantity$\pm 2e_s$ |
| | | $2e_s =$ | (units) |

**Calibration Error Worksheet (page 2)**
*Uncertainty Analysis for Engineers and Scientists: A Practical Guide*
Faith A. Morrison

Table 1: Tolerances for Volumetric Glassware [from Fritz and Schenk, *Quantitative Analytical Chemistry* (Boston: Allyn and Bacon,1987) or www.thomassci.com]

| Capacity, ml | Maximum error allowable, $2e_s$ | | | |
|---|---|---|---|---|
| | Pycnometers (Thomas Scientific) | Volumetric flasks | Volumetric Pipets | Burets |
| 5 | 0.03 | - | 0.01 | 0.01 |
| 10 | 0.04 | - | 0.02 | 0.02 |
| 25 | 0.05 | 0.03 | 0.03 | 0.03 |
| 50 | 0.08 | 0.05 | 0.05 | 0.05 |
| 100 | - | 0.08 | 0.08 | 0.10 |
| 500 | - | 0.15 | - | - |
| 1,000 | | 0.30 | - | - |

Table 2: Tolerances for Laboratory Meters

| meter | Maximum error allowable, $2e_s$ | reference |
|---|---|---|
| Thermocouple, type J or K, standard limits | 2.2°C | www.omega.com /techref/colorcodes.html |
| Thermocouple, type J or K, special limits | 1.1°C | www.omega.com /techref/colorcodes.html |
| RTD (resistance temperature detectors) | Up to 0.001°C with proper calibration | IEC751 Standard |
| Honeywell STD924 DP meter, 0–1,000 mbar | 0.075% of calibrated span | ST 3000 Smart Pressure Transmitter Models Specifications 34-ST-03-65 |

# Error Propagation Worksheet

Faith A. Morrison
*Uncertainty Analysis for Engineers and Scientists: A Practical Guide*
(Cambridge University Press, 2021)

This worksheet guides the user through the determination of the standard error $e_{s_\phi}$ of a quantity $\phi(x_1, x_2, x_3, x_4, x_5)$ that is calculated from measured quantities $x_1, x_2, x_3, x_4,$ and $x_5$. The $x_i$ are subject to uncertainties. The standard error $e_{s,x_i}$ (replicate, reading, calibration; combine in quadrature, if present) for each variable $x_i$ is determined first, and these uncertainties are propagated to determine $e_{s_\phi}$ using the relationship given below.

| $\phi(x_1, x_2, x_3, x_4, x_5)$: | Formula for $\phi$:* | Representative value of $\phi$: (include units) | 95% C.I. of $\phi$: $(\phi \pm 2e_{s,\phi})$ (include units) |
|---|---|---|---|

*Note: units must work as written.

## Measured quantities, $x_i$

| $x_i$ | Symbol | Representative value | $\dfrac{\partial \phi}{\partial x_i}$ | $e_{s,x_i}$ | $\left(\dfrac{\partial \phi}{\partial x_i}\right)^2 e_{s,x_i}^2$ |
|---|---|---|---|---|---|
| $x_1$ | | units | | | |
| $x_2$ | | units | | | |
| $x_3$ | | units | | | |
| $x_4$ | | units | | | |
| $x_5$ | | units | | | |

$$e_{s_\phi}^2 = \left(\frac{\partial \phi}{\partial x_1}\right)^2 e_{x_1}^2 + \left(\frac{\partial \phi}{\partial x_2}\right)^2 e_{x_2}^2 + \left(\frac{\partial \phi}{\partial x_3}\right)^2 e_{x_3}^2 + \left(\frac{\partial \phi}{\partial x_4}\right)^2 e_{x_4}^2 + \left(\frac{\partial \phi}{\partial x_5}\right)^2 e_{x_5}^2$$

* All variables $x_i$ must be independent for this equation to hold.

| | |
|---|---|
| $e_{s,\phi}^2 =$ | |
| $e_{s,\phi} =$ units | Standard error of calculated quantity, $\phi$ |

Note: For some quantities, you will look up the uncertainty; for example, the volume of a volumetric flask may be given as $100.00 \pm 0.04$ ml. In these circumstances it is reasonable to assume that the reported uncertainty is $\pm 2e_s$. For example, if volume is given as $100.00 \pm 0.04$ ml, then $2e_s = 0.04$ ml. [Fritz and Schenk, *Quantitative Analytical Chemistry* (Boston: Allyn & Bacon, 1987), p. 564.]

# Appendix B    Significant Figures

As discussed in the main text, rigorous error analysis is a reliable way to determine the quality of numbers that are touched by measurements. If an error analysis is performed, then the uncertainty analysis indicates how many digits to retain in a final answer. There may be occasions when a complete error analysis is not performed; in this case we follow the rules of *significant figures*.

The significant figures convention (*sig figs*) and its associated calculation rules (presented here) provide a rough way to estimate a lower bound on the uncertainty in a number. In the absence of a complete uncertainty analysis, following the *sig-figs* rules will, at the very least, prevent a result from being reported with grossly inflated precision.

## B.1  Definition of Significant Figures

The significant figures of a number are those digits that are known with assurance plus one digit that is known with less confidence. The implication of the sig-figs convention is that the error in the number is not smaller than $\pm 1$ of the smallest digit.

$$4 \text{ sig-figs answer: } 1.756 \text{ g/cm}^3 \qquad (\text{B.1})$$

$$\text{assumed significance: } 1.756 \pm 0.001 \text{ g/cm}^3 \qquad (\text{B.2})$$

A number written in scientific notation contains only significant digits. Numbers written in more casual notation may contain insignificant digits by error or by design.

In general, all digits in a properly reported number are significant, with the following exceptions:

1. **All trailing and leading zeros are not significant** when they are serving as placeholders to indicate the magnitude of the number. For example, with the number 4,500 the two zeros are not significant figures, and with the number 0.0056 all three zeros are insignificant. Note that for these two numbers written in scientific notation, neither of these problems exist: $4.5 \times 10^3$ and $5.6 \times 10^{-3}$. Scientific notation also allows us to indicate when a trailing zero may be

321

significant; for example, if a sphere is very close to 20 cm in diameter, the scientific notation for this value is $2.0 \times 10^1$ cm.

2. **Spurious digits mistakenly introduced are not significant.** The usual source of such spurious digits is the failure to truncate extra digits used in intermediate calculations. An opportunity to introduce this type of insignificance is generated every time you use a calculator. If you take two numbers such as 34 and 12 and divide them, your calculator will give you an answer with a number of digits that depends on its display, perhaps in this case 2.8333333. Because we had only two digits in the two numbers we divided, only the 2 and the 8 are significant in the answer (we give the general sig-figs rule for dividing numbers later). Note that lack of significance is not a mandate to truncate digits in intermediate calculations. The true value of 34/12 is much closer to 2.8333333 than to 2.8. We truncate to the sig figs only when reporting the final result of a calculation, not during intermediate steps (computers retain all digits at all times to reduce round-off error).

    A second common source of insignificant digits is reporting original data to a precision that cannot be supported by the quality of the equipment or process that produced the data. A common source of this type of insignificance is a digital readout. A temperature indicator may read something like 24.5°C even when the actual measuring device is not accurate to more than ±1.1°C (this is an issue of *calibration error*; see Chapter 4).

3. **Some numbers are infinitely precise.** An example of an infinitely precise number is 2 in the relationship between diameter and radius: diameter = (2)(radius). There is no uncertainty in the number 2 in this case. Another instance of infinite precision is the number 9 in the phrase "9 bricks." There are certainly not 9.5 bricks or 8.5 bricks. We are called on to make appropriate determinations of certainty versus uncertainty in these cases. A related topic is assigning sig figs to a measurement when little is known about the instrument that provided the number. An example would be if we are told the temperature in the room is 300 K, which is approximately room temperature. As written, it has one sig fig (the trailing zeros are not significant) but certainly we know room temperature better than ±100 K. This is a case where common sense should prevail.

Knowing the sig-figs rules allows us to both communicate our answers and to understand the scientific communications of others.

## B.2 Effect of Addition, Subtraction, Multiplication, and Division on Sig Figs

When performing calculations with measured quantities, we face the challenge of determining the appropriate number of sig figs to report with our answer. The basic principle is that the answer cannot be more precise than the *least* precise measurement used in the calculation. To ensure this is the case, we follow two rules – one for when two numbers are multiplied or divided, and a second for when two numbers are added or subtracted. These rules only approximately propagate the uncertainty in calculations; quite often results have more uncertainty in them than suggested by these simple rules.

For this reason we should remember that the sig-figs rules only allow us to estimate a lower bound on the uncertainty in a number. If a more accurate estimation of uncertainty is needed, the full error analysis should be performed.

When determining the number of sig figs we should report with a calculation, we follow two these **Shortcut Rules for Significance Arithmetic:**

1. When **multiplying or dividing two numbers**, the result may have no more significant figures than the least number of sig figs in the numbers that were multiplied or divided. We followed this rule earlier:

$$\frac{34}{12} = 2.8333333 = 2.8 \qquad \text{2 sig figs} \qquad \text{(B.3)}$$

Both numbers had two sig figs, and thus the answer also had two sig figs. In this calculation,

$$\frac{178}{14} = 12.7142857 = 13 \qquad \text{2 sig figs} \qquad \text{(B.4)}$$

Although the numerator has three sig figs, because the denominator has only two sig figs, the answer is rounded to two sig figs.

2. When **adding or subtracting numbers**, the number of decimal places in the numbers determines the number of significant figures in the final answer. The answer cannot contain more places after the decimal point than the smallest number of decimal places in the numbers being added or subtracted.

$$123 - 12 = 111 \qquad \text{0 decimal places, 3 sig figs}$$
$$14.36 - 0.123 = 14.237 = 14.14 \qquad \text{2 decimal places, 4 sig figs}$$
$$1.2 \times 10^2 - 0.12 \times 10^2 = 1.08 \times 10^2 = 1.1 \times 10^2 \text{ 1 decimal place, 2 sig figs}$$
$$1.2 \times 10^5 - 1.2 \times 10^3 = 1.188 \times 10^5 = 1.2 \times 10^5 \text{ 1 decimal place, 2 sig figs}$$

## B.3 Logarithms and Sig Figs

When you take the logarithm of a number with $N$ significant figures, the result should have $N$ decimal places. The number in front of the decimal place indicates only the order of magnitude. It does not enter into the count of significant figures; it simply allows us to reconstruct a number from its logarithm.

Conversely, when raising 10 to a power, if the power of 10 has $N$ decimal places, the result should have $N$ significant figures.

## B.4 Limitations of the Sig-Figs Convention

The sig-figs rules allow a quick estimation of the most optimistic uncertainty level of a quantity. As we see in this book, quite often the uncertainty in the last digit is plus/minus *more* than "1" of that digit. To determine the true uncertainty, the correct method is to follow the processes outlined in this book.

Another shortcoming of relying solely on sig figs to convey uncertainty is that there is no way to quantify the confidence level associated with sig-figs uncertainty. The statistical methods of Chapter 2 allow us to assert a 95% level of confidence, giving a far more meaningful expression of uncertainty than the estimates provided by the sig-figs rules.

# Appendix C   Microsoft Excel Functions
## for Error Analysis

- AVERAGE(range): Arithmetic average of a data range.

$$\bar{x} = \frac{1}{n} \sum_{i=1}^{n} x_i \qquad \text{(Equation 2.8)}$$

- COUNT(range): In the chosen range, the number of cells that contain numbers.
- DEVSQ($x$-range): Sum of squares of deviations of data points $x_i$ from their mean $\bar{x}$.

$$SS_{xx} = \sum_{i=1}^{n} (x_i - \bar{x})^2$$

- ERF(value): The error function is defined in Equation 2.15 (repeated here):

$$\text{Error function (defined): } \int e^{-u^2} du \equiv \frac{\sqrt{\pi}}{2} \text{erf}\,(u) + \text{constant}$$

The error function appears in some probability calculations; see Example 2.3.
- GROWTH(known-$y$'s,known-$x$'s,$x_p$): For the ordinary least squares exponential fit of the dataset $(x_i, y_i)$, the $y_p$ values on the best-fit line at the points $x_p$ (may be a range of $x_p$ values).
- INDEX(arrayname, index1, index2): Excel array function values may be accessed with this command.
- INTERCEPT($y$-range,$x$-range): For the ordinary least squares linear fit of the data set $(x_i, y_i)$, the $y$-intercept of the best-fit line.
- LINEST($y$-range,$x$-range, $y$-intercept-calc-logical, stats-calc-logical): An array function that returns parameters for an ordinary least squares linear fit. Optional logical parameters determine if the $y$-intercept is calculated ("1" indicates yes, "0" instructs to set the $y$-intercept to zero) and whether statistical quantities will be calculated ("1" indicates yes, "0" indicates no). Returns a $5 \times 2$ (or larger) array $A$ with the following quantities:

$A_{11} =$ slope, $m$

$A_{12} =$ intercept, $b$

$A_{21} =$ standard deviation of slope, $s_m$

$A_{22} =$ standard deviation if intercept, $s_b$

$A_{31} =$ coefficient of determination, $R^2 = \dfrac{SS_R}{SS_T}$

$A_{32} =$ standard deviation of $y$ at a chosen value of $x$, $s_{y,x}$

$A_{41} =$ Fisher $F$ statistic

$A_{42} =$ degrees of freedom, $\nu$

$A_{51} =$ regression sum of squares, $SS_R = SS_T - SS_E$

$A_{52} =$ residual sum of squares, $SS_E = \displaystyle\sum_{i=1}^{n} (y_i - \widehat{y}_i)^2$

Note: Total sum of squares $SS_T$ is the summation of the squared differences between the $y_i$ and the average of the $y$-data: $SS_T = SS_{yy} = \sum_{i=1}^{n}(y_i - \bar{y})^2$. $SS_T$ may be calculated from DEVSQ($y$-range).

LINEST is an array function. Array function commands may be invoked in an Excel spreadsheet in a three-step sequence: (1) select a range of the appropriate size for the results; (2) type the matrix-related command into the command line; (3) simultaneously press the keys "control," "shift," and "return." The output array will appear in the range selected in step 1.

LINEST may be used to calculate fits for polynomials of higher order, as well as for exponential functions. See Section 6.3 and Table 6.7 for more on these calculations.

- LOGEST($y$-range,$x$-range,$y$-intercept-calc-logical, stats-calc-logical): An array function that returns parameters for an ordinary least squares fit to an exponential function.
- SLOPE($y$-range,$x$-range): For the ordinary least squares linear fit of the dataset $(x_i, y_i)$, the slope of the best-fit line.
- SQRT(value): $\sqrt{(\text{value})}$
- STDEV.S(range): Standard deviation of a sample set given by range; equal to the square root of the sample variance, VAR.S() (see Equation 2.10).

$$\text{Sample standard deviation } s = \sqrt{s^2}$$

See also VAR.S.

- STEYX($y$-range,$x$-range): Standard deviation of $y(x)$; square root of the variance $s_{y,x}^2$ of $y_i(x_i)$. Used in constructing confidence and prediction intervals for values of $y_p$ at $x_p$ for an ordinary least squares fit.

$$s_{y,x}^2 = \left(\frac{1}{n-2}\right) \sum_{i=1}^{n} (y_i - \widehat{y}_i)^2 = \frac{SS_E}{(n-2)}$$

- T.DIST.2T($t_{limit}, \nu$): Integral of the Student's $t$ distribution probability density function $f(t, \nu)$ ($\nu$ degrees of freedom, Equation 2.29) underneath the two tails $-\infty \le t \le -t_{limit}$ and $t_{limit} \le t \le \infty$. Gives the total probability associated with outcomes in the tails outside of $\pm t_{limit}$.

$$1 - \text{T.DIST.2T}(t_{limit}, v) = \int_{-t_{limit}}^{t_{limit}} f(t', v)dt' \qquad \text{(Equation 2.36)}$$

- T.INV.2T($\alpha, v$): Inverse of the Student's $t$ distribution $f(t, v)$ for a given significance level $\alpha$ and degrees of freedom $v$. The function returns the value of $t_{limit}$ such that the area under the probability density function of the Student's $t$ distribution in the tails above and below $\pm t_{limit}$ is equal to the probability $1 - \alpha$ (expressed as a fraction).

$$\alpha = 1 - \int_{-t_{limit}}^{t_{limit}} f(t', v)dt'$$

$$t_{limit} = t_{\frac{\alpha}{2}, v} = \text{T.INV.2T}(\alpha, v) \qquad \text{(Equation 2.48)}$$

- TREND(known-$y$'s,known-$x$'s,$x_p$): For the ordinary least squares linear fit of the data set $(x_i, y_i)$, the $y_p$ values on the best-fit line at the points $x_p$ (may be a range of $x_p$ values).
- VAR.S(range): Variance of a sample set given by range (see Equation 2.9).

$$\text{Sample variance } s^2 \equiv \left( \frac{\sum_{i=1}^{n} (x_i - \bar{x})^2}{n - 1} \right)$$

When the variance is calculated on an entire population, rather than on a sample set, the $(n - 1)$ in the denominator becomes $n$. See also STDEV.S().

# Appendix D   MATLAB Functions for Error Analysis

## D.1  MATLAB–Excel Table

Table D.1. *Equivalent commands in Excel and MATLAB for simple statistical and arithmetic calculations.*

| Operation | Excel call | MATLAB call |
|---|---|---|
| Mean | AVERAGE(range) | mean(array) |
| Variance | VAR.S(range) | var(array) |
| Standard deviation | STDEV.S(range) | std(array) |
| Sum of numbers | SUM(range) | sum(array) |
| $\Pr[-t_{limit} \le t \le t_{limit}]$ | $1 - \text{T.DIST.2T}(t_{limit}, \nu)$ | $\text{tcdf}(t_{limit}, \nu)$ $- \text{tcdf}(-t_{limit}, \nu)$ or $\text{tcdf}(t_{limit}, \nu)$ $- \text{tcdf}(t_{limit}, \nu, \text{'upper'})$ |
| $t_{limit} = t_{\frac{\alpha}{2}, \nu}$ | $\text{T.INV.2T}(\alpha, \nu)$ | $-\text{tinv}\left(\frac{\alpha}{2}, \nu\right)$ |
| **Least Squares** (Section 6.3) | | |
| Number of model parameters, $p$ | *as desired* | *as desired* |
| Number of points, $n$ | COUNT(range) | length(array) |
| Matrix $A$, $p = 2, 3$ | *not applicable* | $A_{p=2} = [\text{ones(n,1),xdata}]$ $A_{p=3} =$ $[\text{ones(n,1),xdata,(xdata.^2)}]$ |
| Vector bhat, $p = 2, 3$ | *not applicable* | bhat = A\ydata |
| Matrix Cov, $p = 2, 3$ | *not applicable* | Cov = $(s_{yx}^2)$*inv(A.'*A) |
| $SS_E$ | STEYX(y-range,x-range)^2 *(n − p) | sum((ydata- A*(A\ydata)).^2) |
| $s_{y,x}$ (RMSE) | STEYX(y-range,x-range) | sqrt($SS_E$/(n − p)) |
| $SS_{xx}$ | DEVSQ(x-range) | sum((xdata- mean(xdata)).^2) |

Table D.1. *(cont.)*

| Operation | Excel call | MATLAB call |
|---|---|---|
| $SS_{yy} = SS_T$ | DEVSQ(y-range) | sum((ydata-mean(ydata)).^2) |
| $SS_{xy}$ | COUNT(y-range) *COVARIANCE(y-range,x-range) | sum( (xdata-mean(xdata)) *(ydata-mean(ydata)) ) |
| Intercept, $b$, $p = 2$ | INTERCEPT(y-range,x-range) | bhat(1) |
| Slope, $m$, $p = 2$ | SLOPE(y-range,x-range) | bhat(2) |
| $b_j$, $j = 0, 1, 2$, $p = 3$ | INDEX(LINEST(y-range,x-range^{1,2},1,1),1,3-j) | bhat($j + 1$) |
| $s_b$ | INDEX(LINEST(y-range,x-range,1,1),2,2) | sqrt(Cov(1,1)) |
| $s_m$ | INDEX(LINEST(y-range,x-range,1,1),2,1) | sqrt(Cov(2,2)) |
| $s_{b_j}$, $j = 0, 1, 2$, $p = 3$ | INDEX(LINEST(y-range,x-range^{1,2},1,1),2,3-j) | sqrt(Cov(j+1,j+1)) |
| $R^2 \equiv 1 - \frac{SS_E}{SS_T}$ | RSQ(y-range,x-range) | 1-$SS_E$/$SS_T$ |
| $R^2_{adj} \equiv 1 - \frac{(n-1)}{(n-p)} \frac{SS_E}{SS_T}$ | 1-$(n - 1)$/$(n - p)$*$SS_E$/$SS_T$ | 1-$(n - 1)$/$(n - p)$*$SS_E$/$SS_T$ |

## D.2 MATLAB Code for Selected Examples

In this section we provide MATLAB code for several examples, some of the code may be used for end-of-chapter problems as well. All of the code is presented as MATLAB functions, although most (the *run* functions) are better characterized as macros – sets of code that could be entered manually into the MATLAB command line. The output of the functions has been suppressed from being displayed; to access the results, type the name of the variable of interest and the values will print in the Workspace.

1. **Function** *error prop*

The MATLAB function *error prop* creates an error propagation table for the function *fcn* provided in the call. This code includes the internal function *dphidx*(), used to calculate the derivatives.

```
function results = error_prop(fcn,nominal,names,SE,dx)
%
%  function results = error_prop(fcn,nominal,names,SE,dx)
%
%  A function to perform the error propagation worksheet calculations
%  from Morrison, "Uncertainty Analysis for Engineers and Scientists:
%  A Practical Guide," Cambridge University Press
%  ============================================================
%     where fcn:  user-supplied function
%            nominal: representative values of all variables
%            names: names of variables
%            SE: standard errors of variables
%            dx (optional):  user-supplied differentials for determining
%                            partial derivatives (default:
%                            1e-8 times nominal values)
%
%  (c) 2019 by Faith Morrison and Tomas Co
%  @ Michigan Technological University

% Make inputs to be columns.

    nominal = nominal(:);
    names = names(:);
    SE = SE(:);

% Set the default dx for calculating the partial derivatives;
% protect it from negative values of dx and values of zero.

    if nargin<5
        dx = max((1e-8)*abs(nominal),1e-18);
    end

% Calculate partial derivatives; function dphidx is below.

    z = dphidx(fcn,nominal,dx);
    partial_derivatives = z;

% Build the table.

    weighted_SE_squared = (z.*SE).^2;
    tab = table(nominal,partial_derivatives,SE,weighted_SE_squared);
    tab.Properties.RowNames = names; %remove semicolon to echo table
```

```
% Obtain the final result.

    SEfcn = sqrt(sum(weighted_SE_squared));
%    disp(['Propagated Error : ', num2str(SEfcn)]); %displays SE result
%    disp(' ============================= ')

% Build a structure array to save all results together.

    results.table = tab;
    results.SEfcn = SEfcn;
    results.fcn = fcn;

end

function z = dphidx(fcn,x,dx)

    n = length(x);

    for i=1:n
        nx = x;
        nx(i) = nx(i)+dx(i);
        z(i) = (feval(fcn,nx)-feval(fcn,x))/dx(i);
    end
    z = z(:);

end
```

## 2. Example 5.4: Blue Fluid Density Determination

The function *runEP density* provides a series of MATLAB commands that lead up to invoking *error prop* to determine the error propagation table for the Blue Fluid density determination with pycnometers.

```
% runEP_density.m
%
% A set of instructions to calculate propagated error from function
% @phi_density (separate file, density from mass by difference).
% We call the external function @error_prop to perform the error
% propagation worksheet calculations with the inputs specified below.
%
% ========================================================

% Provide variable names.
% To concatinate names (strings), use curly brackets.

    names={'Mass full','Mass empty','Volume'};
    n = length(names);       % n is the number of variables

% We input nominal values and SE for each variable. Units must be
% consistent.  Include unit conversions, if necessary, explicitly.

    nominal=[30.800, 13.410, 10.00];
    SE=[ 5.0e-5,  5.0e-5,  0.02];

%  Call the function "error_prop" to create the output.

    results=error_prop(@phi_density,nominal,names,SE);

%  The output of this set of instructions is "results", which is a
%  MATLAB structure array containing three fields:
%
%  results.table = n x 5 table equivalent to the error prop worksheet
%  results.SEfcn = standard error of the function
%  results.fcn = function for which error was propagated
```

3. **Function** *phi density*
   The function *phi density* provides the function that calculates density from masses and the volume of a pycnometer.

```
%   function  y = phi_density(x)
%
%   A function to evaluate the density from mass by difference
%   ====================================================
%      where x(1)  :   mass of full pycnometer
%            x(2)  :   mass of empty pycnometer
%            x(3)  :   volume of full pycnometer

    M_full  = x(1);
    M_empty = x(2);
    volume  = x(3);

    y = (M_full - M_empty)/volume;

end
```

4. **Example 5.6: Error Propagation Single DP Meter Calibration Point**
   The functions *runEP DP* and *runEP multipleDP* provide a series of MATLAB
   commands that lead up to invoking *error prop* to determine the error propagation
   table for DP meter calibration data with an air-over-Blue-Fluid manometer. This
   problem is solved first for a single call of the function *error prop* and
   subsequently (next page) for multiple calls of the function for different values of
   the standard error for manometer fluid height.

```
% runEP_DP.m
%
% A set of instructions to calculate propagated error from function
% @phi_DP (separate file, differential pressure from a manometer).
% We call the external function @error_prop to perform the error
% propagation worksheet calculations with the inputs specified below.
%
%   ========================================================

% Provide variable names.
% To concatinate names (strings), use curly brackets.

    names={'Density','Gravity','Height_1','Height_2','Unit_Conversion'};
    n = length(names);        % n is the number of variables

% We input nominal values and SE for each variable. Units must be
% consistent.  Include unit conversions, if necessary, explicitly.

    nominal=[1.734,980.66,29.8,46.2,1.4504e-5];
    SE=[ 1.5e-3,  0,   .0577,  .0577, 0];

%  Call the function "error_prop" to create the output.

    results=error_prop(@phi_DP,nominal,names,SE);

%  The output of this set of instructions is "results", which is a
%  MATLAB structure array containing three fields:
%
%  results.table = n x 5 table equivalent to the error prop worksheet
%  results.SEfcn = standard error of the function
%  results.fcn = function for which error was propagated
```

```
% runEP_multipleDP.m
%
% A set of instructions to calculate propagated error for function @phi_DP
% (separate file, differential pressure from a manometer) for seven
% values of the standard error for height.  We call the external function
% @error_prop to perform the error propagation worksheet calculations with
% the inputs specified below.
%
% ========================================================

% Provide variable names.
% To concatinate names (strings), use curly brackets.

    names={'Density','Gravity','Height_1','Height_2','Unit_Conversion'};
    n = length(names);        % n is the number of variables

% We input nominal values and SE for each variable. Units must be
% consistent.  Include unit conversions, if necessary, explicitly.

    nominal=[1.734,980.66,29.8,46.2,1.4504e-5];
    SE = [ 1.5e-3,   0, .0577, .0577, 0  ]

% Make the first call of the function "error_prop" to create the output.
% Place the results into the first element of the array of
% structure arrays, "results".

results(1)=error_prop(@phi_DP,nominal,names,SE);

% Try out different values of the standard error on gravity.
% Replace SE(3:4) with new values to obtain different combined errors.
% Place the results into the appropriate element of the array of
% structure arrays, "results".

SE_height = [.0577, 8e-2, 9e-2, 10e-2, 15e-2, 20e-2, 50e-2];
SE(3:4)=SE_height(2) , results(2) = error_prop(@phi_DP,nominal,names,SE);
SE(3:4)=SE_height(3) , results(3) = error_prop(@phi_DP,nominal,names,SE);
SE(3:4)=SE_height(4) , results(4) = error_prop(@phi_DP,nominal,names,SE);
SE(3:4)=SE_height(5) , results(5) = error_prop(@phi_DP,nominal,names,SE);
SE(3:4)=SE_height(6) , results(6) = error_prop(@phi_DP,nominal,names,SE);
SE(3:4)=SE_height(7) , results(7) = error_prop(@phi_DP,nominal,names,SE);

% The output of this set of instructions is "results", which is a
% structure array containing 6 MATLAB structure arrays, each of which
% (for i:1=6) contains three fields:
%
% results(i).table = n x 5 table equivalent to the error prop worksheet
% results(i).SEfcn = standard error of the function
% results(i).fcn = function for which error was propagated
```

5. **Function** *phi DP*

The function *phi DP* provides the function that calculates differential pressures from manometer fluid heights.

```
function    y = phi_DP(x)
%
%  function  y = phi_DP(x)
%
%  A function to evaluate the differential pressure from a manometer
%  =====================================================
%     where x(1) :   fluid density
%           x(2) :   acceleration due to gravity
%           x(3) :   height 1
%           x(4) :   height 2
%           x(5) :   unit conversion from dynes/cm^2 to psi
%

  rho = x(1);
  g = x(2);
  h1 = x(3);
  h2 = x(4);
  alpha0 = x(5);

  y = alpha0*rho*g*(h1+h2);

end
```

6. **Example 5.7: Error Propagation on Set of DP Meter Calibration Points**

The function *runEP DPall* provides the series of MATLAB commands that lead up to calling *error prop* to determine the error propagation tables for the complete set of DP meter calibration data with an air-over-Blue-Fluid manometer. Code to produce the example's figure is also included.

```
% runEP_DPall.m
%
% Example 5.7 (LINKED)
% Calculate the values and error limits on the DP meter calibration data
% from Example 4.8.
%
% This is a set of instructions to calculate propagated error from
% function @phi_DP (separate file, differential pressure from a manometer).
% We call the external function @error_prop to perform the error
% propagation worksheet calculations with the inputs specified below.
%
% ========================================================
%
% Import or enter the data from Table 4.4 (15 by 3)
% (manometer fluid heights h1 and h2 versus calibration DP meter readings).

    data = importdata('dp.csv');
    calDP_I = data(:, 1);  % signal from DP meter, current mA
    h1 = data(:, 2);        % height up from reference level, cm
    h2 = data(:, 3);        % height down from reference level, cm
    j = length(h1);         %j is the number of data points

% Provide  names of the variables in error propagation;
% to concatinate names (strings), use curly brackets.

    names={'Density','Gravity','Height_1','Height_2','Unit_Conversion'};
    n = length(names);       % n is the number of variables

% The dataset has various values of Height_1 and Height_2 but
% several variables are the same for every calculation.  Provide
% the stationary variables here.

    alpha0 = 14.696/1.01325e6;    %psi/(dynes/cm^2)
    rho = 1.734;                  %g/cm3
    g = 980.66;                   %cm/s^2

% We designate  SE for each variable.

    SE=[ 1.5e-3,  0,  0.15,  0.15,  0];

% We calculate the value of deltaP for each data point in the dataset
% and calculate the standard errors for each data point (SEdeltaP) with
% error propagation. Call function @error_prop(fcn,nominal,names,SE,dx).
% Use "for" loop to step through individual data points.

    for i=1:j
        nominal=[rho, g, h1(i), h2(i), alpha0];
        results(i)=error_prop(@phi_DP,nominal,names,SE);
        SEdeltaP(i) = results(i).SEfcn;
        deltaP(i) = feval(@phi_DP,nominal);   %feval evaluates function
    end
    SEdeltaP = SEdeltaP';
    deltaP = deltaP';
```

```
%  For each data point (i=1:15) we have stored a MATLAB structure array
%  containing three fields:
%  results(i).table = n x 5 table equivalent to the error prop worksheet
%  results(i).SEfcn = standard error of the function
%  results(i).fcn = function for which error was propagated

%  Plot the results.

   hold on

   xdata = deltaP;
   ydata = calDP_I;
   err = 2*SEdeltaP;      % 95% level of confidence error limits
   errorbar(xdata,ydata,err,'horizontal','.k'); %plot with x error bars
   axis([0 4 4 20])
   xtickformat('%.2f')    % force two decimal places on x-axis labels
   ytickformat('%.1f')    % force two decimal places on y-axis labels
   set(gca,'XMinorTick','on','YMinorTick','on') % turn on minor ticks
   xlabel('Differential pressure from manometer standard, psi')
   ylabel('DP meter signal, mA')

%  Add a fit line using fit obtained in Chapter 6 (see text).

   slope = 4.326143;
   intercept = 3.995616;
   yfit = slope * xdata + intercept;
   plot(xdata,yfit,'-k'); %plot fit line from Ch6
   box on

   hold off
```

7. **Example 6.1: Ordinary Least Squares on DP Meter Calibration**

Shown are the series of MATLAB commands that perform ordinary least squares on the fit on the DP meter calibration data.

```
% Example 6.1:  (LINKED)
% Calculate the ordinary least squares model fit on the DP meter
% calibration data from Examples 4.8 and 5.7.
% To obtain exactly the same answer as Excel, we need to include all
% the precision we have (large number of digits) to avoid round-off error.
% The xdata and ydata are from runEP_DPall.
%

% Enter or read in the data.

xdata =[   0.256497670137178    0.577119757808651    0.813886837935278
0.961866263014419    1.159172163119941    1.410737185754481    1.677100150896936
1.874406051002458    2.254219908705587    2.476189046324299    2.668562298927183
3.181557639201540    3.378863539307062    3.576169439412583    3.660024446957431
];
ydata = [5.20    6.60    7.50    8.20    9.00    10.00   11.20    12.00    13.70
14.60    15.50    17.80    18.70    19.50    19.90];

% Perform the ordinary least squares calculation.

ydata = ydata';
n = length(xdata); %number of data points
p = 2; %number of parameters in the least squares model
A = [ones(n,1),xdata']; %column of ones concatinated with xdata
bhat = A\ydata;    % coefficients of the fit; see Section 6.3
%                    b = bhat(1)
%                    m = bhat(2)

% Calculate R-Squared (coefficient of determination)

yhat = A*bhat;    % model predicted values of y
deviations = ydata - yhat;
sqdeviations = deviations.^2;
SSE = sum(sqdeviations);
ymean = mean(ydata);
SST = sum((ydata-ymean).^2);
Rsquared = 1-SSE/SST;
format long % show all digits in results
%
% The results for Example 6.1 are:
% The interceopt is bhat(1)
% The slope is bhat(2)
% R^2 is Rsquared
```

8. **Example 6.3: Ordinary Least Squares on a Single Rheological Pressure Drop versus $L/R$ Plot (Bagley Plot)**

Shown are the series of MATLAB commands that perform an ordinary least squares fit on a single dataset of rheological data of driving pressure drop versus capillary $L/R$ (Bagley plots; one value of apparent shear rate).

```
% Example 6.3:  Bagley fit to DeltaP vs. L/R rheological data
% for one apparent shear rate.
% Import or type in the data from Example 6.3 (driving pressure
%  versus L/R, 8 data points)
% See Section 6.3 for matrix method.
%
%
xdata =[0   1.8400e+00   2.8800e+00   3.7400e+00   4.7800e+00   9.2000e+00
1.8420e+01   2.9340e+01];
ydata = [8.5200e+01   1.1720e+02   1.4480e+02   1.6070e+02   1.8260e+02
2.7390e+02   4.6080e+02   6.7360e+02];
ydata = ydata';
n = length(xdata); %number of data points
p = 2; %number of parameters in the least squares model
A = [ones(n,1),xdata']; %column of ones concatinated with xdata
bhat = A\ydata; %coefficients of the least squares fit
%
% Calculate syx (root mean squared error).
%
yhat = A*bhat;
deviations = ydata - yhat;
sqdeviations = deviations.^2;
SSE = sum(sqdeviations);
% Alternatively:  SSE_newway=sum((ydata-A*(A\ydata)).^2);
syx = sqrt(SSE/(n-p)); %Table D.1
%
% Calculate covariance matrix; the diagonal is var of coefficients bhat
% (matches Excel result); see Section 6.3.3.
%
ATransA = A.'*A;
C = inv(ATransA);
CovMatrix = syx^2*C;
stdev_b = [sqrt(CovMatrix(1,1)),sqrt(CovMatrix(2,2))];
%
% Calculate R-squared
%
ymean = mean(ydata);
SST = sum((ydata-ymean).^2);
Rsquared = 1-SSE/SST;
%
% The results for Example 6.3 are:
% The slope is bhat(2)
% The standard error of slope is stdev_b(2)
% R^2 is Rsquared
```

9. **Example 6.4: Ordinary Least Squares on a Set of Rheological Pressure Drop versus $L/R$ Plots (Bagley Plot)**
   Shown are the series of MATLAB commands that perform several ordinary least squares fits on several datasets of rheological data of driving pressure drop versus capillary $L/R$ (Bagley plots; several values of apparent shear rate).

```
% Example 6.4:  Bagley fit to DeltaP vs. L/R rheological data
% for five apparent shear rates.
% Import or type in the data from Example 6.3 (driving pressure
%  versus L/R, 5 vectors of 8 data points.
% See Section 6.3 for matrix method).
%
%
data = importdata('examp6x4.csv');
xdata = data(2:9,1);
ydata5 = data(2:9,2);
ydata4 = data(2:9,3);
ydata3 = data(2:9,4);
ydata2 = data(2:9,5);
ydata1 = data(2:9,6);
labels = data(1,2:6);
%
% Concatenate the data so that it will work in a "for" loop.
%
ydataconcat = [ydata1,ydata2,ydata3,ydata4,ydata5];
n = length(xdata); %number of data points
p = 2; %number of parameters in the least squares
%
% For index=1:5 perform the regression.
%
for index = 1:5
    ydata = ydataconcat(:,index);
    A = [ones(n,1),xdata]; %column of ones concatenated with xdata
    bhat = A\ydata;  % coefficients of the regression
%
% Calculate syx (root mean squared error) from the definition.
%
    yhat = A*bhat;
    deviations = ydata - yhat;
    sqdeviations = deviations.^2;
    SSE = sum(sqdeviations);
    syx = sqrt(SSE/(n-p));  %Not passed out of function, but could be.
%
% Calculate covariance matrix; the diagonal is varience of
% coefficients slope and intercept (in bhat vector).
%
    ATransA = A.'*A;
    C = inv(A.'*A);
    CovMatrix = syx^2*C;
    stdev_b = [sqrt(CovMatrix(1,1)),sqrt(CovMatrix(2,2))];
%
% Calculate R-squared
%                              ;
    ymean = mean(ydata);
    SST = sum((ydata-ymean).^2);
    Rsquared = 1-SSE/SST;
%
% Choose what to pass out of the function.
%
    bhats(:,index) = bhat;
    stdev_bs(:,index) = stdev_b;
    Rsquareds(:,index) = Rsquared;

end
%
% The results of Example 6.4 (Table 6.4) are:
% Deltap_entrance (5 values) = bhats(1,:)
% standard error of intercepts - stdev_bs(1,:)
% R^2 for each fit:  Rsquareds
```

10. **Example 6.5: Ordinary Least Squares on an Overall Heat Transfer Coefficient versus Steam Flow Rate Plot**

Shown are the series of MATLAB commands that perform an ordinary least squares fit on a dataset of heat exchanger overall heat transfer coefficient versus steam mass flow rate (for 10 similar double-pipe heat exchangers). Code to produce the accompanying figure is included.

```
% Example 6.5:  Linear fit to heat exchanger data.
% Import or type in the data from Example 6.5 (overall heat transfer
% coefficient versus steam mass flow rate).
% See Section 6.3 for matrix method.
%
% Import or enter the data from Table 6.5(86 by 2).
%
data = importdata('he.csv');
xdata = data(:, 1);
ydata = data(:, 2);
%
% We need the t statistic.
%
n = length(xdata); %number of data points
p = 2; %number of parameters in the least squares model
tstat = -tinv(0.025,n-p);
%
% Least squares (see Section 6.3 for matrix method).
%
A = [ones(n,1),xdata]; %column of ones concatinated with xdata
bhat = A\ydata;  % coefficients of the least squares fit
%
% Calculate syx (root mean squared error).
%
yhat = A*bhat;  %predicted y's for xdata values
deviations = ydata - yhat;
sqdeviations = deviations.^2;
SSE = sum(sqdeviations);
syx = sqrt(SSE/(n-p));
%
% Calculate covariance matrix; the diagonal is var of coefficients bhat.
% Store in stdev_b.
%
ATransA = A.'*A;
C = inv(A.'*A);
CovMatrix = syx^2*C;
stdev_b = [sqrt(CovMatrix(1,1)),sqrt(CovMatrix(2,2))];
%
% Calculate R-squared
%
ymean = mean(ydata);
SST = sum((ydata-ymean).^2);
Rsquared = 1-SSE/SST;
%
% Calculate the intervals for the 95% CI and PI regions.
% Use arbitrary xvec that spans the range of the data.
%
xvec = ((1:91)-1)*0.055+1;
xvec = xvec';
xmean = mean(xdata);
SSxx = dot(xdata-xmean,xdata-xmean);
hold = 1/n + ((xvec-xmean).^2)/SSxx;
CIinterval = tstat*syx*sqrt(hold);
PIinterval = tstat*syx*sqrt(1+hold);
%
```

```
% Create vectors for plotting:  the model fit plus/minus the intervals
% Reproduce Figure 6.12.
%
x0 = [ones(91,1),xvec];
fitmodel = x0*bhat;
PIplus = fitmodel+PIinterval;
PIminus = fitmodel-PIinterval;
CIplus = fitmodel+CIinterval;
CIminus = fitmodel-CIinterval;
plot(xdata,ydata,'k+',xvec,fitmodel,'k',xvec,PIplus,'r:',xvec,PIminus,'r:',xv
ec,CIplus,'g--',xvec,CIminus,'g--');
title('Example 6.5 (MATLAB)')
xlabel('steam mass flow rate, g/s')
ylabel('overall heat transfer coefficient, kW/m2K')
%
% The results for Example 6.5 are:
% The intercept is bhat(1)
% The slope is bhat(2)
% syx is syx
% SSxx is SSxx
% xmean is the mean of the xdata
% The standard error of slope is stdev_b(2)
% The standard error of intercept is stdev_b(1)
% R^2 is Rsquared
```

11. **Examples 6.9 and 6.10: Ordinary Least Squares Polynomial Fit Including Error on Parameters**

Shown are the series of MATLAB commands that perform a second-order polynomial fit using ordinary least squares. The data are rheological entrance pressure losses as a function of apparent shear rate. The standard deviations of the three model parameters are also calculated.

```
% Examples 6.9 and 6.10:  Polynomial fit to rheological data of entrance
% pressure loss versus apparent shear rate (5 data points) including
% error on the regression coefficients.
%
% Import or type in the data from Example 6.4 (entrance pressure
% losses versus apparent shear rate, 5 data points).
% Place in column matrices.

xdata =[40 60 90 120 250];
xdata = xdata';
ydata = [37.54 66.87 85.44 106.13 164.07];
ydata = ydata';

% Create the A matrix for polynomial fit (p=3 coefficients).

p = 3;  % p= number of parameters in the least-squares model
n = length(xdata);
A = [ones(n,1),xdata,(xdata.^2)]; %column of ones concatenated with xdata

%  Calculate the fit (bhat).

bhat = A\ydata;  %the coefficients

% Calculate syx (root mean squared error).

yhat = A*bhat;
deviations = ydata - yhat;
sqdeviations = deviations.^2;
SSE = sum(sqdeviations);
% SSE_newway=sum((ydata-A*(A\ydata)).^2);
syx = sqrt(SSE/(n-p));

% Calculate covariance matrix; the diagonal is var of coefficients bhat.

ATransA = A.'*A;
C = inv(ATransA);
CovMatrix = syx^2*C;
stdev_b = [sqrt(CovMatrix(1,1)),sqrt(CovMatrix(2,2)),sqrt(CovMatrix(3,3))];
stdev_b = stdev_b';

% Calculate R-squared

ymean = mean(ydata);
SST = sum((ydata-ymean).^2);
% SST_newway = sum((ydata-mean(ydata)).^2);
Rsquared = 1-SSE/SST;  %matches
Rsquared_adjusted = 1-(n-1)/(n-p)*SSE/SST; %matches

% The results for this example are:
% Regression coefficients:  3x1 matrix bhat
% Standard deviation of coefficients:  3x1 matrix stdev_b
```

12. **Example 6.11: Ordinary Least Squares Polynomial Fit Including 95% CI and PI Regions**

Shown are the series of MATLAB commands that perform a second-order polynomial fit using ordinary least squares. The data are rheological entrance pressure losses as a function of apparent shear rate. The 95% confidence interval and the 95% prediction interval of the fit are both calculated and Figure 6.19 is created.

```
% Example 6.11:  Polynomial fit to rheological data of entrance pressure
% loss versus apparent shear rate with % 95% CI and PI on polynomial fit
% shown and Figure 6.19 created.
%
% Load results from Example 6.9 and 6.10 calculations.

ls_MATLAB_Ex6x9;

% Choose the values of x0 and create a 3x126 matrix
% as discussed in the text.

x0 = importdata('X0.csv');
x0 = x0';

% We need the t statistic.

tstat = -tinv(0.025,n-p);

%As noted in text, calculate 126 by 3 matrix 'hold'.

hold = x0'*C;

%As noted in text, vector dot the rows of hold with the columns of x0.
%This creates hold2, which is x0' C x0, a quantity under the sqrt.
%From hold2 we calculate the intervals for the CI and the PI
% (see equations in text).

for index = 1:length(x0)
    hold2(index) = hold(index,:)*x0(:,index);
end
CIinterval = tstat*syx*sqrt(hold2');
PIinterval = tstat*syx*sqrt(1+hold2');

%Create vectors for plotting:  the model fit plus/minus the intervals.
%Produces Figure 6.19.

fitmodel = x0'*bhat;
PIplus = fitmodel+PIinterval;
PIminus = fitmodel-PIinterval;
CIplus = fitmodel+CIinterval;
CIminus = fitmodel-CIinterval;
xvec = x0(2,:);
plot(A(:,2),ydata,'ko',xvec,fitmodel,'k',xvec,PIplus,'r:',xvec,PIminus,'r:',xvec,CIplus,'
g--',xvec,CIminus,'g--');
title('Example 6.11 (MATLAB)')
xlabel('apparent shear rate, 1/s')
ylabel('entrance pressure loss, bar')
```

# Appendix E  Statistical Topics

This appendix includes discussion of selected topics in statistics related to uncertainty analysis. For more information, see the literature [4, 17, 18, 20, 23, 27, 28, 32, 56].

## E.1  Properties of Probability Density Functions

The probabilities associated with continuous stochastic variables are expressed as integrals over their probability density functions (pdf) [38].

Definition of $f$:
probability is expressed
as an integral of a
probability density function (pdf)
(continuous stochastic variable)

$$\Pr[a \leq w \leq b] \equiv \int_a^b f(w')dw' \quad \text{(Equation 2.12)}$$

where $f(w)$ is the probability density function for continuous stochastic variable $w$. The quantity $f(w')dw'$ is the probability that $w$ takes on a value between $w'$ and $w' + dw'$. The integral represents the result of adding up all the probabilities of observations between $a$ and $b$. If we let $a = -\infty$ and $b = \infty$, then the probability is 1.

Integrating over
all probabilities
equals 1

$$\int_{-\infty}^{\infty} f(w')dw' = 1 \quad \text{(E.1)}$$

For all other intervals $[a, b]$, the probability is less than 1 and is calculated by integrating the pdf $f(w)$ between $a$ and $b$.

For a continuous stochastic variable, the mean $\bar{w}$ of a variable $w$ characterized by probability density distribution $f(w)$ is calculated as follows:

346

| Mean of a continuous stochastic variable | $$\bar{w} = \mu = \int_{-\infty}^{\infty} f(w')w'dw'$$ | (Equation 3.9) |

For a continuous stochastic variable characterized by probability distribution $f(w)$, the variance is calculated as follows:

| Variance of a continuous stochastic variable | $$\text{Var}(w) = \sigma^2 = \int_{-\infty}^{\infty} (w' - \bar{w})^2 f(w')dw'$$ |

(Equation 3.10)

where $\bar{w}$ is the mean from Equation 3.9.

**Example E.1: A variance identity.** *For a continuous stochastic variable x of mean $\bar{x} = \mu$ and variance $\text{Var}(x) = \sigma^2$, show that the following identity holds:*

$$\text{Var}(x) = \sigma^2 = \left( \int_{-\infty}^{\infty} x'^2 f(x')dx' \right) - \mu^2 \qquad \text{(E.2)}$$

where $f(x)$ is the probability density function for $x$.

**Solution:** We calculate variance of a continuous stochastic variable with Equation 3.10.

$$\sigma^2 = \int_{-\infty}^{\infty} (x' - \bar{x})^2 f(x')dx'$$

$$= \int_{-\infty}^{\infty} \left( x'^2 - 2x'\bar{x} + \bar{x}^2 \right) f(x')dx'$$

$$= \int_{-\infty}^{\infty} x'^2 f(x')dx' - 2\bar{x} \int_{-\infty}^{\infty} x' f(x')dx' + \bar{x}^2 \int_{-\infty}^{\infty} f(x')dx'$$

For all continuous stochastic variables $\int_{-\infty}^{\infty} f(x')dx' = 1$, which we use to simplify the last term. Substituting Equation 3.9 into the middle term we obtain

$$\sigma^2 = \int_{-\infty}^{\infty} x'^2 f(x')dx' - 2\mu^2 + \mu^2$$

$$\sigma^2 = \int_{-\infty}^{\infty} x'^2 f(x')dx' - \mu^2$$

## E.2 Classic Distributions

### Normal Distribution

It turns out that we can quite often reasonably assume that the stochastic effects in measurements are *normally distributed*; that is, their pdf has the shape of the normal distribution (Figure 2.6).

Normal
probability
distribution
(pdf)

$$f(x) = \frac{1}{\sqrt{2\sigma^2\pi}} e^{-\frac{(x-\mu)^2}{2\sigma^2}}$$

(Equation 2.17)

The normal distribution (the bell curve) is a symmetric distribution with two parameters: a mean $\mu$, which specifies the location of the center of the distribution, and the standard deviation $\sigma$, which specifies the spread of the distribution. With the assumption that the random effects are normally distributed, we reduce the problem of determining the system's pdf to determining these two parameters.

The probability density of the normal distribution is such that 68% of the probability density is located within $\pm$ one standard deviation of the mean ($\mu \pm \sigma$). Within two standard deviations ($\mu \pm 2\sigma$), this amount rises to 95% probability; within $\mu \pm 3\sigma$, the probability is 99.7% (see Problem 2.37). Many random processes produce stochastic results that are normal or approximately normal.

The sampling (of size $n$) of a normal distribution of known standard deviation $\sigma$ results in a normal distribution of variance $\sigma/\sqrt{n}$. The sampling of a normal distribution of unknown standard deviation produces another common statistical distribution, the Student's $t$ distribution.

## Student's $t$ Distribution

The Student's $t$ distribution is the sampling distribution of a normal distribution with unknown standard deviation. The Student's $t$ distribution is written in terms of the scaled deviation of sample mean, $t$:

$$\text{scaled deviation } t \equiv \frac{(\bar{x} - \mu)}{s/\sqrt{n}}$$

(Equation 2.25)

where $n$ is the number of observations in the sample and $s$ is the sample standard deviation. The Student's $t$ distribution is a function of $t$ and $v$, where $v = n - 1$ is called the degrees of freedom.

Student's $t$
distribution (pdf)

$$f(t, v) = \frac{\Gamma\left(\frac{v+1}{2}\right)}{\sqrt{v\pi}\,\Gamma\left(\frac{v}{2}\right)} \left(1 + \frac{t^2}{v}\right)^{-\left(\frac{v+1}{2}\right)}$$

(Equation 2.29)

$\Gamma()$ is a standard mathematical function called the gamma function [54]. The Student's $t$ probability distribution function $f(t, v)$ is plotted in Figure 2.9.

The Student's $t$ distribution has a mean of zero and a standard deviation that varies with the number of degrees of freedom:

$$v > 2, \quad \text{standard deviation} = \frac{v}{v-2}$$
$$1 < v \leq 2, \quad \text{standard deviation} = \infty$$

When the degrees of freedom are large, the Student's $t$ distribution becomes the standard normal distribution (the normal distribution with $\mu = 0$ and $\sigma = 1$).

The proof that the Student's $t$ distribution is the sampling distribution of a normal distribution with unknown standard deviation may be found in the literature [34].

# E.3  Combined Uncertainty and Error Limits (GUM)

In Chapter 2 we discuss the statistics behind the 95% confidence interval of the mean, which is used to obtain a good value of a measured quantity along with appropriate error limits. Errors associated with two common systematic aspects of measurements – the need to read a value in some way (reading error) and the role of device calibration in the value obtained (calibration error) – are addressed in Chapters 3 and 4. In Chapter 5, we see how to combine error contributions such as replicate, reading, and calibration errors to obtain a combined standard uncertainty $e_{s,cmbd}$, both for a measured quantity subject to multiple error sources and for a derived quantity arrived at by mathematical manipulations of stochastic variables. Including the reading and calibration errors in the combined standard uncertainty is correct, but it complicates the assessment of what the appropriate multiplier is on $e_{s,cmbd}$ when error limits are constructed.

In this section, we discuss in greater detail the justification behind our recommendation that final answers arrived at through combining errors be expressed with error limits $\pm k_{95} e_{s,cmbd} \approx \pm 2 e_{s,cmbd}$, an interval called the *expanded uncertainty*. The material in this section is largely drawn from the *Guide to the Expression of Uncertainty in Measurement*, known by the acronym GUM [21]. GUM is a product of an international working group in the field of metrology, the science of measurement.

## Types of Standard Errors

The standard errors associated with measurements can be divided into two types based on the method used to determine the standard errors [21]:

*Type A:* Method of evaluation of uncertainty by the statistical analysis of a series of observations. Replicate error $e_{s,random} = \frac{s}{\sqrt{n}}$ is a Type A standard error.

*Type B:* Method of evaluation of uncertainty by means other than the statistical analysis of series of observations. Reading error $e_{s,reading} = \frac{e_R}{\sqrt{3}}$ and calibration error $e_{s,cal}$ are Type B standard errors. Evaluating a Type B error requires exercising scientific judgment and sometimes making estimates, as discussed in Chapters 3 and 4. Types of relevant information that go into determining a Type B uncertainty include [21]:

- Previous measurement data
- Experience with or general knowledge of the behavior and property of relevant materials and instruments
- Manufacturer's specifications
- Data provided in calibration and other reports
- Uncertainties assigned to reference data taken from the literature

Methods of arriving at replicate, reading, and calibration standard errors, and of combining them into a combined standard uncertainty $e_{s,cmbd}$, are explained in the main text [21]. After obtaining $e_{s,cmbd}$, and following the logic of the 95% confidence interval of the mean of replicates, we seek to establish an interval based on $e_{s,cmbd}$ that will, with 95% confidence, capture the true value of the variable. Since $e_{s,cmbd}$ contains Type B standard errors, however, we cannot develop the desired interval using

the same statistical process that leads from a known pdf to the 95% confidence interval.
We must follow a different path.

## Expanded Uncertainty

The *expanded uncertainty* is defined as

$$\text{Expanded uncertainty at a level of confidence of 95\%} \equiv \pm k_{95} e_{s,cmbd} \qquad \text{(E.3)}$$

where $k_{95}$ is called the *coverage factor* for a level of confidence of 95%. The coverage
factor typically takes on values between $k_{95} = 2$ (approximately what is obtained for
95% CI based on the normal distribution or on the Student's $t$ distribution with $v \longrightarrow$
$\infty$) and $k_{95} = 3$, a number that broadens error limits when small sample sizes reduce
our ability to estimate the variance (such as when using data replicates to estimate the
mean of a stochastic variable).

The determination of the precise value of $k_{95}$ requires considerable reflection. The
convolution of pdfs, when they are known, can lead to a rigorous determination of $k_{95}$
(see GUM Annex G, sections G.1.3–G.1.6 [21]). These convolutions are rarely, if ever,
carried out either due to our not knowing the pdfs or due to satisfaction with easier
methods involving estimation (discussed later in this appendix).

Estimations of $k_{95}$ are based on the central limit theorem [21, 38]. The features
of the $k_{95}$ derivations are summarized in GUM (see sections G.2–G.3 [21]). The
result of the consideration of many possible standard error types and distributions is
the recommendation that $k_{95}$ be calculated from the Student's $t$ distribution using an
effective number of degrees of freedom $v_{eff}$. The GUM-recommended expression for
$v_{eff}$ is that given by the Welch–Satterthwaite formula [10, 51, 57]:

$$v_{eff} = \frac{e_{s,i}^4}{\sum_{i=1}^{N} \frac{e_{s,i}^4}{v_i}} \qquad \text{(E.4)}$$

where $e_{s,i}$ is the ith standard error used in the error propagation or combination of
errors and $v_i$ is the number of degrees of freedom associated with the ith contribution.
If the contribution is a Type A standard error, $v_i$ will be readily determined from the
statistics; if the contribution is a Type B standard error, the discussion in GUM guides
how $v_i$ is to be determined (section G.4 [21]).

We do not give the details of the determination of the Type B $v_i$ that are needed to
obtain $v_{eff}$ because we are adopting an even simpler heuristic, also described in GUM
(section G.6.6 [21]). As stated there:

*For many practical measurements in a broad range of fields, the following conditions
prevail:*

*– The estimate y of the measurand Y is obtained from estimates $x_i$ of a significant
  number of input quantities $X_i$ that are describable by well-behaved probability
  distributions, such as the normal and rectangular distributions;*

- *The standard uncertainties $[e_{s,i}]$ of these estimates, which may be obtained from either Type A or Type B evaluations, contribute comparable amounts to the combined standard uncertainty $[e_{s,cmbd}]$ of the measurement result $y$;*
- *The linear approximation implied by the law of propagation of uncertainty is adequate [This is the error propagation rule presented in Chapter 5]...;*
- *The uncertainty of $[e_{s,cmbd}]$ is reasonably small because its effective degrees of freedom $v_{eff}$ has a significant magnitude, say greater than 10.*

*Under these circumstances, ... [and] based on the discussion given in this annex, including that emphasizing the approximate nature of the uncertainty evaluation process and the impracticality of trying to distinguish between intervals having levels of confidence that differ by one or two percent, one may do the following:*

- *adopt $[k_{95}] = 2$ and assume that $[\pm 2e_{s,cmbd}]$ defines an interval having a level of confidence of approximately 95 percent;*
- *or, for more critical applications, adopt $[k_{95}] = 3$ and assume that $[\pm 3e_{s,cmbd}]$ defines an interval having a level of confidence of approximately 99 percent.*

We thus arrive at a simple expression for the expanded uncertainty:

$$\text{Expanded uncertainty at a level of confidence of 95\%} \equiv \boxed{\pm 2e_{s,cmbd}} \tag{E.5}$$

This version of error limits for nonrandom errors appeared in the main text as Equation 2.56.

Note that we are drawing a distinction between a 95% confidence interval, which is derived from an associated probability distribution function such as the Student's $t$ distribution (Type A), and a reasoned interval such as the expanded uncertainty (Type B), which is assigned an estimated level of confidence through less statistically direct methods. Members of the scientific field concerned with measurements recommend maintaining the distinction between these two types of intervals, so as to represent more precisely the degree of statistical rigor associated with error limits.

There are circumstances for which an expanded uncertainty of $\pm 2e_{s,cmbd}$ is less reliable. The details are in GUM [21]; the cases most likely to be problematic for us are listed next.

### Problematic Cases for $k_{95} = 2$ Approximation

1. A replicate standard error $s/\sqrt{n}$ with small $n$ ($n < 6$) dominates the combination. In this case, the reading and calibration errors should be neglected and $k_{95}$ set to $t_{0.025,n-1}$. The reasoning for this is that when $n < 6$, $t_{0.025,n-1}$ is greater than 2.5; using the replicate $t$ will be more representative than the data than using $k_{95} = 2$ (GUM section G.2.3 [21]).
2. If reading error based on resolution dominates, it is recommended to neglect the replicate and calibration errors and determine a 95% confidence interval from the reading error pdf, the rectangular distribution (GUM sections G.2.3 and G.6.5 [21]).

In summary, for errors with "well-behaved distributions," where replicate, reading, and calibration errors "contribute comparable amounts," and are based on "the approximate nature of the uncertainty evaluation process and the impracticality of trying to distinguish between intervals having levels of confidence that differ by one or two percent" (all quotes from GUM), it is quite reasonable to create error limits (called expanded uncertainty) with a level of confidence of 95% using $\pm 2e_{s,cmbd}$ (Equation E.5).

## E.4  Additional Statistics Topics

### Mean and Variance of the Sampling Distribution of a Normal Distribution, $\sigma$ Known

When the standard deviation of the underlying normal distribution is known, we can derive the standard deviation of the sampling distribution.

For a population with a normal distribution of standard deviation $\sigma$, we take a sample of $n$ measurements of the variable $x$. The mean of the sample is the mean of the underlying normal distribution.

$$\mu = \bar{x} = \frac{\sum_{i=1}^{n} x_i}{n}$$

As $n \to \infty$, the estimate of the mean becomes very accurate. The variance of the mean is variance of this expression.

$$\text{Var}(\bar{x}) = \text{Var}\left(\sum_{i=1}^{n} \frac{x_i}{n}\right)$$

The variance of a function $y = \phi(x_i)$ may be calculated from Equation 5.16 (shown for the case of three variables) [38].

$$\sigma_y^2 = \left(\frac{\partial \phi}{\partial x_1}\right)^2 \sigma_{x_1}^2 + \left(\frac{\partial \phi}{\partial x_2}\right)^2 \sigma_{x_2}^2 + \left(\frac{\partial \phi}{\partial x_3}\right)^2 \sigma_{x_3}^2 + 2\left(\frac{\partial \phi}{\partial x_1}\right)\left(\frac{\partial \phi}{\partial x_2}\right)\text{Cov}(x_1, x_2)$$
$$+ 2\left(\frac{\partial \phi}{\partial x_1}\right)\left(\frac{\partial \phi}{\partial x_3}\right)\text{Cov}(x_1, x_3) + 2\left(\frac{\partial \phi}{\partial x_2}\right)\left(\frac{\partial \phi}{\partial x_3}\right)\text{Cov}(x_2, x_3)$$

(Equation 5.16)

Thus we can calculate $\text{Var}(\bar{x}) = \sigma_y^2$ for $y = \phi(x_i) = \sum_{i=1}^{n} x_i/n$. If we take all the individual measurements to be uncorrelated, then all the covariance terms are zero. Carrying out the derivatives with respect to each $x_i$ and substituting, we obtain the variance of $\bar{x}$:

$$\text{Var}(\bar{x}) = \text{Var}\left(\sum_{i=1}^{n} \frac{x_i}{n}\right)$$
$$= \sum_{i=1}^{n} \left(\frac{1}{n}\right)^2 \text{Var}(x_i)$$

$$= \sum_{i=1}^{n} \left(\frac{1}{n^2}\right) \sigma^2$$

$$= \frac{1}{n^2} \left(n\sigma^2\right)$$

$$= \frac{\sigma^2}{n}$$

## Covariance of Least-Squares Slope and Intercept

We can calculate the covariance of the least-squares slope and intercept by taking advantage of the linearity properties of covariance.

In Chapter 6, we calculate the least-squares fit to a straight line for dataset $(x_i, y_i)$, with $i = 1, 2, \ldots n$. Beginning with Equations 6.5 and 6.6 for slope and intercept, we first show that these may be written in terms of sums over the appropriate coefficient $(\alpha_i, \beta_i$, see below) multiplied by the values of $y_i$ [55]:

Least-squares slope

$$m = \frac{SS_{yx}}{SS_{xx}} = \frac{\sum_{i=1}^{n}(x_i - \bar{x})(y_i - \bar{y})}{SS_{xx}}$$

$$= \sum_{i=1}^{n} \frac{(x_i - \bar{x})}{SS_{xx}} y_i - \frac{\sum_{i=1}^{n}(x_i - \bar{x})\bar{y}}{SS_{xx}}$$

$$= \sum_{i=1}^{n} \frac{(x_i - \bar{x})}{SS_{xx}} y_i - \frac{\bar{y}\left(\sum_{i=1}^{n} x_i - n\bar{x}\right)}{SS_{xx}}$$

$$= \sum_{i=1}^{n} \left[\frac{(x_i - \bar{x})}{SS_{xx}}\right] y_i$$

$$= \sum_{i=1}^{n} \alpha_i y_i \qquad (E.6)$$

Least-squares intercept

$$b = \bar{y} - m\bar{x}$$

$$= \sum_{i=1}^{n} \frac{y_i}{n} - \frac{\bar{x}\sum_{i=1}^{n}(x_i - \bar{x})y_i}{SS_{xx}}$$

$$= \sum_{i=1}^{n} \left[\frac{1}{n} - \frac{\bar{x}}{SS_{xx}}(x_i - \bar{x})\right] y_i$$

$$= \sum_{i=1}^{n} \beta_i y_i \qquad (E.7)$$

We seek the covariance of $m$ and $b$.

$$\mathrm{Cov}(m, b) = \mathrm{Cov}\left(\sum_{i=1}^{n} \alpha_i y_i, \sum_{i=1}^{n} \beta_i y_i\right) \qquad (E.8)$$

Note that from the linearity properties of covariance:

$$\text{Cov}((aU + bW), (cU + dW)) = ac\text{Cov}(U, U) + ad\text{Cov}(U, W)$$
$$+bc\text{Cov}(W, U) + bd\text{Cov}(W, W) \qquad \text{(E.9)}$$

For the least-squares slope $m$ and intercept $b$, we write:

$$\text{Cov}(m, b) = \text{Cov}\left(\sum_{i=1}^{n} \alpha_i y_i, \sum_{i=1}^{n} \beta_i y_i\right)$$
$$= \alpha_1\beta_1\text{Cov}(y_1, y_1) + \alpha_1\beta_2\text{Cov}(y_1, y_2) + \ldots \alpha_1\beta_n\text{Cov}(y_1, y_n)$$
$$+\alpha_2\beta_1\text{Cov}(y_2, y_1) + \alpha_2\beta_2\text{Cov}(y_2, y_2) + \ldots \alpha_2\beta_n\text{Cov}(y_2, y_n)$$
$$+\ldots$$
$$+\alpha_n\beta_1\text{Cov}(y_n, y_1) + \alpha_n\beta_2\text{Cov}(y_n, y_2) \ldots + \alpha_n\beta_n\text{Cov}(y_n, y_n) \qquad \text{(E.10)}$$

The $y_i$ are independent and homoscedastic; therefore

$$\begin{array}{ll} \text{for } i \neq k & \text{Cov}(y_i, y_k) = 0 \\ \text{for } i = k & \text{Cov}(y_i, y_k) = s_{y,x}^2 \end{array} \qquad \text{(E.11)}$$

This leads to the following simplification:

$$\text{Cov}(m, b) = \sum_{i=1}^{n} s_{y,x}^2 \left(\frac{(x_i - \bar{x})}{SS_{xx}}\right)\left(\frac{1}{n} - \frac{\bar{x}(x_i - \bar{x})}{SS_{xx}}\right)$$

$$= s_{y,x}^2 \sum_{i=1}^{n} \left(\frac{(x_i - \bar{x})}{SS_{xx}}\right)\left(\frac{SS_{xx} - n\bar{x}(x_i - \bar{x})}{nSS_{xx}}\right)$$

$$= \frac{s_{y,x}^2}{n(SS_{xx})^2} \sum_{i=1}^{n} (x_i - \bar{x})\left[SS_{xx} - n\bar{x}(x_i - \bar{x})\right]$$

$$= \frac{s_{y,x}^2}{n(SS_{xx})^2} \left[SS_{xx}\sum_{i=1}^{n} x_i - \bar{x}SS_{xx}\sum_{i=1}^{n} 1 - n\bar{x}\sum_{i=1}^{n}(x_i - \bar{x})^2\right]$$

$$= \frac{s_{y,x}^2}{n(SS_{xx})^2} [n\bar{x}SS_{xx} - n\bar{x}SS_{xx} - n\bar{x}SS_{xx}]$$

$$= -\frac{s_{y,x}^2\bar{x}}{(SS_{xx})}$$

<table>
<tr><td>Covariance of<br>least-squares<br>slope and intercept</td><td>$\text{Cov}(m, b) = -\dfrac{s_{y,x}^2\bar{x}}{(SS_{xx})}$</td><td>(E.12)</td></tr>
</table>

# Appendix F   Choosing an Empirical Model

Empirical modeling connects an equation to data. When trying to understand new data, we typically begin by plotting the data along with error bars. If we decide to model the data, the next step is to choose a model equation that behaves like the data behave. In this section, we summarize some common model choices and describe how the equations behave. This section serves as a short catalog of empirical models.

## F.1  Straight Lines, Parabolas, and Cubics

Straight lines, parabolas, and cubics are three types of polynomials. The simplest polynomial is the straight line (polynomial of order 1), which may be written as

$$
\begin{array}{c}
\text{Equation for a straight line} \\
(y\text{-intercept version}) \\
\text{2 parameters: } m, b
\end{array}
\qquad
\boxed{y = mx + b}
\qquad (\text{F.1})
$$

$$(y - b) = mx \qquad (\text{F.2})$$

where $m$ is the slope of the line and $b$ is the $y$-intercept. An alternative form that uses the $x$-intercept, $c = -b/m$, is

$$
\begin{array}{c}
\text{Equation for a straight line} \\
(x\text{-intercept version})
\end{array}
\qquad
\boxed{\begin{array}{c} y = m(x - c) \\ c = -b/m \end{array}}
\qquad (\text{F.3})
$$

Note that plotting the data either as $(y - b)$ versus $x$ or as $y$ versus $(x - c)$ produces a line that passes through the origin $(0, 0)$ for the axes chosen (Figure F.1).

If the relationship shown by the data has some curvature, the next simplest functional relationship to use as an empirical model is the quadratic equation (polynomial of order 2), which has the shape of a parabola (Figure F.2, top):

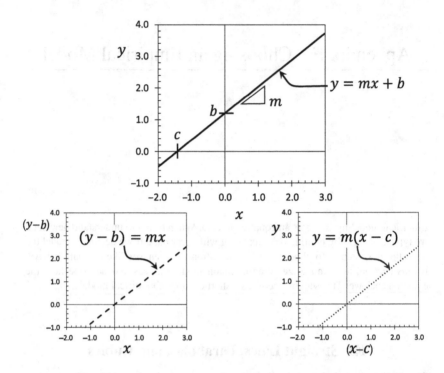

Figure F.1 Shown at the top is the straightforward $y(x)$ plot of a straight-line model, $y = mx + b$. Note that plotting the data either as $(y - b)$ versus $x$ or as $y$ versus $(x - c)$ produces a line that passes through the origin $(0, 0)$ for the axes chosen ($c = b/m$).

Quadratic equation
3 parameters: $a_0, a_1, a_2$

$$y = a_0 + a_1x + a_2x^2 \qquad \text{(F.4)}$$

For the case in which model parameter $a_2$ is positive, the parabola opens upward; for $a_2$ negative, the parabola opens downward. Parabolas are symmetric with respect to the vertical line through the *vertex* of the parabola, the point at either the minimum of the curve (for $a_2 > 0$) or the maximum of the curve (for $a_2 < 0$). The coordinates $(x_0, y_0)$ of the vertex are obtained by finding the minimum (or maximum) of the quadratic equation:

Find the $x$ location
of the minimum/maximum:

$$\frac{dy}{dx} = a_1 + 2a_2x = 0$$

$$x_0 = \left(-\frac{a_1}{2a_2}\right) \qquad \text{(F.5)}$$

## Order 2 polynomials: Parabolas

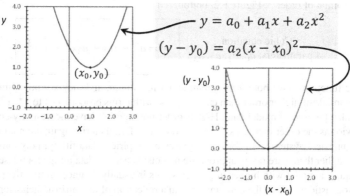

## Higher-order polynomials:

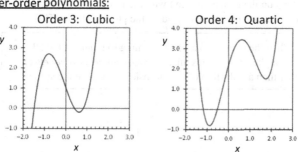

Figure F.2 A second-order polynomial (parabola) has a single extremum, shown here as the vertex point $x_0, y_0$. Increasing the order of a polynomial increases the number of minima/maxima that may be present. Note that a plot of $(y - y_0)$ versus $(x - x_0)$ is a centered parabola with vertex at the origin of the chosen axes, and the curve is symmetric with respect to the $(y - y_0)$-axis.

Find the $y$ location
of the minimum/maximum:

$$y_0 = a_0 + a_1 x_0 + a_2 x_0^2$$

$$y_0 = \left( a_0 - \frac{a_1^2}{4a_2} \right) \tag{F.6}$$

Note that a plot of $(y - y_0)$ versus $(x - x_0)$ is a centered parabola with vertex at the origin of the chosen axes, and the curve is symmetric with respect to the $(y - y_0)$-axis (Figure F.2, top right).

Quadratic equation
(centered version)

$$
\begin{aligned}
(y - y_0) &= a_2 (x - x_0)^2 \\
x_0 &= \left( -\frac{a_1}{2a_2} \right) \\
y_0 &= \left( a_0 - \frac{a_1^2}{4a_2} \right)
\end{aligned}
\tag{F.7}
$$

The addition of an $x^3$ term to a quadratic equation produces a cubic equation (polynomial of order 3; Figure F.2, bottom left).

Cubic equation
4 parameters: $a_0, a_1, a_2, a_3$

$$y = a_0 + a_1 x + a_2 x^2 + a_3 x^3 \qquad \text{(F.8)}$$

Cubic functions have both a minimum and a maximum, and they are not symmetric. Polynomials of higher order than cubic – those with terms proportional to $x^4$, $x^5$, etc. – may also be used to model data. Higher-order polynomials have increasingly complex behavior as the order increases. Data that rise, then fall, then rise again are not unheard of in physical systems, but for the purposes of empirical data fitting, polynomials of order higher than 3 are rarely appropriate models: When the data jump around, seeming to require a model with higher-order terms, this is usually an uncertainty effect due to the stochastic nature of the data, rather than a reflection of the intrinsic behavior of the system. Error bars on the data associated with the fit can clarify if variations observed in data are due to uncertainty or are caused by the physics (Figure F.3).

> When the data jump around, seeming to require a model
> with higher-order terms, this is usually *an uncertainty effect*
> rather than a reflection of the intrinsic behavior of the system.

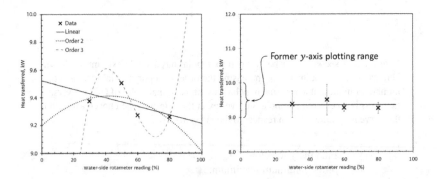

Figure F.3  When error bars are omitted, scattered data may give the impression of a complex data response, particularly when plotting software is allowed to select the $x$- and $y$-axis scales. For the heat-transfer data shown, if we are guided by guesswork and the $R^2$ of the fit, we would find $R^2$ values for linear ($R^2 = 0.32$), quadratic ($R^2 = 0.45$), and cubic ($R^2 = 1$) fits, that would lead, perhaps, to choosing a cubic model as the "best fit." When error bars are present, however, it becomes clear that the data variation is due to uncertainty and that the data are, in fact, independent of $x$ (data credit: N. Benipal [2]). A coefficient of determination $R^2 = 1$ is a danger sign – in the present case, $R^2 = 1$ for the cubic because four data points is the minimum amount of data to specify a cubic. This is an inappropriate "fit," because it ignores the stochasticity of the data. For more on coefficient of determination $R^2$, see the discussion around Equation 6.17.

Note that two points define a line, three points define a parabola, and, in general, $(p + 1)$ points define a polynomial of order $p$. For polynomial fits to observations of stochastic variables, the number of data points $n$ to be used to determine model parameters must be greater than $p$. The difference between the available number of points $n$ and the required minimum number of points $(p + 1)$ to determine the model is called the number of *degrees of freedom* associated with the fit. Ideally the number of observations $n$, and hence the degrees of freedom, would greatly exceed $(p + 1)$, giving enough information to separate stochastic variation from the true trend, which is to be captured by the model. We return to the discussion of the concept of degrees of freedom later in the appendix.

## F.2 Exponentials

Polynomials can capture a wide variety of physical behavior, but they fail to describe *exponential* behavior, which is a very strongly varying type of response. Exponential growth and decay are quite common in physical systems; for example, bacteria grow exponentially and chemical radioactivity decays exponentially. Exponential growth occurs when the quantity's growth rate is proportional to the quantity's current value. For such a system, if $z$ is the quantity of something that changes with time, and $dz/dt$ is its growth rate with respect to time, then requiring the value of $z$ to be proportional to growth rate gives

$$\frac{dz}{dt} = \alpha z \tag{F.9}$$

where the value of model parameter $\alpha$ determines how strongly the growth rate is tied to the current value of $z$. For the case in which $\alpha$ is constant, we can integrate Equation F.9 to obtain a function $z(t)$ that gives the values of $z$ at all times.

$$\frac{dz}{z} = \alpha dt$$

$$\int \frac{dz}{z} = \int \alpha dt \tag{F.10}$$

$$\ln z = \alpha t + C_1$$

$$z(t) = e^{\alpha t} e^{C_1}$$

where $C_1$ is the integration constant. If $z_0$ is the value of $z$ at $t = 0$, we can solve for the integration constant and obtain the final equation for $z(t)$.

$$\begin{array}{c} \text{Equation for exponential} \\ \text{2 parameters: } \alpha, z_0 \end{array} \qquad \boxed{z(t) = z_0 e^{\alpha t}} \tag{F.11}$$

Exponential functions change quite rapidly, and conventional $z$ versus $t$ plots of Equation F.11 may be difficult to interpret (Figure F.4, top left). At short times, the function appears to be unchanging and has a value near zero, and at long times it grows rapidly, in an accelerating fashion, and soon leaves the field of view. When rapidly

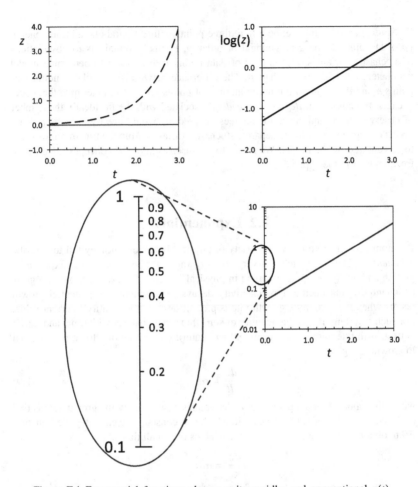

Figure F.4 Exponential functions change quite rapidly, and conventional $z(t)$ plots (upper left) of equations such as Equation F.11 are unable to convey much about the process. Taking the logarithm of the $z$ data (plotted on the ordinate or $y$-axis, upper right) produces a graphical representation that is much clearer. Logarithmic plotting with a logarithmic scale (lower right) is one in which the label of the scale displays the antilogarithm, making it easier to interpret a log-scaling graph in terms of the untransformed variable.

changing data are plotted in a linear–linear fashion such as this, it is difficult to tell the nature of the growth rate – is it exponential? The sum of two exponentials? Does it follow some other functional form?

More can be discerned about the functional form of the behavior of rapidly or extensively changing data if instead of plotting $z(t)$ on a linear–linear plot, we plot the common logarithm $\log() = \log_{10}()$ of the data versus the dependent variable (Figure

F.4, top right). The same data shown at top left in Figure F.4 have been transformed by computing the $\log_{10}$ of the $z$ data. The transformed values $\log z$ are plotted versus the $t$ data. When plotted this way, the systematic variations and trends in the data at short times become visible. Because of the nature of the act of taking the logarithm of a quantity, when we plot $\log z$ versus $t$ rather than $z$ versus $t$, data over a very wide range of values (many decades) may be displayed and considered on one compact graph. On such a graph, an exponential function such as Equation F.11 would appear as a straight line, and we can obtain the parameter $\alpha$ from the slope.

$$z(t) = z_0 e^{\alpha t} \tag{F.12}$$

$$\log z = (\log z_0) + (\alpha t) \log e$$

$$\log z = \underbrace{(\log z_0)}_{\text{intercept}} + \underbrace{(\alpha \log e)}_{\text{slope}} t \tag{F.13}$$

A disadvantage of plots of the common logarithm of a quantity is that, except for values equal to powers of 10, whose logarithms are whole numbers (for example, $\log(1) = 0$, $\log(10) = 1$, and $\log(100) = 2$), all other numbers have hard-to-recognize values when transformed to $\log y$ (for example, $\log 3 \approx 0.477$). We can make plots of $\log y$ versus $x$ more informative by marking, for example, 0.477 as "3," and $\log 2 \approx 0.301$ as "2." A plot marked in this way is said to use *logarithmic scaling* on the axes so marked (Figure F.4, bottom).

The logarithmic scale is a hybrid representation of data in which the data on that scale have been transformed by taking their base-10 logarithm. When the logarithmic scale is used, the convention is to replace the ordinate axis-label "$\log y$," and its associated evenly spaced values of $\log y$, with the label "$y$" and unevenly spaced tick marks on the axis placed where the *antilogarithmic* values are whole numbers (or whole numbers multiplied by a power of 10).

Antilogarithm examples 

$$\log 3 = 0.47712125472$$

$$3 = \text{antilog}(0.47712125472) = 10^{0.47712125472}$$

$$\log 0.3 = -0.528787453$$

$$3 \times 10^{-1} = \text{antilog}(-0.528787453) = 10^{-0.528787453}$$

On a logarithmic-scale axis, the placement of tick marks labeled with whole-number values of $y$ allows the user of the graph to think in terms of the untransformed variable, while benefiting from the visual advantages of the transformation of $y$ to $\log y$ scaling. A graph that uses logarithmic scaling on one axis is called a *semilog* plot; when both axes are logarithmically scaled, it is called a *log–log* or *double-log* plot. The logarithmic scaling on a graph may be recognized by the uneven placement of the tick marks.

As mentioned earlier, when a variable is changing exponentially with respect to its dependent variable, the semilog plot of the variable is a straight line. When the variable's behavior is not purely exponential, a semilog plot can still be revealing and can help us to deduce an appropriate model for the data. Consider, for example, the data at the top of Figure F.5, which is the time-dependent response of a thermocouple signal when the thermocouple was switched from an ice-water bath to a bath of boiling water (Example 6.12). On this linear–linear plot, we observe the signal rapidly transition from the starting temperature to the final temperature. During the transition, the difference

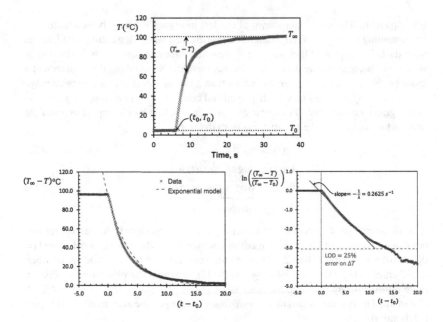

Figure F.5 The time-dependent response of a thermocouple signal when the thermocouple was switched from an ice-water bath to a bath of boiling water displays first-order response behavior. Plotting $(T_\infty - T)$ versus time shows the region of exponential decay. A semilog plot of these variables allows us to fit the exponential region to a straight line. The limit of determination was calculated (LOD; see Section 4.3 and Example 6.12). Data below the LOD (with more than 14% error) were omitted during calculation of the fit, as were data recorded before the transfer of the thermocouple ($t < t_0$).

between the signal's ultimate value $T_\infty$ and the signal's time-dependent value $T(t)$ is decreasing rapidly with time. This suggests that the temperature-difference data may be represented by a model such as

$$\text{(difference from ultimate value)} = \text{(exponential decay)} \qquad \text{(F.14)}$$

$$(T_\infty - T) = Ae^{\alpha t}$$

More generally, for such behavior we can write

$$(y_\infty - y) = Ae^{-t/\lambda} \qquad \text{(F.15)}$$

where we have written the equation in terms of the generic variable $y$ and have used $\alpha \equiv -1/\lambda$ both to emphasize that the data *decays* (rather than grows) and to use a parameter ($\lambda$) with units of time. If we label the point when the response begins as $(t_0, y_0)$, we can incorporate these coordinates into the model. First, we substitute the starting point into the model.

$$(y_\infty - y) = Ae^{-t/\lambda}$$
$$(y_\infty - y_0) = Ae^{-t_0/\lambda}$$
$$A = (y_\infty - y_0)e^{t_0/\lambda} \qquad (F.16)$$

Now, we use Equation F.16 to eliminate $A$ from Equation F.15:

$$(y_\infty - y) = (y_\infty - y_0)e^{-(t-t_0)/\lambda}$$
$$= y_\infty e^{-(t-t_0)/\lambda} - y_0 e^{-(t-t_0)/\lambda}$$
$$y = y_\infty - y_\infty e^{-(t-t_0)/\lambda} + y_0 e^{-(t-t_0)/\lambda}$$

We obtain the standard form for this model by subtracting $y_0$ from both sides and rearranging.

$$y = y_\infty - y_\infty e^{-(t-t_0)/\lambda} + y_0 e^{-(t-t_0)/\lambda}$$
$$(y - y_0) = y_\infty\left(1 - e^{-(t-t_0)/\lambda}\right) - y_0\left(1 - e^{-(t-t_0)/\lambda}\right)$$
$$= (y_\infty - y_0)\left(1 - e^{-(t-t_0)/\lambda}\right)$$

| | |
|---|---|
| First-order response model<br>4 parameters: $y_0$, $y_\infty$, $\lambda$, $t_0$ | $(y - y_0) = (y_\infty - y_0)\left(1 - e^{-(t-t_0)/\lambda}\right)$    (F.17) |

This model, called *first-order response*, displays a rising behavior in a transitioning region beginning at $(t_0, y_0)$, but settles out to long-time behavior $y = y_\infty$ after a characteristic amount of time related to the value of the parameter $\lambda$. The value of $\lambda$ may be determined from the data by creating a plot designed to give a straight line. We can rearrange Equation F.17 as follows:

$$1 - \left(\frac{y - y_0}{y_\infty - y_0}\right) = e^{-(t-t_0)/\lambda}$$
$$\left(\frac{y_\infty - y}{y_\infty - y_0}\right) = e^{-(t-t_0)/\lambda}$$
$$\ln\left(\frac{y_\infty - y}{y_\infty - y_0}\right) = -\frac{1}{\lambda}(t - t_0) \qquad (F.18)$$

which shows that a semilog plot of $(y_\infty - y/(y_\infty - y_0)$ versus $(t - t_0)$ would display a straight line of slope $(-1/\lambda)$ and an intercept of zero. A related fit using the thermocouple data is shown at the bottom right of Figure F.5.

# F.3 Power Laws

Power-law functions are also useful for empirical model fitting. For dependent variable $y$ and independent variable $x$, the power-law function has the form

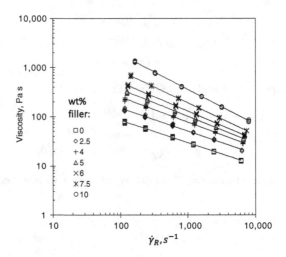

Figure F.6 The viscosity of carbon-filled liquid crystal polymer composites is a power-law function of the shear rate at the wall, $\dot{\gamma}_R$. Shown are the viscosity functions of composites of Vectra with carbon black (Ketjenblack) [22].

Power-law equation
2 parameters: $A$, $B$

$$y = Ax^B \tag{F.19}$$

where $B$ is the power-law exponent and $A$ is a scalar pre-factor. If we take the logarithm of both sides of Equation F.19, we obtain

$$\log y = \underbrace{(\log A)}_{intercept} + \underbrace{(B)}_{slope} \log x \tag{F.20}$$

From this result, if we plot the data as $\log y$ versus $\log x$, we will see a straight line of slope equal to $B$ and intercept of $\log A$. Thus, when we wish to test if data are following a power-law function, we plot them with logarithmic scaling on both axes (Figure F.6).

Even when data are not expected to represent true power laws, it is often useful to view the data on a double-log plot as an early diagnostic. If there is a region of power-law behavior in the data, it will readily be apparent as a linear region on such a graph. Also, log–log graphs can help us to identify boundaries between regions in which different physics governs the behavior of a system [33]. For example, the friction-factor versus Reynolds-number data that characterize steady flow in a circular tube show three distinct regions, which are readily visible in a log–log plot (Figure F.7). At Re < 2,100, friction factor follows a true power law with $B = -1$ and $A = 16$. Above a Reynolds number of 2,100, no power-law behavior is seen, and two distinct regions are observed: the transitional flow region for $2,100 \leq \mathrm{Re} \leq 4,000$ and the turbulent flow region for $4,000 \leq \mathrm{Re} \leq 10^6$, in which the friction factor depends also on pipe roughness $\epsilon$. Plotting the data on a double-log plot helps guide the investigation into this flow problem. The problem of pipe flow was investigated in the 1800s by Osborne Reynolds.

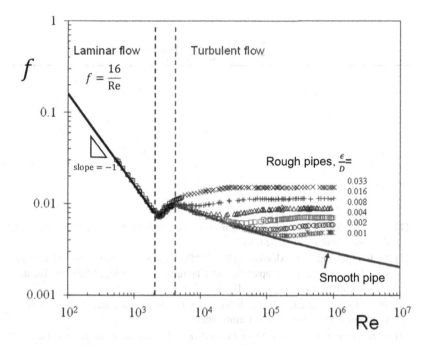

Figure F.7 The Fanning friction factor $f$ (dimensionless fluid wall drag) versus Reynolds number Re (dimensionless flow rate) data that characterize flow in a circular tube show three distinct regions: laminar, which shows power-law behavior; transitional; and turbulent [41]. Plotting the data on a double-log plot conveys the variation of the data. Also shown are the data of J. Nikuradse [47], whose data on sand-roughened pipes established the quantitative effect of pipe roughness $\epsilon$ on friction factor in pipes of diameter $D$.

The existence of a power-law region at low Reynolds number, including the values of its parameters $A$ and $B$, can be derived from the physics of slow laminar flow [41]. The more complex responses of the transitional and turbulent regions have been subject of intense study since Reynolds' early research, and research.

Researchers and engineers routinely use the models presented in this section to capture the behavior of experimental data. In Section 6.2, we discuss how to perform least-squares linear regression to determine good values of model parameters for a given dataset.

# Bibliography

[1] E. B. Bagley. (1957). "End corrections in the capillary flow of polyethylene." *J. Applied Physics,* 28(5), 624–627.

[2] N. Benipal and F. A. Morrison. (2013). "Data associated with CM3215 Laboratory, December 2013." Department of Chemical Engineering, Michigan Technological University, Houghton, MI.

[3] H. J. C. Berendsen. (2011). *A Student's Guide to Data and Error Analysis.* Cambridge University Press, Cambridge.

[4] P. R. Bevington. (1969). *Data Reduction and Error Analysis for the Physical Sciences.* McGraw-Hill, New York.

[5] P. R. Bevington and D. K. Robinson. (2003). *Data Reduction and Error Analysis for the Physical Sciences*, 3rd ed. McGraw-Hill Higher Education, New York.

[6] H. S. Carslaw and J. C. Jaeger. (1959). *Conduction of Heat in Solids,* 2nd ed. Oxford University Press, Oxford.

[7] T. B. Co. (2013). *Methods of Applied Mathematics for Engineers and Scientists.* Cambridge University Press, New York.

[8] Copper Development Association. (2019). *Copper Tube Handbook.* CDA Publication A4015-14/19. Available at www.copper.org/publications/pub_list/pdf/copper_tube_handbook.pdf, accessed August 9, 2019.

[9] E. O. Doebelin. (1990). *Measurement Systems: Application and Design,* 4th ed., McGraw-Hill, New York.

[10] H. Fairfield-Smith. (1936). "The problem of comparing the results of two experiments with unequal errors." *J. Couns. Sci. Indust. Res. (Australia)* 9(3), 211.

[11] Fluke Corporation. "Accuracy, resolution, range, counts, digits, and precision." Available at en-us.fluke.com/training/training-library/test-tools/digital-multimeters/accuracy-resolution-range-counts-digits-and-precision.html, accessed January 25, 2017.

[12] J. S. Fritz and G. H. Schenk. (1987). *Quantitative Analytical Chemistry.* Allyn and Bacon, Boston.

[13] W. A. Fuller. (1987). *Measurement Error Models.* Wiley, New York.

[14] C. J. Geankoplis. (2003). *Transport Processes and Separation Process Principles: Includes Unit Operations,* 4th ed. Prentice Hall, New York.

[15] L. Gonick and W. Smith. (1993). *The Cartoon Guide to Statistics*. HarperCollins, New York.

[16] W. S. Gosset. (1908). "The probable error of a mean." *Biometrika*, VI(1), 1–25.

[17] I. Guttman, S. S. Wilkes, and J. S. Hunter. (1982). *Introductory Engineering Statistics*, 3rd ed. Wiley, New York.

[18] A. J. Hayter. (2002). *Probability and Statistics for Engineers and Scientists*, 2nd ed. Wadsworth Group, Duxbury.

[19] D. B. Hibbert and J. J. Gooding. (2006). *Data Analysis for Chemistry*. Oxford University Press, Oxford.

[20] I. G. Hughes and T. P. A. Hase. (2010). *Measurements and Their Uncertainties: A Practical Guide to Modern Error Analysis*. Oxford University Press, Oxford.

[21] Joint Committee for Guides in Metrology (JCGM/WG 1). (2008) "Evaluation of measurement data Guide to the expression of uncertainty in measurement," 1st ed., JCGM 100:2008 GUM 1995 with minor corrections, document produced by Working Group 1. Especially Section 5, Determining combined standard uncertainty (pp. 18–23), and Annex G, Degrees of freedom and levels of confidence (pp. 70–78). Available at www.bipm.org/utils/common/documents/jcgm/JCGM_100_2008_E.pdf.

[22] J. A. King, F. A. Morrison, J. M. Keith, M. G. Miller, R. C. Smith, M. Cruz, A. M. Neuhalfen, and R. L. Barton. (2006). "Electrical conductivity and rheology of carbon-filled liquid crystal polymer composites." *J. Applied Polymer Sci.*, 101, 2680–2688.

[23] D. G. Kleinbaum, L. L. Kupper, K. E. Muller, and A. Nizam. (1998). *Applied Regression Analysis and Other Multivariable Methods*. Duxbury Press, Pacific Grove.

[24] G. Kraus and J. T. Gruver. (1965). "Rheological properties of multichain polybutadienes." *J. Polymer Sci. A*, 3, 105–122.

[25] R. LeBrell and S. LeRolland-Wagner. (2013). "Data from CM3215 Laboratory, calibration of Honeywell differential pressure meter at Laboratory Bench 1." Department of Chemical Engineering, Michigan Technological University, Houghton, MI.

[26] D. R. Lide, ed. (2004). *CRC Handbook of Chemistry and Physics*, 88th ed. CRC Press, New York.

[27] C. Lipson and N. J. Sheth. (1973). *Statistical Design and Analysis of Engineering Experiments*. McGraw-Hill, New York.

[28] A. J. Lyon. (1970). *Dealing with Data*. Pergamon Press, Oxford.

[29] L. Lyons. (1991). *A Practical Guide to Data Analysis for Physical Science Students*. Cambridge University Press, Cambridge.

[30] S. E. Manahan. (1986) *Quantitative Chemical Analysis*. Wadsworth, Monterey, CA.

[31] D. L. Massart, B. G. M. Vandeginste, S. N. Deming, Y. Michotte, and L. Kaufman. (1998). *Chemometrics: A Textbook*. Elsevier, Amsterdam.

[32] R. H. McCuen. (1985). *Statistical Methods for Engineers*. Prentice Hall, Englewood Cliffs, NJ.

[33] T. A. McMahon and J. T. Bonner. (1983). *On Size and Life*. Scientific American Books, New York.

[34] S. Midorikawa. "Derivation of the $t$-distribution," Professor, Department of Information Science, Aomori, Japan University. Available at https://shoichimidorikawa.github.io/Lec/ProbDistr/t-e.pdf, accessed July 22, 2019.

[35] J. N. Miller and J. C. Miller. (2000). *Statistics and Chemometrics for Analytical Chemistry,* 4th ed. Prentice Hall, New Jersey.

[36] J. N. Miller and J. C. Miller. (2005). *Statistics and Chemometrics for Analytical Chemistry,* 5th ed. Prentice Hall, New Jersey.

[37] J. C. Miller and J. N. Miller. (1984). *Statistics for Analytical Chemistry.* Wiley, New York.

[38] D. C. Montgomery and G. C. Runger. (2011). *Applied Statistics and Probability for Engineers,* 5th ed. Wiley, New York.

[39] F. A. Morrison. (2000). *Understanding Rheology.* Oxford University Press, New York.

[40] F. A. Morrison. (2013). "CM3215 Laboratory data, Fall 2011," Department of Chemical Engineering, Michigan Technological University, Houghton, MI. Data were taken by Samantha Armstrong, Ryan Barrette, Christina Basso, Ellesse Bess, Ellesse Bess, Tyler Boyea, Alexander Bray, Chris Bush, Jeff Caspary, Michelle Chiodi, Neeraj Chouhan, Ben Clemmer, Alex Culy, Courtney David, Stephen Doemer, Erik Drake, Henry Eckert, Ian Gaffney, Jeffrey Graves, Connor Gregg, Tyler Gygi, Paul Hagadone, Kim Hammer, Peter Heinonen, Brian Howard, Amber Johnson, Amber Johnson, Brian Kaufman, Kerry King, Joshua Kurdziel, Hwiyong Lee, Carissa Lindeman, Krista Lindquist, Leandra Londo, Kelsey Maijala, Ben Markel, Tristan McKay, David Mellon, Jordan Meyers, Ainslie Miller, Caroline Minkebige, Adam Moritz, Zach Newmeyer, Kevin Osentoski, William Paddock, Robert Parker, Morgan Parr, Josh Patton, Nick Phelan, James Podges, Carlos Prado, Mark Preston, Mitchell Redman, Alex Reese, Brian Ricchi, Timothy Rossetto, Samolewski, Cory Schafer, Tom Schneider, Lindsay Seefeldt, Chris Shocknesse, James Sikarskie, Stephanie Stevens, Katrina Swanson, Zach Tanghetti, Dillon Verhaeghe, Alex Wegner, Ethan Weydemeyer, Kelly-Anne Zayan, and Long Zhang.

[41] F. A. Morrison. (2013). *An Introduction to Fluid Mechanics.* Cambridge University Press, New York.

[42] F. A. Morrison. (January 15, 2014). "How to add 95% confidence interval error bars in Excel 2010," unpublished course handout, Department of Chemical Engineering, Michigan Technological University, Houghton, MI. Available at www.chem.mtu.edu/%7Efmorriso/cm3215/2014WordFigureErrorBars95CI.pdf, accessed June 21, 2018.

[43] F. A. Morrison. (April 12, 2005). "Using the Solver add-in in Microsoft Excel." unpublished course handout CM4650 Polymer Rheology, Department of Chemical Engineering, Michigan Technological University, Houghton, MI. Available at pages.mtu.edu/~fmorriso/cm4650/Using_Solver_in_Excel.pdf, accessed July 4, 2017.

[44] F. A. Morrison. (April 4, 2016). "Unsteady heat transfer to a sphere: measuring the heat transfer coefficient (fitting PDE solution," unpublished course lecture slides, CM3215 Fundamentals of Chemical Engineering Laboratory, Department of Chemical Engineering, Michigan Technological University,

Houghton, MI. Available at https://pages.mtu.edu/~fmorriso/cm3215/Lectures/ CM3215_Lecture10HeatConductSphere_2019, accessed May 21, 2020.

[45] F. A. Morrison. (2017). "YouTube channel DrMorrisonMTU," instructional videos on chemical engineering and other topics. Available at www.youtube.com/ user/DrMorrisonMTU, accessed 4 July 2017.

[46] F. A. Morrison. (July 25, 2018). Unpublished data, Department of Chemical Engineering, Michigan Technological University, Houghton, MI.

[47] J. Nikuradse. (1933). "Stromungsgesetze in Rauhen Rohren." *VDI Forschungsh*, 361; English translation, NACA Technical Memorandum 1292.

[48] Omega Corporation. (2018). "Omega Web Technical Temperature Reference." Available at www.omega.com/techref/Z-section.html, accessed July 17, 2018.

[49] R. H. Perry, D. W. Green, and J. O. Maloney. (1973). *Perry's Chemical Engineers' Handbook,* 6th ed. McGraw-Hill, New York.

[50] B. C. Reed. (1990). "Linear least-squares fits with errors in both coordinates." *Am. J. Phys.* 57(7), 642–646; also erratum *Am. J. Phys.* 58(2), 189.

[51] F. E. Satterthwaite. (1941). "Synthesis of variance." *Psychometrika* 6, 309–316; (1946). "An approximate distribution of estimates of variance components." *Biometrics Bull.* 2(6), 110–114.

[52] J. R. Taylor. (1997). *An Introduction to Error Analysis: The Study of Uncertainties in Physical Measurements*, 2nd ed. University Science Books, Herndon, VA.

[53] Thermometrics Corporation, (2018). "RTD sensor accuracy and tolerance standards, Class B RTD." Available at www.thermometricscorp.com/acstan.html, accessed August 1, 2018.

[54] G. B. Thomas, Jr., and R. L. Finney. (1984). *Calculus and Analytic Geometry,* 6th ed. Addison-Wesley, Boston.

[55] Tutor Web. (2018). "Covariance between estimates of slope and intercept." Available at tutor-web.net/stats/stats3103simplereg/lecture70/sl00, accessed February 2, 2018.

[56] D. D. Wackerly, W. Mendenhall III, and R. L. Scheaffer. (2002). *Mathematical Statistics with Applications,* 6th ed., Wadsworth Group, Duxbury.

[57] B. L. Welch. (1936). "The specification of rules for rejecting too variable a product, with particular reference to an electric lamp problem." *J. R. Stat. Soc. Suppl.* 3, 29–48; (1938). "The significance of the difference between two means when the population variances are unequal." *Biometrika* 29, 350–362; (1947). "The generalization of 'Student's *t*' problem when several different population variances are involved." *Biometrika* 34, 28–35.

# Index

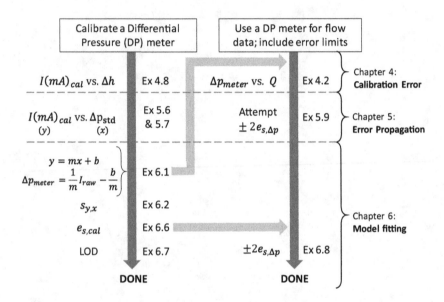

Figure 1.3 Ten linked examples show how the techniques of error analysis of this text are applied to a practical example, the calibration and use of a differential-pressure (DP) meter. Also addressed is the lowest value that can be accurately measured by a device, the limit of determination (LOD).

| Replicate errors and error basics | Eqn. | Type A errors |
|---|---|---|
| $\Pr[a \le x \le b] \equiv \int_a^b f(x')dx'$ | 2.12 | Probability as an integral of probability density function (pdf) |
| $t \equiv \dfrac{(\bar{x} - \mu)}{s/\sqrt{n}}$ | 2.25 | Scaled deviation |
| $\Pr\left[-t_{limit} \le x \le t_{limit}\right] \equiv$ $\displaystyle\int_{-t_{limit}}^{t_{limit}} f(t'; n-1)dt'$ | 2.34 | Probability that scaled deviation $t$ is between $-t_{limit}$ and $t_{limit}$ |
| $(1 - \alpha) = \displaystyle\int_{-t_{limit}}^{t_{limit}} f(t'; v)dt'$ | 2.47 | $\begin{aligned} t_{limit} &= \text{T.INV.2T}(\alpha, v) \\ &= -\text{tinv}\left(\tfrac{\alpha}{2}, v\right) \end{aligned}$ |
| $\mu = \bar{x} \pm t_{0.025, n-1}\dfrac{s}{\sqrt{n}}$ | 2.52 | 95% CI of the mean (replicate error only) $\begin{aligned} t_{0.025, n-1} &= \text{T.INV.2T}(0.05, n-1) \\ &= -tinv(0.025, n-1) \end{aligned}$ |
| $x_{n+1} = \bar{x} \pm t_{0.025, n-1} s\sqrt{1 + \dfrac{1}{n}}$ | 2.58 | 95% PI of $x_{n+1}$ (replicate error only) |

| Reading error | Eqn. | Type B errors |
|---|---|---|
| Sensitivity | Figure 3.5 | $1 \le \text{ROT} \le 15$ times last digit, *or* Obtain from the manufacturer |
| Resolution (subdivisions) | 3.3 | $\left(\dfrac{\text{smallest division}}{2}\right)$ |
| Fluctuations | 3.4 | $\left(\dfrac{\text{max} - \text{min}}{2}\right)$ |
| $e_R = \text{max(sensitivity, resolution, fluctuations)}$ | 3.5 | Nonstandard reading error |
| $e_{s, reading} = \dfrac{e_R}{\sqrt{3}}$ | 3.12 | Standard reading error |

| Calibration error | Eqn. | Type B errors |
|---|---|---|
| Digits provided | 4.2 | ROT = 1 times last digit |
| Manufacturer maximum error | | $e_{s,est} = \dfrac{(\text{error limit})}{2}$ |
| User calibration | 4.3 | $e_{s,est} = \dfrac{(\text{max value} - \text{min value})}{4}$ <br> *or* carry out rigorous calibration |
| $e_{s,cal} = \max(\text{ROT, look up, user})$ | 4.5 | Standard calibration error |
| $\text{LOD} = 8e_{s,cal}$ | 4.8 | ROT = 25% error, maximum |

| Error Propagation | Eqn. | Type B errors |
|---|---|---|
| $e_{s,cmbd}^2 = e_{s,random}^2 + e_{s,reading}^2 + e_{s,cal}^2$ | 2.55 | Measurement errors combine in quadrature |
| $e_{s,y}^2 = \left(\dfrac{\partial\phi}{\partial x_1}\right)^2 e_{s,x_1}^2 + \left(\dfrac{\partial\phi}{\partial x_2}\right)^2 e_{s,x_2}^2 + \left(\dfrac{\partial\phi}{\partial x_3}\right)^2 e_{s,x_3}^2$ | 5.18 | Error propagation, 3 variables, uncorrelated |
| $e_{s,y}^2 = \displaystyle\sum_{i=1}^{n}\left(\dfrac{\partial\phi}{\partial x_i}\right)^2 e_{s,x_i}^2$ | 5.20 | Error propagation, $n$ variables, uncorrelated |
| $\sigma_y^2 = \left(\dfrac{\partial\phi}{\partial x_1}\right)^2 \sigma_{x_1}^2 + \left(\dfrac{\partial\phi}{\partial x_2}\right)^2 \sigma_{x_2}^2 + \left(\dfrac{\partial\phi}{\partial x_3}\right)^2 \sigma_{x_3}^2 + 2\left(\dfrac{\partial\phi}{\partial x_1}\right)\left(\dfrac{\partial\phi}{\partial x_2}\right)\text{Cov}(x_1,x_2) + 2\left(\dfrac{\partial\phi}{\partial x_1}\right)\left(\dfrac{\partial\phi}{\partial x_3}\right)\text{Cov}(x_1,x_3) + 2\left(\dfrac{\partial\phi}{\partial x_2}\right)\left(\dfrac{\partial\phi}{\partial x_3}\right)\text{Cov}(x_2,x_3)$ | 5.16 | Error propagation, 3 variables, correlated |
| error limits $= \pm 2e_{s,cmbd}$ | 2.56, E.5 | Expanded uncertainty at a level of confidence of 95% (Type B errors) |

| Ordinary Least Squares (LS) | Eqn. | Type A errors |
|---|---|---|
| $m = \dfrac{SS_{xy}}{SS_{xx}}$ | 6.5 | SLOPE(y-range,x-range) or INDEX(LINEST(y-range,x-range),1,1) |
| $b = \bar{y} - m\bar{x}$ | 6.6 | INTERCEPT(y-range,x-range) or INDEX(LINEST(y-range,x-range),1,2) |
| $s_{y,x} = \sqrt{\dfrac{1}{(n-2)} \displaystyle\sum_{i=1}^{n} (y_i - \widehat{y}_i)^2}$ | 6.25 | STEYX(y-range,x-range) |
| $y_{i,true} = y_i \pm t_{0.025,n-2} s_{y,x}$ | 6.27 | 95% CI on $y_i$ values in the LS dataset |
| $s_m = \sqrt{s_{y,x}^2 \left(\dfrac{1}{SS_{xx}}\right)}$ | 6.32 | INDEX(LINEST (y-range,x-range,1,1),2,1) |
| $m_{true} = m \pm t_{0.025,n-2} s_m$ | 6.37 | 95% CI on LS slope |
| $s_b = \sqrt{s_{y,x}^2 \left(\dfrac{1}{n} + \dfrac{\bar{x}^2}{SS_{xx}}\right)}$ | 6.33 | INDEX(LINEST (y-range,x-range,1,1),2,2) |
| $b_{true} = b \pm t_{0.025,n-2} s_b$ | 6.38 | 95% CI on LS intercept |
| $s_{\widehat{y}_0}^2 = s_{y,x}^2 \left(\dfrac{1}{n} + \dfrac{(x_0 - \bar{x})^2}{SS_{xx}}\right)$ | 6.62 | Variance of the LS model fit |
| $\widehat{y}_{0,true} = (mx_0 + b) \pm t_{0.025,n-2} s_{\widehat{y}_0}$ | 6.63 | 95% CI on LS fit |
| $s_{y_{0,n+1}}^2 = s_{y,x}^2 \left(1 + \dfrac{1}{n} + \dfrac{(x_0 - \bar{x})^2}{SS_{xx}}\right)$ | 6.73 | Variance of the next value extending the LS dataset |
| $y_{0,n+1,true} = (mx_0 + b) \pm t_{0.025,n-2} s_{y_{0,n+1}}$ | 6.65 | 95% PI on LS fit |

Printed in the United States
By Bookmasters